Wackenheim · Braun The Veins of the Posterior Fossa

A. Wackenheim J. P. Braun

The Veins of the Posterior Fossa

Normal and Pathologic Findings

With a Foreword by J. Bull

With 171 Figures

Springer-Verlag Berlin Heidelberg New York 1978

Auguste Wackenheim

Professeur agrégé
Head of the Department of Neuroradiology, University Hospital
F-67005 Strasbourg

Jean Paul Braun

Docteur de médecine
Head of the Department of Neuroradiology, Pasteur Hospital
F-68000 Colmar

The cover picture is an x-ray of meningioma of right cerebellopontine angle which is shown in more detail in Figure 162 on page 137.

ISBN-13: 978-3-642-66714-5 e-ISBN-13: 978-3-642-66712-1
DOI: 10.1007/978-3-642-66712-1

Library of Congress Cataloging in Publication Data. Wackenheim Auguste. The veins of the posterior fossa. Bibliography: p. Includes index. 1. Brain—Blood-vessels—Abnormalities. 2. Cranial fossa, Posterior—Blood-vessels. 3. Veins. I. Braun, Jean Paul, joint author. II. Title. [DNLM: 1. Skull—Bloody supply. 2. Cerebral angiography. WL141 W115v]. RC388.5.W32 616.8′1 77-23953

Softcover reprint of the hardcover 1st edition 1978

Foreword

It is a privilege and pleasure to write a foreword to this work. Who, twenty years ago, would have thought that a whole book could be devoted to the veins of the posterior fossa? Even today the standard text-books of anatomy only give a paragraph to the subject. Thanks to the dedicated work of rather a few neuroradiologists our knowledge has now reached a very sophisticated level. The careful correlation of anatomical dissection with the angiographic image has achieved this, but without the aid of subtraction (discovered by a world figure in neuroradiology—Professor ZIEDSES DES PLANTES) the subject would not be as advanced as it is.

The authors are to be congratulated not only for presenting the most up to date anatomical information but also for showing how wide the applications are in the clinical field by the careful analysis of displacements of veins, many of which were quite unrecognised only a very few years ago. Drs WACKENHEIM and BRAUN are in the forefront of this discipline to which they and the French school have contributed so much. Also it was their great enthusiasm and drive which led to the birth of the European Society of Neuroradiology, the first meeting of which was held in Colmar in September 1969. The Society has flourished under their guidance and very successful annual meetings have been held for nearly a decade.

The schools of neuroradiology at Strasbourg and Colmar have attracted dedicated assistants who have contributed in no small way to the teaching and research which is so active. It is pleasant to see that the authors pay tribute to these men and women who have assisted in the production of this book.

All those involved in the neurological sciences will benefit by carefully studying the text and excellent illustrations.

January 1978 JAMES BULL

Acknowledgments

This book is the fruit of a regular collaboration between the medical staffs of the Neuroradiology Departments in Strasbourg and Colmar. This collaboration was made possible by the understanding and support of the authors' families, especially their wives, who bore their necessary absences in loving stride. We owe them our devoted thanks.

We thank Dr. Y.P. Huang, pioneer in the field of posterior fossa veins, for his valuable suggestions. We are grateful to our medical collaborators, Dr. Elisabeth Babin and Dr. Alain Tournade, for their efficient participation in our work, and to our technical staffs, especially Mrs. Eleonore Bromhorst and Miss Marie-Paule Heitz, whose radiographic documents illustrate this book. Our special thanks go to Mr. Auguste Frey for his competent work in subtraction and photography, and to Mrs. Rosemarie Kipper and Miss Francine Burgy, our secretaries, for their capable work.

This book could not have been written without the kindness and friendship of our colleague, Dr. Nicole Heldt, neuropathologist, who prepared specimens with our study in mind. To her we express our sincere gratitude. To our many other colleagues who remain unnamed, we express our thanks.

To our translator, Mrs. Lillian Braun, our last, but not least, thanks.

A. Wackenheim J.P. Braun

Contents

A. Introduction

The relatively restricted and often little known anatomic studies of the veins of the posterior fossa are now unadaptable to the extraordinary development of cerebral angiography in neuroradiology, especially with the use of subtraction, angiotomography, and magnification. If anatomists seem to have attributed a somewhat limited importance to the multiple veins of the posterior fossa, this has not been due to negligence or deliberate disregard on their part. It is rather the expansion of cerebral angiography and the evolution of neuroradiology in general which creates this impression. Medicine is being oriented by cerebral angiography toward structures heretofore invisible and now subjected to scientific analysis. Under this drive, a detailed anatomic study of the arterial and venous cerebral system has been undertaken. This elaboration is warranted by imperious clinical and therapeutic reasons. Arterial and venous structures, brought to light by the use of cerebral vascular investigation, can now be rapidly analyzed. Of course the approach to this study differs, depending on whether this knowledge is applied to the establishment of a rapid diagnosis intended for the neurosurgeon, or rather, to the teaching of general anatomy, whose applications are deferred. Let us not forget that neuroradiology is a kind of "highly specialized artificial anatomy" and that its interests, methods, and aims place this speciality on a different level than that of gross anatomy.

A perfect structural knowledge of the normal anatomy of the posterior fossa is indispensable to the study of its veins and arteries and their radiologic applications. This demands intimate collaboration with anatomists interested in the matter.

We are fortunate enough to benefit of the friendship and active collaboration of a dynamic research team, directed by Professor J.G. Koritke and assisted by his collaborators, Dr. C. Maillot and Dr. A. Tournade. But it was principally Dr. N. Heldt who oriented our research and helped us correct our errors, and it was at the Institute of Pathologic Anatomy in Professor Y. Legal's department that we made extensive research and anatomic comparisons of our phlebographic data, especially concerning pathologic findings.

B. Historical Background

It is impossible to give a detailed history of the literature on the venous system of the posterior fossa. One would have to comment at length on the bibliography which appears at the end of this study to arrive at a complete analysis. We shall nevertheless try to recall to mind the dates and principal papers referring to the radiologic approach of this study.

The first general study of the venous system of the posterior fossa was made in 1950 by GOMEZ OLIVEIROS [58]. NAMIN's thesis in 1955 added no new element to the preceding work [95]. In 1964, LINDGREN's study [92] divided the venous system of the posterior fossa into two groups. From 1965 on, HUANG [67–75] has published a series of detailed studies which have led to the radiologic and clinical exploitation of the phlebogram in vertebral angiography. VIALE and ROSA [137–147] published in 1968 a radiologic and anatomic review. BULL and KOSLOWSKI [29] made a general study of the petrosal veins in 1969.

In the past years, more and more radiologic-phlebographic studies of the posterior fossa have been made; our own since 1968 [5, 11–13, 21–25] and those of our pupils LOPEZ [93b], BRADAC [16–20], BEN AMOR [9–13], and BILLEWICZ [14, 15].

More recently two radiologic-anatomic studies of the basal vein have been published: one by SIGNARGOUT [126b] and the other by BABIN [6]. Recently (1974) DUVERNOY published a monography on the superficial veins of the human brain [42].

C. Practical Methodology

Is it necessary to prove the diagnostic importance of the phlebogram of the posterior fossa? We are accustomed to interpreting arteriograms rapidly. To analyze the veins as well is tantamount to doubling the intellectual effort of the physician, both as to the interpretation of radiograms proper as to the physiologic and anatomic identification. Subtraction and/or magnification, which are often indispensable, add to the amount of work necessary to the rendering of a correct interpretation of the phlebogram.

The terminology of the veins remains confused. We shall try to simplify matters as much as possible, bearing in mind the terminology used during the *International Course in Neuroradiology held in Strasbourg in 1972.*

Arteries are often removed from the surface of the nervous parenchyma, whereas veins, due to their long superficial course, *trace* its contours quite well, especially at the level of the cerebral trunk. The cisterns are often delimited in like manner.

Veins are precious reference points in those cerebral segments not marked by arteries. This is the case at the level of the pontocerebellar angle, the cerebellar vermis, the brain stem, and the isthmus of the brain, regions which are *much better localized by veins than by arteries.*

Veins are *less resistant to mechanical pressure* than arteries, and will, by this very fact, endure readily and precociously those displacements useful in a neuroradiologic diagnosis (thinner walls, lower blood pressure, absence of pulse). Veins are *not subject to angiosclerosis,* whereas arteries are deformed and elongated by this disorder, becoming more and more resistant to pressure. This state is a source of difficulty in the reading of an arteriogram. At the level of the posterior fossa there are as many *anatomic variants* of the arteries as of the veins. It was believed at one time that only the complementary procedure of pneumography, for example, established the presence of a perianeurysmal hematoma. Today, the veins complement this information. In a lateral projection of the posterior fossa, veins are perhaps easier to analyze than in a frontal one, as the former is more distinct from the photographic point of view. But as a rule both frontal and lateral projections are necessary.

Considering the conspicuous wealth of information furnished by the arteries, physicians have a tendency to neglect the cerebral veins. However, conscious as we are of the inadequacies of arterial semeiology, we have come to use the phlebogram and its complementary information, especially as it does not impose any supplementary burden on the patient.

In any event, we have progressed beyond our predecessors, who, like NAMIN in 1955, formulated an opinion which now belongs to the history of neuroradiology when he said, "Cerebellar veins have a disconcerting aspect which defies all description."

With practical neuroradiology in mind we are led to describe the venous system in accordance with the main goal, i.e., the clinical diagnosis. It would certainly be tempting to follow the inspiration of embryology or physiology, which clarify certain aspects, but which at the same time introduce a wealth of useless information in practical and immediate phlebography. We have thus [67] followed the system outlined by HUANG which divides the venous system into three groups: *the superior group* (drainage by the great vein of Galen), *the anterior group* (drainage by the superficial petrosal vein), and the *posterior group* (drainage by the sinuses of the tentorium). However this system proves to be too hazardous and *does not provide for our daily neuroradiologic needs.*

This situation led us to divide the phlebogram of the posterior fossa into five parts. Each division refers to a group of veins having anatomic-physiologic relations, but having above all a radiologic context and sufficient anatomic invariability so as to be a reference value whose study adduces a useful "radiologic sign"; i.e., establishing a diagnosis. These five venous groups are:

1. The basal veins and their tributaries
2. The superior and inferior vermian veins
3. The precentral vein
4. The peduncular and pontine veins
5. The superior petrosal veins and their tributaries.

1. The Basal Veins and Their Tributaries

(see also page 9)

In practice the observer analyzes the basal vein by identifying its general morphologic characteristics (page 25) and by applying certain simple geometrical guidemarks.

a) In frontal suboccipital (Towne) projections, *the basal vein* always presents two well-defined segments converging toward point *L*, an easy guidemark as it is the most lateral in the vein's course. *This point corresponds* in fact to the lateropeduncular angle of the vein. It indicates the change in direction of the basal vein and delimits the cerebral peduncle from the temporal lobe. Point *L* has the advantage of being visible both in frontal and lateral projections (Figs. 1–3, 21, 22). In a frontal projection the anterior segment of the basal vein appears more or less shortened, depending on the obliquity of the central ray with respect to the head, hence this visualized sinuosity depends on the aforementioned factors. It is almost always possible to measure, on the one hand, the distance from point *L* to the midline, called *A*, and, on the other hand, to measure the distance from the midline to the most internal point of the curve, called *B* (Fig. 1). Because of the variations in the morphology of the venous course, these two measurements only retain our attention in this type of projection. We have found an average distance (not corrected) of 25 mm for *A* and of 17 mm for *B*. These figures approach those arrived at by SIGNARGOUT and coworkers [126b].

The average relation A/B measures 1.45, that is, a little less than 1.5. In consequence, it is possible to arrive at an approximative evaluation of the normal venous course by dividing distance *A* into three segments and by placing the vein's anterior curve tangent to the line which passes through the second third.

b) *In a lateral projection* the basal veins are more comparable from one examination to another. On the one hand we have tried to locate the direction, and on the other hand the origin, point *L*, and the terminating point of the vein with reference to classical cranial bony guidemarks (Fig. 3). *The direction of the anterior segment* of the basal vein from its origin to point *L* has been the object of our attention (Fig. 3a). Most often it presents a parallel course in its general direction, apart from a few slight curves, either at the nasio-internal occipital protuberance line, or at the tuberculum sellae-nasion line. When it is not parallel to either of these two lines (10% of cases), it generally takes an ascending direction backward. We have evaluated *the distance from the tuberculum sellae to the internal occipital protuberance with reference to the three principal guidemarks of the vein:* its anterior extremity marked by the confluence of the strial and insular veins, point *L*, and its confluence with the internal cerebral vein (Fig. 3b), and have called these distances α, β, γ. They average respectively, 13 mm, 15 mm, and 31 mm. It is interesting to note that the confluence of the basal vein and the internal cerebral vein is approximately twice removed from the internal occipital protuberance-tuberculum line than point *L*. This constitutes a simple and rapid radiologic guidemark of point *L* in lateral projections ($\gamma = 2\beta$).

We have further tried *to mark the position of point L with reference to the line joining the tuberculum sellae and the lambda* when these two elements are visualized simultaneously on the film (Fig. 3c). This was obtained by evaluating on the one hand, the distance from point *L* to this line, and on the other hand, the relation of the distance *a*, which represents the segment from the tuberculum to point *L* (or its projection on the line, and *b* the segment which represents the distance point *L*-lambda. Point *L* is often found on the line tuberculum-lambda, that is to say, the curve traced by the basal vein is tangent to the line, coursing at most 4 mm above and 8–9 mm below. As to the relationship *a–b*, it averages 1–3 and constitutes a third simple radiologic guidemark of point *L*: *slightly before the first third of the line tuberculum lambda, on the line or a few mm* below ($b = 2a$).

The distance from the basal vein to the internal cerebral vein is interesting as these two venous trunks course, the former in the medial section, and the latter in the lateral section of the choroid fissure, on both sides of the thalamus (Fig. 3d). In order to proceed, we have measured the distance separating the two vessels on the vertical line, which from the external auditory canal course through point L or toward the most removed point of the skull, that is the vertex. This distance averages 2.50 cm.

The point of origin of the basal vein, on the contrary, is difficult to locate due to its anterior position with relation to the sella turcica. It is usually found 1 cm above the extremity of the dorsum. We have checked this and found an average distance of 9.6 mm. The distance which separates the origin of the basal vein from the tuberculum sellae is a radiologic guidemark more constant in nature (an average of 16.6 mm), but this figure is distorted by variations in the length of the anterior segment of the vein.

The relationship between the basal vein and the calcified pineal body was studied in 1970 in our monograph [154] on the angiography of the mesencephalon. Let us reiterate that calcification of the pineal body remains generally distant from the posterior segment of the basal vein. Pathologic displacement, either of the basal vein or of the pineal body, brings them into contact.

2. The Superior and Inferior Vermian Veins

(pages 12 and 20)

At the opposite pole of the posterior fossa, the inferior vermian vein provides information on the depth of the cisterna magna. Neuroradiologists know that care must be taken when examining this cistern during pneumoencephalography. Hence it is justifiable that we outline the vein on the phlebogram. It delimits the anterior wall of the cisterna magna, thus permitting the evaluation of the latter's depth with respect to the occipital squama.

In consequence, the inferior vermian vein indicates indirectly hypertension or a malformation of the inferior pole of the cerebellum. In contrast to the postero-inferior cerebellar artery, which might ramify independently in the cerebellum up to and into the cervical canal, the inferior vermian vein embraces the vermian parenchyme proper. Before concluding that its course is displaced in a postero-inferior direction, we firmly recommend that its medial or slightly paramedial position be checked in a frontal projection. In fact, its plainly paramedial or lateral localization (frontal view), that is, on the posterior aspect of the cerebellar hemisphere, excludes its validity as a guidemark in a lateral view (Figs. 4, 64–66). The superior vermian vein is indispensable in outlining the subdural supravermian space. This space (Figs. 35–38), is occupied in cases of hypertension (page 78) when there is a herniation of the superior vermis in the subdural space [152].

3. The Precentral Vein

(page 10)

After a study of the veins of the upper pole (basal vein) and the lower pole (inferior vermian vein), we shall examine the *precentral vein* in a lateral projection. This vein furnishes us primary information, since the "precentral region" is visualized neither in pneumoencephalograms nor in arteriograms. The precentral vein is usually invariable and is thus a perfect guidemark on the midline. Its elementary backward displacements, due to anterior and anterolateral masses, as well as forward displacements by posterior and posterolateral masses, are basic and established venous signs.

On the contrary, normal morphology and topography of this vein exclude mass expansions. As described by Huang [68], the precentral vein indicates the *collicular point,* hence constituting a precious guidemark, as illustrated in various figures, especially in Figures 5 and 30.

4. The Peduncular and Pontine Veins

(page 13)

The fourth venous complex which we recommend be taken systematically into consideration is the pontopeduncular network. It is perhaps the most difficult to understand. The prepontine venous group touches both Dandy's veins, the basal veins, the veins of the fourth ventricle, and the bulbospinal veins. It represents the transversal connection between the right and left sides of these venous groups. The variants of the pontopeduncular group do not exceed those of any other arterial group in the brain. This group has the main advantage of being intimately associated with the pons, whereas the basilar artery is the most removed from it (Figs. 2, 28, 54). It is in this light and at this level that we emphasize the arteriovenous dissociation (Fig. 156).

5. The Superior Petrosal Veins and Their Tributaries

(page 13)

These veins have been known for some time [43] as Dandy's veins and are the most useful ones in a frontal projection. Under satisfactory technical conditions, they permit the localization of the external part of the pontocerebellar cisterns, and are opacified on both sides during vertebral angiography.

We shall see the systematization of their tributaries. The neuroradiologist through his experience in pneumoencephalography can recognize the normal position of the superior petrosal vein, above the superior edge of the petrosal bone and beyond the porus, in the external part of the pontocerebellar cistern.

In our present study we shall outline in detail the normal radiographic anatomy of all the veins of the posterior fossa. In actual fact, an analysis of the above-mentioned five venous groups will permit the determination of whether the area under practical routine study is normal or abnormal.

D. Normal Findings

I. The Venous Sinuses

This subject will not be dealt with. A radiograph of the sinuses is roughly reviewed in Figures 6–11. The venous sinuses are often poorly opacified in vertebral angiography. They can be studied better in retrograde jugulography, which can easily be performed in frontal, lateral, and axial projections. This method is of practical interest in diagnosing jugular glomus tumors and venous thromboses. Numerous hemodynamic phenomena can disturb the opacification of the sinuses, so that care must be exercised in the interpretation of the radiographs. When in doubt, it would be well to perform direct and retrograde phlebography of all ingoing vessels.

The venous sinuses are fed by the veins of the posterior fossa, which coalesce directly after having described a small segment parallel to the sinus, both at the level of the superior petrosal sinus and at that of the lateral sinus (Fig. 11). As a variant one may see veins which course in the tentorium and especially in its free margin (Fig. 67).

II. The Emissary Veins

The intracranial venous system and the extracranial venous system communicate through the emissary veins which are rarely visualized in angiography. Besides, they are of no great practical usefulness. They can be localized by their characteristic bony canal, bordered by a clearly visible margin.

Three main emissary veins can be found at the level of the posterior fossa:

1. *The Occipital Emissary Veins,* which are quite variable, are localized on both sides of the midline, above the posterior edge of the foramen magnum.
2. *The Anterior and Posterior Condylar Veins* appear often in angiography and in jugulography (Figs. 6, 74, 75).
3. *The Mastoid Emissary Vein,* which can be of high caliber, and whose bony canal is well known by radiologists (Fig. 78). This vein is clearly visualized in cases of thrombosis of the lateral sinus (Fig. 83).

The anterior temporal emissary veins emerge in the sphenotemporal region. We observed an abnormal dilatation of the temporal emissary vein in two cases where the anterior segment of the superior longitudinal sinus was impermeable (Fig. 10).

III. The Great Vein of Galen and Its Tributaries

1. The Internal Cerebral Vein

The topography and morphology of the internal cerebral vein are precisely known. It has no real place in the vertebrobasilar systematization, hence we shall cover this point only briefly, illustrating the principal tributaries of the vein. All the venous tributaries of the internal cerebral vein are seen in a carotid phlebogram. Under certain conditions, they can appear on a vertebral phlebogram (Fig. 24).

a) The Pellucidal Vein

This vein follows a rather straight course and its direction is the same as that of the internal cerebral vein in a lateral projection (Figs. 17 and 18).

b) The Vein of the Caudate Nucleus

The caudate veins are sometimes three in number:

The vein of the head of the caudate nucleus,
The vein of the body of the caudate nucleus,
The vein of the tail of the caudate nucleus.

These veins are relatively well distinguished in angiograms, especially the vein of the head of the caudate nucleus. This particular vein usually runs into the striothalamic vein; it rarely courses directly into the internal cerebral vein (Fig. 17).

c) The Superior Choroidal Vein

The superior choroidal vein runs into the internal cerebral vein. Its caliber varies quite often (it is frequently inferior to that of the thalamostriate vein) and it originates in the choroidal plexuses of the lateral ventricle (several small veins run together into the trunk of the superior choroidal vein). The choroidal plexuses of the lateral ventricle can be visualized in a frontal projection (Fig. 15a) of a vertebral angiography, as we demonstrated in 1973 [151], or in cases of selective opacification of the anterior choroidal artery (Fig. 15b, c).

This vein can have a tortuous course, as demonstrated in the specimen in Figure 13, and ends in the thalamostriate and pellucidal vein to form the origin of the internal cerebral vein. The superior choroidal vein opacifies better than the inferior choroidal vein, which is a tributary of the basal vein. In carotid angiography, the superior choroidal vein is always badly distinguished (Fig. 14b), due to the strong dilution of the contrast medium. On the contrary, in vertebral angiography, it is generally possible to distinguish the trunk of the superior choroidal vein in a variable length (Fig. 14a).

In carotid angiography, the internal cerebral vein is well opacified since its three original tributaries carry along the contrast medium (pellucidal, striothalamic, and superior choroidal veins).

In vertebral angiography, on the other hand, the internal cerebral vein is badly, or not at all, opacified, as only one of its tributaries carries the contrast medium.

The terminal segment of the internal vein is opacified, as it receives the contrast medium from other tributaries (superior thalamic vein and vein of the choroidal plexus of the third ventricle). Angiographic observations lead us to believe that the superior choroidal vein has a considerable output.

In a lateral angiographic view three veins (Figs. 14, 16) run in an anteroposterior direction and are parallel-projected. The upper vein is the *superior choroidal vein* (which, in fact, has a very lateral origin and an oblique course, as demonstrated in Fig. 12), the middle vein is the posterior part of the internal cerebral vein, and the lower vein is the superior thalamic vein (Fig. 16). During the course of vertebral angiography, deep venous opacification is not simultaneous since in early phlebogram the veins of the plexus of the lateral and third ventricle, the superior thalamic veins and the posterior segment of the basal vein are visible. The internal cerebral vein appears only shortly in connection with the superior choroidal vein. Lastly, the opacification of the superior choroidal vein disappears before that of the thalamic veins whose opacification, as well as that of the posterior segment of the basal vein, persists (Fig. 16).

d) The Ventricular or Subependymal Veins

These do not belong to the vertebrobasal system, thus we shall review them only briefly.

Four veins can be distinguished (Figs. 17, 18, 20):

α) The medial atrial vein or *vein of the internal wall of the trigonum.* The number of tributaries varies (two or three) and these converge to form a trunk which lies extraventricularly and courses into the internal cerebral vein. We have already published the gross morphology of this vein in our previous monograph [154a]. In our department, Billewicz and Ben Amor devoted papers to the systematization of the subependymal veins [14, 15].

β) The direct medial vein or *vein of the internal wall of the body of the lateral ventricle:* it is smaller and lies more anteriorly than the previous vein.

γ) The lateral atrial vein or *vein of the external wall of the trigonum.* It usually terminates in the basal vein but can be a tributary of the internal cerebral vein.

δ) The direct lateral vein or *vein of the external wall of the body of the lateral ventricle.* During ordinary carotid angiography, the veins of the internal wall of the trigonum and those of the external wall of the ventricular body opacify with greater frequency.

e) The Anterior Thalamic Vein

This vein fills during the early venous phase and is seen as a small vein coursing more or less vertically in the anterior thalamic area and ending in the internal cerebral vein. On rare occasions the single anterior thalamic vein is replaced by several very tiny venous vessels.

f) The Striothalamic Vein

This vein courses at the junction of the upper and lateral aspect of the thalamus and is of very great importance in practical angiography, especially in diagnosing thalamic tumors (Figs. 17, 18).

2. The Basal Vein

Although the basal vein has no real part in a vertebral phlebogram, we shall outline in brief its topography, morphology, and connections with the veins of the posterior fossa. This systematization (Fig. 24) has been taken from BABIN.

a) Origin of the Basal Vein (Figs. 21 and 22)

The vein is formed by the confluence of small veins. In a lateral view the point of origin is situated slightly above the dorsum sellae. It is seen as a stellar burst formed by:

1. *The Anterior Cerebral Vein* (anterior limbic vein, anterior pericallosa vein). It anastomoses with the controlateral vein through an anterior communicating vein.
2. *The Medial and Inferior Frontal Veins,* among which is found the olfactory vein, coursing in the olfactory groove.
3. *The Deep Sylvian Vein* which collects blood from insular convolutions.
4. *The Inferior Striate Veins,* satellite veins of the anterior striate veins.

b) Trunk of the Basal Vein (Fig. 23)

The lateral topography of the anterior segment can be delimited by point *L*, as illustrated in several figures.

In a frontal view, the projection of the basal vein depends on the angle of the ray beam. However, one can distinguish a sigmoid course formed by 3 segments (Fig. 1):

An anterior segment which obliques upward to the midline,

A middle segment which courses laterally up to point *L*,

And finally a posterior segment which courses toward the midline to coalesce with the great vein of Galen.

c) Tributaries of the Basal Vein

The internal tributaries are made up of the veins of the chiasma, of the optical tract, of the tuber cireneum, of the mammillary tubercles, of the cerebral peduncles, and the posterior communicating veins.

The external tributaries are made up of the vein of the hippocampus, the inferior choroidal vein, or the inferior ventricular vein (Figs. 25 and 26).

The posterior tributaries are made up of the internal occipital veins which usually join the great vein of Galen but which can also drain into the basal vein (Fig. 34).

The superior tributaries are formed by the posterior thalamic veins and/or the pineal veins (Figs. 12, 41).

The inferior tributaries are made up of the peduncular veins (pages 5 and 13), the lateral mesencephalic veins (page 14), the pontine veins (pages 5 and 13), the precentral vein (pages 5 and 10), and the quadrigeminal veins (page 12).

d) Veins of the Cerebral Peduncles

The veins of the cerebral peduncles, still called "posterior communicating veins", originate at the level of the lateral mesencephalic sulcus and course forward and downward around the anterior face of the cerebral peduncle. They bend their course toward the midline, at which point they join. The posterior communicating veins connect with the longitudinal veins from the pons at the level of the posterior perforated space in the interpeduncular cistern. They link transversally the two basal veins and are well visible in a frontal projection, but badly and superimposed in a lateral projection (Fig. 28). Usually the trunk of the veins of the cerebral peduncles outlines the inferior aspect of the cerebral peduncle so that they participate in the general venous outlining of the upper brainstem, as depicted in Figure 27.

e) Variations in the Drainage of the Basal Vein

These variants are described in our department by BABIN

The sinuous courses of the vein's trunk in front of point *L* or behind it, or the whole of the trunk, are rather frequent (Fig. 22).

Megadolichobasal veins can be observed (Fig. 23).

The basal vein can drain the internal cerebral vein instead of the great vein of Galen.

A more or less long intratentorial course and direct drainage in the straight sinus can be observed.
Drainage can also occur in the lateral sinus.
At times the anterior segment of the basal vein forms a part of the *lateral anastomotic mesencephalic vein* (Fig. 29a).
In a lateral view the course can be divided into two segments:

An anterior or anteroposterior segment in front of point *L* corresponding to the anterior segment of the basal vein;
A posterior segment in a craniocaudal direction corresponding to the lateral mesencephalic vein.

Drainage can also take place by means of an incomplete lateral anastomotic mesencephalic vein (Fig. 29b).

When the anterior segment of the basal vein is absent, the lateral mesencephalic vein drains directly into the posterior segment of the basal vein.

3. The Posterior Mesencephalic Vein

The posterior mesencephalic vein remains the object of continued angiographic study. The differences in interpretation of this vein are due to the fact that two different veins can course parallel to the posterior segment of the basal vein (sometimes replacing it). So as to avoid further confusion, we suggest maintaining the term posterior mesencephalic vein in the two cases, as follows:

a) The posterior mesencephalic vein described by HUANG in 1965 can exist at the same time as the basal vein in which case it drains the optopeduncular region (Figs. 32 and 52). In the two variations of drainage described by HUANG, the anterior cerebellar vein and the anastomotic lateral mesencephalic vein, the posterior segment of the basal vein is represented by a venous segment "called" the posterior mesencephalic vein.

b) The inferior ventricular vein or choroidoventricular vein of Lazorthes originates in the choroidal plexus of the lateral ventricle and flows into the basal vein near point "*L*". Along this course it flows above the basal vein. Instead of flowing directly into the basal vein, it can have a recurrent course backward in direction of the great vein of Galen. This segment, parallel to the basal vein, corresponds to the so-called posterior mesencephalic vein.

In a frontal projection, the three veins can be seen: *the posterior mesencephalic vein* (according to HUANG), *the posterior segment of the basal vein*, and *the choroidoventricular vein* which courses obliquely upward and toward the medial line, interiorly to the posterior segment of the basal vein. It corresponds to the external aspect of the cerebral peduncle.

4. The Precentral, Paracentral and Postcentral Veins

a) The Precentral Vein

The denomination of precentral vein can be replaced by that of anterior vermian vein, a more adequate anatomic nomenclature. The angiographic study presented by HUANG and WOLFF in 1966 gives the principal normal radioanatomic characteristics of this vein. The authors especially describe the collicular point (Figs. 5 and 30). The precentral vein arises in the groove which separates the lingula and the lobulus centralis in its lateral right and left parts by two veins whose confluence forms the trunk of the precentral vein.

In this original course, above the superior cerebellar peduncle (brachium conjunctivum), the precentral vein may be anastomosed with the superior petrosal vein through thin veins joining the precentral and the lateral mesencephalic vein (see the drawing of Fig. 42). Some authors have adopted the term of "brachial vein" for the original segment, a term which is confusing since it was also used for the vein which joins the lateral mesencephalic vein to the pontocerebellar angle. In 1971 we proposed [159] maintaining the term "brachial vein" only for the brachial segment of the precentral vein and abandoning this term for the inferior segment of the lateral mesencephalic vein. The two original veins of the precentral vein can remain independent along a certain course and thus form two precentral paramedian trunks. In a frontal projection the precentral vein describes an inverted "Y" whose two arms delimit the roof of the fourth ventricle (Fig. 30). The medial branch corresponds to the principal venous trunk which flows into the great vein of Galen or into the superior vermian vein. This angiographic image is often badly visible in a frontal projection because of superimpositions and optic telescoping of the vein (Fig. 30).

Laterally, the precentral vein has a characteristic bisegmentary form (originating segment or brachial vein and main trunk). The inferior segment corresponds to the course in the groove between the lobulus centralis and the lingula. The superior segment corresponds to the superior cerebellar vermis. The inferior and superior segments form an angle opening backward, the peak of which corresponds to the summit of the inferior quadrigeminal tubercules (Figs. 5 and 30). This collicular point of HUANG is a very useful landmark. The precentral vein, when it is unique, localizes on the other hand the midline of the posterior fossa. Furthermore, the precentral vein in a lateral view divides the posterior fossa into two areas (anterior and posterior to the vein), so that the anteroposterior displacement, as we shall see, is very useful in practical radiology.

b) The Paracentral Vein

At times one can observe a small satellite vessel of the precentral vein which courses laterally and exteriorly called the paracentral vein. This vessel is of no practical interest in angiography.

c) The Postcentral Vein

The course of this vein lies in the groove between the lobulus centralis and the culmen. It flows into the precentral vein, thus forming the postcentral vein, or it can drain in the superior vermian vein thus forming the preculminary vein. The postcentral or preculminary vein can split into two trunks (Figs. 31 and 40).

5. The Anterosuperior Cerebellar Hemispheric Vein

This vein was named superior internal (superocisternal cerebellar hemispheric vein) by GREITZ and LINDGREN (in 1961), anterosuperior cerebellar vein by ROSA and VIALE (in 1968), and is still named great cerebellar vein by HUANG (1963).

We feel that this vein is of slight importance, as it is infrequent, very variable in size and course, and tends to confuse the reader, as seen in Figures 29 and 31. This vein may have a large diameter and drains into the great vein of Galen. It thus differs from the other hemispheric veins which flow into the lateral sinus, the superior petrosal vein, or the inferior vermian vein. When hypertrophied the superointernal cerebellar hemispheric vein is in fact a drainage variant of the basal vein and corresponds in its vertical segment to the lateral mesencephalic vein and in its anteroposterior segment to the posterior segment of the basal vein (Figs. 32 and 52). Another variant is illustrated in Figure 32b where the equivalent of the lateral mesencephalic vein flows in front into the optopeduncular system.

6. The Internal Occipital Veins

The internal occipital veins drain the inferior internal side of the occipital lobe toward the great vein of Galen. In accordance with the general nomenclature, we propose to call internal occipital veins the entire system which includes the precunian, cunian, and calcarine veins.

These veins are visible when the posterior cerebral artery is well opacified either via the carotid or the vertebral artery.

a) The precunian vein is the most anterior one. It drains the territory of the precuneus and runs in a lateral projection upward and forward to join Galen's vein. It rarely drains directly into the posterior cerebral vein or limbic vein to form the cuneus limbic vein (TESTUT).

b) The cuneocalcarine vein or internal occipital vein originates in the junction of the cunian and calcarine veins and joins the great vein of Galen. It joins more rarely the basal vein. These veins are generally well opacified, they are clearly identified on lateral projections (Fig. 34), and less on frontal projections.

7. The Posterior Pericallosal Vein (Posterior Cerebral Vein)

The posterior cerebral vein, called also the posterior marginal vein of the corpus callosum, originates on the dorsal aspect of the truncus corporis callosi, surrounds the splenium, and terminates in the great vein of Galen. The vein of the quadrate lobe joins it and the two veins form the cuneolimbic vein. This vein drains the posterior part of the corpus callosum, the splenium, and the quadrate lobe. The normal radiologic anatomy has been studied by JOHANSON.

The posterior cerebral vein comprises two segments:

The anteroposterior segment, parallel to the inferior longitudinal sinus, is frequently split in two.

The terminal segment courses around the splenium under the form of a single trunk and flows into the great vein of Galen.

In some cases the vein continues forward to join the anterior cerebral vein (Fig. 33). We have to determine as a test of normality the distance between the course of the internal cerebral vein and the upper segment of the posterior cerebral vein (average 1.5–2 cm). When deep cerebral and callosal tumors increase this distance, it indicates infiltration of the splenium.

8. The Superior Thalamic Veins

Thalamic veins are invisible in a frontal view, whereas easily identifiable in a lateral view. The superior thalamic vein courses below the internal cerebral vein and joins the great vein of Galen, close to its confluence with the internal cerebral vein. At its origin, it presents a characteristic thickening due to the confluence of several slender veins. The superior thalamic vein remains quite distant from the internal cerebral vein, approaching it progressively in its posterior course. Quite often the two thalamic veins are recognizable, but for reasons of geometric projection, they are not superimposed. The superior thalamic vein appears early in the venous phase and remains opacified a few seconds longer than the superior choroidal vein (Figs. 12, 16, 34). The superior thalamic vein can be replaced by several small undefined veins in a phlebogram. Beyond the superior thalamic vein, one can distinguish more rarely other small anterior and inferior thalamic veins.

9. The Superior Vermian Vein

The superior vermian vein is formed by a single medial trunk: it is the equivalent of the precentral vein at the culmen and of the postcentral vein at the lobulus centralis. This vessel is sometimes called the supraculminate vein. It could also be called the middle vermian vein, as opposed to the anterior vermian vein, or the precentral vein. This vein is always clearly visible in a lateral view, as there are few bony and venous overlying structures in its area (Fig. 35). It is of great topographic interest as it delimits the anterosuperior vermis and is subject to modifications in cases of intracranial hypertension (Figs. 36–39).

In a frontal view it is hidden by the other medial veins. The superior vermian vein arises from affluents coming from the declive and the culmen, as well as from neighboring areas of the cerebellar hemisphere. These affluents form an arched trunk which runs along the convexity of the culmen from which it also receives affluents.

It drains the great vein of Galen, more rarely Rosenthal's basal vein. In certain cases, the vermian vein joins the precentral vein to form a common trunk (Fig. 30a) which terminates in the great vein of Galen. In other cases, it anastomoses with the inferior vermian vein, thus delimiting the whole of the vermis (Figs. 36, 63). The superior vermian vein delimits, with the straight sinus and the great vein of Galen, a triangular surface defined in a lateral projection. This region represents the subdural supravermian space and is not visualized in any other examinations. In effect, it is not bounded by arteries nor is it airfilled in pneumoencephalography (Figs. 36–39). The height of this space depends on the more or less medial localization of the superior vermian vein (Fig. 38). It is probable that the air-filling of the subdural space which can occur during a pneumoencephalography corresponds to the progressing of the air along the arachnoid orifice which gives passage to the superior vermian vein (Figs. 38c and 39).

10. The Veins of the Quadrigeminal Plate

Several veins drain the quadrigeminal tubercules. These veins flow into the basal vein, the great vein of Galen or into the precentral vein. These veins, which are of slight caliber, are not distinguishable in a frontal view. On the contrary lateral projections permit a good viewing of these veins, especially when the projection is enlarged directly, with or without subtraction. Several veins are very often distinguished in the triangle formed by the basal vein or the posterior mesencephalic vein above and the precentral vein below (Figs. 5, 37, 40b, c).

11. The Veins of the Pineal Gland

These veins are very difficult to distinguish from the posterior thalamic veins. According to their topography, it is probable that they often form a common trunk with the posterior thalamic veins, the trunk of which flows into the vein of Galen as demonstrated on lateral projections (Figs. 12, 37, 41, 52).

IV. The Petrosal Veins

1. The Superior Petrous Vein

The superior petrosal vein has a caliber of 0.1–1 mm and an average length of about 1 cm. It is located in the external region of the cerebellopontine cistern, outside the trigeminal nerve and above the eighth cranial pair, of which it is a constant and trustworthy landmark. It drains a very large area, including the anterior parts of the cerebellum, the posterior part of the medulla and of the mesencephalon, and the anterior part of the pons and medulla oblongata. This very important vein, described by MERKEL in 1885, is also called the superior floccular vein (POIRIER, 1921), and the superior lateral cerebellar vein (LAZORTHES, 1949), but is best known under the name "Dandy's vein" (1929).

The superior petrosal vein is the most important lateral venous confluent and represents ipso facto an important angiographic landmark. The opacification of these veins is obtained regardless of the vertebral artery injected, so that generally both veins, the left and the right, are clearly visible in a frontal projection. The superior petrosal vein drains into the superior petrosal sinus. The topographic importance of this vein is due also to its affluents, which converge forming a characteristic stellar pattern.

These affluents are the:

Transverse pontine veins.

Lateral medullopontine vein together with the vein of the lateral recess of the fourth ventricle.

Superior hemispheric cerebellar veins and especially the vein of the horizontal fissure of the cerebellum, which is considered as the main affluent from the anatomical point of view by M. DUVERNOY (1975).

Brachial vein (vein of the brachium conjonctivum) (Fig. 42).

Lateral mesencephalic vein.

The superior petrosal veins are well set in evidence in a frontosuboccipital projection (Towne's projection). A lateral projection is less interesting, since there is on the one hand a superimposition of the two veins and on the other hand, a petrosal opacity which is not always eliminated by subtraction. The localization of this vein is facilitated by the identification of the point of convergence of its affluents. The method illustrated in Figure 42 shows the superior petrosal vein situated approximately 1.5 cm in front of the midpoint of Twining's line. We shall raise the question of the topographic importance of this vein in the drainage of mass expansions in the cerebellopontine angle.

2. The Inferior Petrous Vein

In adults the petrosal vein appears in only 30% of cases, as its anatomical presence is inconstant. It originates in the confluence of the vein of the great horizontal fissure and the vein of the lateral recess of the fourth ventricle which, instead of flowing into the superior petrosal veins, drains into the inferior one. Their junction occurs at the level of the superior pole of the flocculus in the depth of the great horizontal sulcus of Vicq d'Azyr.

The inferior petrosal vein courses obliquely forward toward the midline, crosses the lateral part of the prepontine cistern, and flows into the inferior petrosal sinus, along the glossopharyngeal nerve through the jugular foramen. At times the inferior petrosal vein is made up of a transversal vessel which loops forward and joins directly the inferior petrosal sinus. In our cases it seems that the inferior petrosal veins are very rarely seen in adults, as opposed to children (Fig. 43c) where the petrous bones are translucent.

V. The Superficial Veins of the Pons

(Figs. 44–49)

These veins form a network which covers the anterior and lateral aspects of the pons, and drains especially the superior petrosal vein (Fig. 2) while also communicating with the basilar plexus.

Two groups of veins are distinguishable:

A transversal or laterolateral distribution.

A longitudinal or craniocaudal distribution.

1. The Transversal Pontine Veins

join the left and right cerebellopontine angles. In general they go in pairs. Among them we distinguish:

a) A superior transversal pontine vein, which can result in a medial emissary vein draining into the clival plexus. This vein corresponds to the vein of the pontopeduncular sulcus and generally drains in the superior petrosal vein or in the optopeduncular veins.

b) An inferior transversal pontine vein of a smaller caliber than in (a), joining the two lateral medullopontine veins which contribute to the formation of the superior petrosal vein. This vein can split and sometimes drain the inferior petrous sinus. It may correspond to the medullopontine sulcus.

c) A medial transversal pontine vein is inconstant and joins with the superior petrous veins in a more or less parallel course to the superior and inferior transversal veins.

2. The Longitudinal Pontine Veins

Longitudinal veins which connect the transversal veins, forming the prepontine venous network.

Longitudinal veins which anastomose downward the inferior transversal pontine vein and the prebulbar veins.

Longitudinal veins which anastomose with the superior transversal pontine vein upward with the optopeduncular venous system.

Prepontine veins are thus represented by a checkered network in which transversal veins dominate what is anatomical and functional. In a minority of specimens (20% for TOURNADE), it is the longitudinal system which predominates, at which point one can observe an important medial longitudinal vein which delimits the anterior aspect of the pons Varolii (Fig. 48) in angiography.

The prepontine venous network is especially visualized in angiographic lateral projections, where one can clearly distinguish longitudinal veins having anterior convexity. They do not always correspond to the most anterior point of the pons, since the view obtained can be due to longitudinal pontine veins in a paramedial position.

Lateral projections also show superior and inferior transversal veins, the inferior transversal vein usually being badly visible since hidden by petrosal bones. Anastomosis of a longitudinal vein with the optopeduncular system is common. Thanks to subtraction, one can witness the confluence of the superior veins with the superior petrosal vein and the inferior transversal pontine vein with the lateral medullopontine vein. Frontally, the venous views are less clear (Fig. 46). They are often characterized by sinuous veins having a superior concavity which join the two superior petrous veins. We shall see that the information obtained by a frontal projection is of lesser interest in pathology, whereas lateral projections are always of great interest.

VI. The Lateral Mesencephalic Veins

The lateral mesencephalic vein courses in the lateral mesencephalic sulcus. It connects the cerebellopontine angle (superior petrous sinus or superior petrosal vein) with the basal vein. In the cerebellopontine angle the origin of the vein generally arises in the confluence of three veins: the superior petrous vein, the cerebellar hemispheric vein, and the transversal pontine vein. At the level of the basal vein the junction with the lateral mesencephalic vein is generally localized behind the point "*L*". In rare cases the lateral mesencephalic vein has an anastomotic function between the basal venous system and the superior petrosal venous system. In these cases the lateral mesencephalic vein has a larger and an invariable caliber (Figs. 50–52).

The lateral mesencephalic vein is a very important lateral landmark because it divides the mesencephalic area into two main parts: the posterosuperior part (quadrigeminal plate) and the anteroinferior part (basis of the cerebral peduncle). This vein permits the determination of the lateral projection of the venous limits of the quadrigeminal plate as demonstrated in Figure 80. The lateral mesencephalic vein marks laterally the anterior margin of the cerebellum in a fashion similar to the precentral vein which marks the limit on the midline.

There is some confusion in the terminology of the inferior segment of the lateral mesencephalic vein because of its connection to the brachium pontis (see below). In order to clarify the terminology, we propose to distinguish two segments of the lateral mesencephalic vein.

Firstly, *the superior segment,* which has already been called the lateral mesencephalic vein by HUANG (1963) and by ROSA (1968). This segment runs in the ambiant cistern and reaches the lateral mesencephalic sulcus where it delineates the tegumentum and the peduncle.

Secondly, *the inferior segment* which runs on the superior and the anterior aspect of the brachium pontis and is therefore called the brachial vein by HUANG. In 1971 we proposed to abandon the denomination of "brachial" vein. It may be simpler to call this part "the inferior segment" of the lateral mesencephalic vein. Normally it is lambda or trident-shaped in frontal as well as in lateral views. The two branches of the lambda are connected respectively with the superior petrous vein and with a cerebellar hemispheric vein, generally the vein of the great horizontal fissure. When there is a third branch, this is connected with the pontine veins. The lateral mesencephalic vein is not evident in all patients but may be seen in about 30% of presumably normal vertebral angiograms. In this chapter we report variations in shape and connections of the lateral mesencephalic vein. Let us first recall the variation which HUANG described in 1963 and which he called the lateral anastomotic mesencephalic vein. We propose adding the following 6 variations (Fig. 50):

1. Strong predominance of the connection with a cerebellar hemispheric vein, whereas the connection with the petrous vein is hypotrophied but still present.
2. Connection with the precentral vein.
3. The lateral mesencephalic vein originates in the posterior communicating vein.
4. Exclusive connection with the pontine veins.
5. The anterior branch of the lambda, instead of joining the petrous vein, is connected to the pontine vein.
6. Trident-shaped lateral mesencephalic vein connected, respectively, to the pontine vein, the petrous vein, and the cerebellar vein.

One of the most important studies on the lateral mesencephalic vein and its practical use in angiographic diagnosis was published by our pupil BILLEWICZ [15b].

Some practical rules he set up follow:

1. In a frontal projection the inferior segments of the lateral mesencephalic veins have a symmetrical position with respect to the midline.
2. The inferior segment of the lateral mesencephalic vein forms an angle with the horizontal level of both petrous upper aspects. This angle is variable, depending on the orientation of the central beam, but is identical on both sides (variations of 0–5°). Pathologic variations due to unilateral hemispheric tumors are responsible for differences of 10–20° between the left and right side.
3. In a frontal projection the distance between the left and right lateral mesencephalic vein (superior segment) allows the determination of the transversal diameter of the midbrain. The average value of this distance is 26 ± 3.7 mm. In brainstem tumors this distance can measure up to 32 mm.
4. In a lateral projection BILLEWICZ measured the distance between the precentral and the lateral mesencephalic vein. The normal average value of this distance is 10 ± 1.2 mm. In cerebellar hemispheric tumors the distance decreases. In cases of brainstem tumors the distance corresponds to the anteroposterior diameter of the quadrigeminal plate, as we demonstrated in Figure 80.

VII. The Cerebellar Hemispheric Veins

The surface veins which course on the cerebellar hemispheres can be divided into 3 groups (Figs. 57 and 58).

1. The Anterior Group

These veins drain toward the superior petrosal vein and the superior petrosal sinus, rarely in the great vein of Galen.

2. The Posterior Group

These veins course into the lateral sinus. The vein of the great horizontal fissure of the cerebellum (Fig. 68) is part of this group. This vein is an affluent of the superior petrosal vein. These veins often follow a para- and intratentorial course parallel to the lateral sinus (Fig. 11a) or have direct communications (Fig. 11b). Some of these veins flow into the inferior vermian vein (Fig. 57).

3. The Medial Group

These veins follow a horizontal course. They drain into the paravermian and inferior vermian veins. This venous network, whose functional importance is slight, is drained by the cerebellar cortex. Phlebographic diagnosis is simple, as these vessels are numerous and generally have a high caliber. In a frontal view, they are located in the external and superior part of the posterior fossa.

In a lateral view, however, they follow an anterior and superior direction.

VIII. The Veins of the Medulla Oblongata and of the Fourth Ventricle

1. Anatomical Study

The study of the veins of the bulb, undertaken within the framework of their systematization, permits us to distinguish anterior, posterior, and lateral bulbar veins.

a) The Anterior Bulbar Veins

The anterior bulbar veins are formed by the anteromedial bulbar vein, the anterolateral bulbar veins, and by the superficial pyramidal network.

Inasmuch as the superficial pyramidal network is made up of venules of low caliber, only the anterior and medial bulbar vein and the lateral bulbar veins are likely to be seen in a phlebography.

α) The anterior and medial bulbar vein (Fig. 56) anastomosed at its lower extremity with the anterior spinal vein, courses in the medial sulcus of the bulb, which it covers, up to the level of the foramen caecum of Vicq d'Azyr, where it separates into the two anterior medullopontine right and left veins (Fig. 56). These anterior medullopontine veins course on the surface of the bulbopontine sulcus to the level of the lower pole of the cerebellopontine angle where they contribute to the formation of the lateral medullopontine vein (Figs. 55, 56).

β) The anterior and lateral bulbar veins, whose characteristic morphology permits their phlebographic identification, are represented by the preolivary veins and by the anterior transversal veins of the medulla oblongata.

The preolivary veins (Fig. 56), on the right as well as on the left, follow the preolivary sulcus, and course along it from the first cervical root up to the lateral fossette where they anastomose with the anterior bulbopontine veins and the corresponding lateral bulbar veins.

The anterior transversal veins of the medulla oblongata (Figs. 55, 56) run along the anterior aspect of the bulbar pyramids in two or three venous trunks, to the right as well as to the left. These veins terminate, certain ones at the middle section of the preolivary vein, others at the level of the origin of the preolivary vein, and then continue as the anterior radical satellite vein of the first cervical root (Fig. 56).

b) The Lateral Bulbar Veins

The lateral bulbar veins are represented by the lateral medullopontine vein, by the retroolivary veins, and by the radical satellites of the mixed nerves.

α) The lateral medullopontine vein (Figs. 55, 56) originates at the level of the lower pole of the pontocerebellar angle.

Each lateral medullopontine vein issues from the junction of the anterior medullopontine vein with the vein of the restiform body and the pre- and retrohomolateral olivary veins. Each voluminous lateral medullopontine vein crosses the pontocerebellar angle to anastomose with the pontine and cerebellar veins.

β) The retroolivary veins (Fig. 55) make up the venous circle of the great hypoglossal nerve. They anastomose generally with the upper and lower extremities of the preolivary veins, coursing along the paraolivary fossette in an upward and outward oblique direction.

γ) The radical satellite veins of the IXth, Xth, and XIth nerves. As a general rule, the radical satellite veins of the IXth, Xth, and XIth nerves course along on the anterior aspect of their corresponding nerve.

The most voluminous of these veins is the satellite vein of the glossopharyngeal nerve which runs into the vein of the restiform body.

Venous drainage of the medulla oblongata follows an ascending course, represented by the lateral medullopontine vein, and a descending course, constituted by the radical satellite vein of the first cervical root.

c) The Posterior Bulbar Veins

Phlebographic interpretation can be applied solely to the posteromedial and the posterolateral bulbar veins. The posterior tributary veins cannot be seen because of their low caliber.

α) The posterior and medial bulbar vein (Fig. 56) at the level of the bulb prolongs the posterior spinal trunk and courses on the surface of the posterior medial sulcus, up to the level of the obex where the posterior and medial bulbar vein divides into two bordering veins of the floor of the fourth ventricle (Fig. 56).

These veins take an upward and outward oblique direction, reaching the bulbar fossette where they anastomose with the lateral medullopontine vein (Fig. 56).

A vein, anastomosed with the posterior occipital sinus through the cisterna magna, is frequently observed around the middle section of the posterior and medial bulbar vein.

β) The posterior and lateral bulbar veins are made up of the two veins of the restiform bodies (Fig. 56) which emerge from the lower section of the posterior and medial bulbar vein. They course along, in an upward and outward oblique direction, behind what appears to be the origin of the spinal, pneumogastric, and glossopharyngian nerves. Subsequently, each vein of the restiform body, curving forward and upward, reaches the lateral fossette of the bulb and anastomoses with the anterior medullopontine vein to form the lateral medullopontine vein.

2. Radiographic Study

Whereas an anatomical study of the bulbar veins provides us with a precise morphology and topography, angiography runs into numerous difficulties. The medulla oblongata is a zone of transition in the circulatory system between medullary and pontine vascularization.

Contrast medium injected into the vertebral system is diluted by the blood flow. In addition to these circulatory conditions, the low caliber of these veins permits only a partial and faint opacification of these vessels. Subtraction is indispensable, since this fine venous network, subject to feeble contrast, is screened by the thickness of the skull in this area.

A frontal projection permits in certain cases differentiation in the venous system of the medulla oblongata, while a lateral projection reveals only the anterior trunks, and rarely the posterior trunks.

In order to interpret a bulbar phlebogram, it is necessary to individualize certain venous elements of sufficiently high caliber and whose topography is known.

a) The Lateral Medullopontine Veins (page 13), formed by the junction of the olivary veins, the restiform bodies, and the marginal veins of the fourth ventricle, course along the lateral border of the medulla oblongata and the pons in a slightly oblique vertical and outward direction. They flow toward the upper petrosal junction, a region which one can almost always locate and which is a venous crossroads of the utmost importance.

b) The Medial Bulbar Vein flows among them in a vertical direction. When this medial vein communicates with the marginal veins of the fourth ventricle, it can be located on the posterior aspect. When it communicates with the transversal medullopontine vein, it courses on the anterior aspect.

c) The Transversal Medullopontine Vein joins the two lateral medullopontine veins following an angular convex downward course or a horizontal course. At times a fine venous network connects the different elements among themselves.

d) The Marginal Veins of the Fourth Ventricle follow an oblique course and join together on the medial line in the posterior medial bulbar vein. They are flanked exteriorly by the veins of the restiform bodies which also join together on the medial line. Drainage of these veins occurs toward the pontocerebellar angle.

IX. The Vein of the Lateral Recess of the Fourth Ventricle

1. Anatomical Study

Classical authors have neglected the vein of the lateral recess of the fourth ventricle, which has been the object of the attention of contemporary authors such as GOMEZ OLIVEIROS [58], HUANG and WOLF [67–75], and ROSA and VIALE [112–120, 137–147]. The study of this vein, which is a cerebellar one, falls into the framework of the general systematization of the veins of the cerebral trunk, since in its final lap it is a tributary of the petrosal venous system, of which it is one of the principal affluents at the level of the cerebellopontine angle. The vein of the lateral recess of the fourth ventricle anastomoses most frequently with its homologue in front of the anterior aspect of the cerebellar nodulus. The venous anastomotic loop thus formed outlines quite regularly the anterior convexity of the nodulus. The two right and left veins thus constituted are symmetrical in course and caliber.

Each vein of the lateral recess follows a concave course forward and oblique upward and outward, thus leading it to the anterior aspect of the middle cerebral peduncle where it terminates by becoming the direct or indirect affluent of the superior or inferior petrosal veins. In the initial course, it runs successively in front of the nodulus, then of the cerebellar tonsil, but remains behind the choroidal plexus of the fourth ventricle.

In the second part of its course, it inserts itself below and behind the posterior velum medullare thus entering a deep groove which separates the peduncle from the flocculus in front, and from the digastric lobulus below. Hence it arrives at the external extremity of the lateral recess of the fourth ventricle where it is entirely covered by the flocculus and the digastric lobulus.

In the third part of its course, this vein reaches the anterior face of the middle cerebellar peduncle, at the level of the internal extremity of the great horizontal fissure. This vein terminates by draining directly into the inferior petrosal vein or by joining the great circumferential sulcus of Vicq d'Azyr. During its course, the vein of the lateral recess of the fourth ventricle increases in caliber. At its origin it measures about 80 microns, while measuring 300–400 microns at its end. Different affluents run into the vein of the lateral recess along its course. At its origin this vein anastomoses on both sides with each of the two inferior vermian veins, later on it receives veins coming from the plexus of the fourth ventricle, as well as vein outlets from the lamina of the digastric and frail lobuli, from the cerebellar tonsil and from the flocculus.

In the terminal part of its course, the vein of the lateral recess of the fourth ventricle sometimes presents a quite different distribution. It can, in fact, bend partially forward and then course along the internal and superior edge of the flocculus, passing behind the statoacoustic nerve, where it joins the lateral medullopontine vein.

The vein of the lateral recess of the fourth ventricle, incorrectly named since it remains constantly outside the fourth ventricle, is a satellite cerebellar vein of the posterior velum medullare. As it terminates in the cerebellopontine angle, it participates most often in the creation of the original trunk of the inferior petrosal vein, and represents one of the principal venous cerebellar affluents at the venous confluence of the cerebellopontine angle.

2. Phlebographic Study

During the phlebographic phase of vertebral angiography, the vein of the lateral recess of the fourth ventricle appears under three different morphologic aspects:

The two veins forming a common trunk by their junction on the medial line are opacified.

The vein appears opacified on one side.

Anastomosis of the vein of the lateral recess of the fourth ventricle with the inferior homolateral vermian vein is usually visible unilaterally.

a) Morphologic and Topographic Study of the Vein of the Lateral Recess of the Fourth Ventricle in a Frontal Projection

α) The opacification of each of the two veins of the lateral recess of the fourth ventricle is characterized by a sinuous vascular course, in a transverse direction to the inferior pole of each of the pontocerebellar angles. Anastomosis of the two veins shows up as a sharp descending angle in front of the anterior aspect of the cerebellar nodulus. In this particular projection, the three segments of the course of the vein of the lateral recess of the fourth ventricle are readily visible. The initial segment takes a generally oblique direction downward and outward; the two other segments form a concave sharp angle upward, corresponding to the drainage of each of the two veins toward the petrosal venous system.

β) When the vein of the lateral recess of the fourth ventricle anastomoses with the homolateral inferior vermian vein, there is generally no opacification of the opposite vein, and its course seems to continue directly through the vein of the lateral recess of the fourth ventricle.

γ) Unilateral opacification of the vein of the lateral recess of the fourth ventricle includes an important injection of the superior and inferior petrosal venous system. In this particular frontal projection, it is important to dissociate the opacification of the veins of the lateral recess of the fourth ventricle from that of the posterior communicating veins of the transverse veins of the pons, the topographic characteristics of which are as follows:

The veins of the posterior communicating system are situated higher than the vein of the lateral recess of the fourth ventricle and have a morphology whose double superior concavity is more regular, while inverted with reference to the course of the vein of the lateral recess of the fourth ventricle.

The transverse veins of the pons are situated lower in an angiogram and their asymmetrical morphology is characterized by a shorter transverse course. On the other hand, their direction tends to a slightly superior concavity.

b) Morphologic and Topographic Study of the Vein of the Lateral Recess of the Fourth Ventricle in Lateral Projection

The vein of the lateral recess of the fourth ventricle is characterized by:

α) Its Morphology: It consists of an anterior and posterior segment. The posterior or initial segment localizes the lateral part of the roof of the fourth ventricle (Fig. 63). The terminal, anterior part prolongs the first segment at an angle of about 30°, then continues in the lateral medullopontine vein which follows a direction quite parallel to the clivus up to the level of the petrosal venous system. In Figure 63 we include a drawing of the main types of morphology encountered in our cases, both in frontal and in lateral projections.

β) Its Topography: The inital part of the vein of the lateral recess of the fourth ventricle, on a lateral view, is localized about 1 cm in front of the midpoint of a line which joins the internal occipital protuberance to the tuberculum sellae (Fig. 55). It is particularly important to demonstrate the initial segment of the vein of the lateral recess of the fourth ventricle, which is often displaced by a mass in the posterior fossa. The point of junction of the vein of the lateral recess of the fourth ventricle and the lateral medullopontine vein lies on this vertical line dropped from the midpoint of Twining's line (Fig. 55). With the idea of establishing a mark, a metal wire was affixed on an anatomical specimen intended to demonstrate the characteristic projection of the vein in frontal and lateral projections, to the right and to the left (Fig. 59).

X. The Lateral Medullopontine Vein

The lateral medullopontine vein rises in the area of the lateral groove of the medulla oblongata at the junction of the anterior medullopontine vein, the pre- and retroolivary veins, and the vein of the restiform body. The vein's course follows an oblique upward, forward, and outward direction, attaining the inferior pole of the cerebellopontine angle. The vein is situated in the pontofloccular groove and passes behind and beyond or between the facial and statoacoustical nerves, reaching the internal and anterior part of the great horizontal cerebellar fissure. It straightens its course upward, arriving at the superior part of the cerebellopontine angle, below the origin of the trigeminal nerve. It is at this level that the lateral medullopontine vein rejoins those other veins which participate in the formation of the superior petrosal vein. The lateral medullopontine vein is related to the vein of the great horizontal cerebellar fissure, which courses on the anterior aspect of the middle cerebellar peduncle. The caliber of the lateral medullopontine veins is irregular, one vein having a third or a fourth of the diameter of the next. The affluents of the lateral medullopontine veins are:

The vein of the lateral recess of the fourth ventricle.
The inferior transversal pontine vein.
The vein of the great horizontal cerebellar fissure.
And the radicular veins, satellites of the facial and statoacoustical nerves.

In an AP phlebographic view, the lateral medullopontine vein follows a lateral, oblique, and outward direction toward the superior petrosal vein. Anastomosis with the vein of the lateral recess of the fourth ventricle is often observed in the lower segment of its course (Fig. 55).

In a side view, the course of the lateral medullopontine vein follows an almost vertical direction and connects the lateral segment of the vein of the lateral recess of the fourth ventricle to the superior petrosal vein. It is impossible to identify the opacified side in this view, as the veins are superposed. This vein is compressed and shifted forward in cases of tumors of the fourth ventricle and of the cerebellopontine angle.

In Figures 55 and 63 we present a drawing of the main types of lateral medullopontine veins demonstrated by subtraction.

XI. The Inferior Vermian Vein

When in a medial position in the posteroinferior region of the posterior fossa, there is a constant single or double vein called the vermian vein. When the vein is situated laterally on the internal face of the cerebellar hemispheres, it is called the paravermian vein (Figs. 57, 68).

The paravermian vein drains the vein of the lateral recess of the fourth ventricle when the latter does not communicate with the contralateral vein. The origin of the inferior vermian and paravermian veins is formed by the peritonsillar network. This network (Fig. 70) is made up of the superior, lateral, and inferior tonsillar veins, as well as the veins of the digastric lobule. In practice one distinguishes the superior and inferior retrotonsillar veins in a lateral projection of the phlebogram in a vertebral angiography. Some variations are visible in the drawing given in Figure 69.

The position of the tonsillar veins is a source of information on the morphology of the cerebellar tonsil, while the topographic data furnished by the copular point, the point of convergence of the tonsillar veins, is hardly reliable. According to Huang, the copular point corresponds to the lamina of white substance which joins the cerebellar pyramid to the digastric lobule at the bottom of the paravermian sulcus in front of and outside of the cerebellar tonsil. The position of this point of convergence of the superior and inferior posterior tonsillar veins varies greatly and has been located up to 2 cm above the inferior pole of the tonsil. We attach little importance to the location of the copular point proper and feel that it is of greater importance to study the angle formed by the superior retrotonsillar vein with the trunk of the inferior vermian vein. This angle is equal to or superior to 90° (Figs. 64, 65, 69, 70) and diminishes in cases of mass expansions. The trunk of the inferior vermian vein courses at some distance of the paravermian sulcus on the posterior face of the small digastric lobulus, then on the semilunar lobulus, terminating at the level of the lateral sinus at a variable distance from the junction of the straight sinus into the torcular. Along this paravermian course the vein follows the internal face of the cerebellar lobe and at times even its posterior face. These veins are generally even. More rarely the vein leaves the internal face of the cerebellar hemisphere to course on the so-called posterior vermis, as we demonstrated earlier (Fig. 4).

It is only in this case that the vein deserves the name of inferior vermian vein and becomes a landmark on the anterior face of the great cistern. The inferior vermian vein anastomoses with the vein of the lateral recess at its origin. Less frequently the inferior vermian vein constitutes a direct anastomosis with the superior vermian vein and describes thus a perivermian circle (Fig. 79). The network of the inferior vermian vein can furnish information on the topography of the anteroinferior vermis owing to its anastomosis with the vein of the lateral recess of the fourth ventricle. Further it delimits the posterior and superior part of the inferior vermis when it is in a medial position. The paravermian veins are not of great use in localizing the nervous structures since they course, as we have seen, either on the internal face or on the posterior face of the cerebellar hemisphere. It is easy to distinguish the so-called vermian localization, or paravermian, of this constant venous trunk. A frontal projection permits the identification of the medial localization of the vermian form and the paramedial localization of the paravermian form.

In spite of its variations in course, the inferior vermian vein may be an interesting landmark in the detection of lateral shiftings by tumors of the cerebellum. We shall refer to this in the chapters on pathology. Both the superior and inferior retrotonsillar veins can be seen only through subtraction in a frontal view. A satisfactory outline of these veins in such a view is rare (Fig. 69a) but very useful, especially in cases of intraparenchymatous tumors of the medulla oblongata.

XII. The Vein of the Tentorial Margin

The vein of the free margin of the tentorium is a variant of the venous systematization and connects the straight sinus to the superior petrosal vein (Fig. 67). While of no special significance, it should not be misinterpreted in a frontal view and should be differentiated from a basal or posterior mesencephalic vein.

XIII. Venous Drainage of the Craniocervical Junction

The transitional cervico-occipital area is covered by a heavy venous network which consists of a highly important functional blood reservoir. The occipital muscular veins are tributaries of the posterior jugular vein. Other drainage channels are situated in depth and flow into the vertebral and internal jugular venous system.

The area of the foramen magnum is drained by three principal channels (Figs. 73–78).

1. The Internal Jugular Vein

which follows the lateral and the inferior and superior petrosal sinuses. This vein collects the affluents of Virchow's retrobasilar venous plexus. On the other hand, the internal jugular vein drains the collateral and anastomotic veins in the cervical region and joins the external and anterior jugular veins in the subclavian vein.

2. The Posterior Jugular Vein

is formed by the mastoid emissary vein through which it anastomoses with the lateral sinus. The bony canal formed by this emissary vein is single or double. Its course and diameter are very variable. Generally it is situated at the level of the curve formed by the lateral sinus. The posterior jugular vein collects the occipital and cervical muscular veins.

3. The Vertebral Vein

is formed by three principal affluents:

a) The anterior condylar vein, which forms the anterior root of the vertebral vein, courses into the anterior condylar canal. According to Stolic, topographic variants can be distinguished in these veins, which are diagrammed in Figure 73:

The inferior petrosal sinus flows into the internal jugular vein.
Absence of the emissary condylar vein.
Anastomosis between the inferior petrosal sinus and the anterior condylar vein.
The inferior petrosal sinus flows into the anterior condylar vein.
Anastomosis between the clivus plexus and the inferior petrosal sinus.

b) The posterior condylar vein is not as easily distinguishable as the anterior condylar vein. It joins the posterior section of the plexus of the foramen magnum to the vertebral vein, thus forming the posterior root. During its course it perforates the atlanto-occipital membrane.

c) The emissary condylar vein, present in 30% of cases, is usually to one side, and connects the internal jugular vein to the vertebral vein while crossing the occipital bone via a small canal.

4. Anastomoses

In the intracranial space, anastomoses can develop along the lines of the diagram of Figure 73d depending on the presence or absence of the emissary condylar mastod vein.

In the exocranial space, anastomoses cause the junction of the three important drainage channels (Figs. 73b, c).

Anastomotic spinal plexuses exist which join the two vertebral veins by transversal anastomoses.

XIV. Venous Landmarks as Observed in Normal Vertebral Angiography

The veins course along the cortex outlining its contours while providing characteristic angiographic landmarks of the nerve surfaces, and by the same token delimiting the cisterns. Thus venous topography delimits the cerebellar vermis, the mesencephalon, the pons and the cerebral peduncle of the quadrigeminal plate, the interpeduncular cistern, and the pontocerebellar angle.

1. The Copular Point of Huang

The cerebellar vermis is surrounded by a venous network (Figs. 79 and 102). The superior vermian vein outlines faithfully the contours of the superior vermis, while the inferior vermian vein, when it courses on the midline, gives the topography of the inferior vermis. It must be remembered that the paravermian veins course along the posterior or internal face of the cerebellar lobes, in consequence they can offer no indication on vermian topography. The superior and inferior posterior tonsillar veins, which are at the origin of the inferior vermian vein, provide us with tonsillar morphology and topography. The point of confluence of the tonsillar network, called copular point by HUANG, has a variable topography and cannot be used as a landmark. On the other hand, the peritonsillar venous network gives us the morphology and topography of the tonsils (Fig. 70). It is important to analyze the angle formed by the venous trunk of the inferior vermian vein and the superior posterior tonsillar vein. This angle measures nearly 90° in normal cases and decreases in cases of dilatation of the fourth ventricle (Figs. 69, 107, 125).

2. The Anteroinferior Cerebellar Vermis

is delimited in a lateral projection by the initial segment of the vein of the lateral recess of the fourth ventricle. This vein projects itself at the level of the lateral part of the roof of the fourth ventricle anastomosing with the supratonsillar vein. This area corresponds to the nodulus of the cerebellum also outlined in gas encephalography.

3. The Isthmus of the Mesencephalon

is limited by the vertical portion of the course of the lateral mesencephalic veins (Fig. 80). These veins, as they follow the superior cerebellar peduncle in their horizontal portion, arrive at the superior petrosal vein, which represents the external and superior boundary of the pons.

4. The Lateral Groove of the Medulla Oblongata

The superior petrosal veins receive the lateral medullopontine vein on both sides. These are formed by the anterior medullopontine veins and the vein of the lateral recess of the fourth ventricle. This point of confluence provides the lateral limits of the lateral groove of the medulla oblongata (Fig. 72).

5. The Cerebral Peduncle

In a lateral projection the cerebral peduncle encroaches the area outlined by the following veins (Fig. 80):

The mean segment of the basal vein above
The superior petrosal vein and the transversal pontine veins
The superior veins of the pons below
The peduncular veins forward
The superior segment of the lateral mesencephalic vein behind.

The distance between the inferior segment of the precentral vein and the most posterior point of the lateral mesencephalic vein corresponds to the peduncular cap and measures 11 mm with a focal distance of 1 m.

6. The Pons

The interpeduncular cistern and the pons seen laterally are outlined by the mean longitudinal venous network, especially when a large anteromedial drainage vein is present.

The superior segment of the mean longitudinal vein of the pons forms the posterior and superior contours of the cistern. The superior ridge of this cistern is given by the peduncular veins.

7. The Quadrigeminal Plate

in a lateral projection appears diamond-shaped and is outlined by the following veins (Fig. 80):

The lateral mesencephalic vein forward.
The posterior segment of the basal vein above.
The precentral vein behind.
The superior peduncular vein below. This vein anastomoses the precentral vein with the lateral mesencephalic vein.

8. The Cerebellopontine Angle

is outlined by the following venous elements (Fig. 81):

The superior pole is given by the superior petrosal vein and its affluents: the lateral mesencephalic vein above, the transversal veins of the pons within, and the cerebellar hemispheric veins without.

The external edge is formed by the lateral medullopontine vein which courses upward and outward, draining the superior petrosal vein. The 7th and 8th pairs of cranial nerves are situated within the course of this vein.

The inferior pole is localized on the lowest point of the course of the vein of the lateral recess of the fourth ventricle. In a lateral view this change in direction of the vein corresponds to the lateral part of the inferior segment of the fourth ventricle on a level with the vermian nodulus.

9. The Anteroposterior Localization

of the junction point between the vein of the lateral recess of the fourth ventricle and the lateral medullopontine vein may be of some interest (Wackenheim and Stoffels). This point is easily located thanks to a vertical line dropped on the midpoint of Twining's line. In normal conditions the point is located on this vertical line (Fig. 55) and may be shifted forward or backward under pathologic conditions.

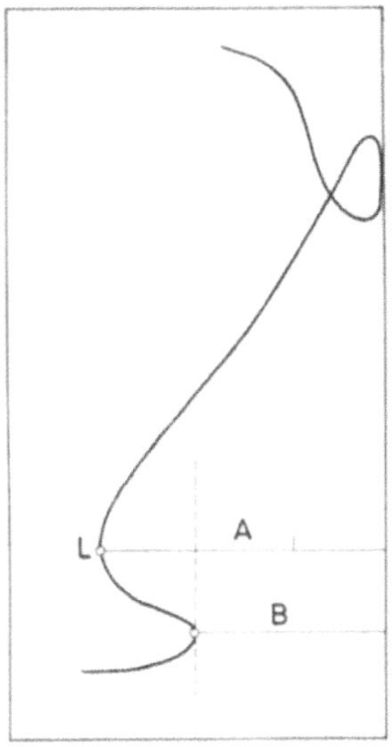

Figure 1

Method of locating the normal basal vein in frontosuboccipital view (according to BABIN and WACKENHEIM).

A distance between point *L* and midline noncorrected average: 25 mm; focal distance 100 cm); *B* distance between most internal point of basal vein with reference to midline noncorrected average: 17 mm; focal distance: 100 cm). $A/B=1.45$ (in practice 1.5)

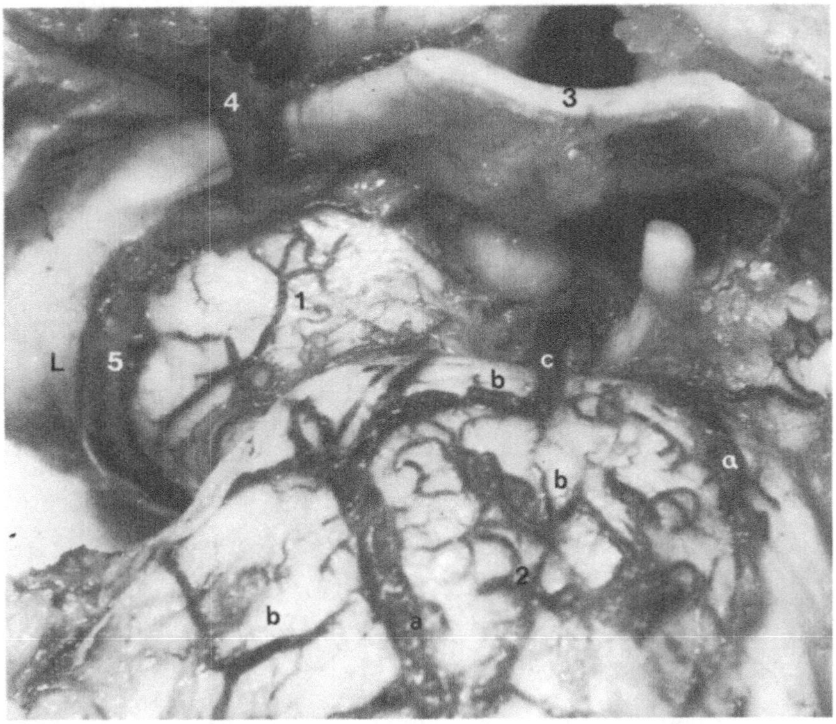

Figure 2

Specimen of ventral aspect of pons and cerebral peduncles

1 peduncle; *2* pons; *3* chiasma; *4* basal vein (subchiasmatic segment); *5* basal vein (peripeduncular segment); *L* most external point of peripeduncular segment of basal vein; *a* longitudinal pontine veins; *b* transversal pontine veins; *c* medial pontine longitudinal veins

Figure 3 A–D

Method of locating basal vein (*BV*) in lateral projection

(according to BABIN and WACKENHEIM)

N nasion; *P* internal occipital protuberance; *L* basal vein's farthest point from peripeduncular segment; *La* lambda; *E* external auditory canal; *V* vertex

A Anterior segment of basal vein is approximately parallel to line *NP* or *NT*.

B Distances separating basal vein from line *TP*: origin of basal vein, $\alpha=13$ mm; point *L*, $\beta=15$ mm; finishing point of basal vein, $\gamma=31$ mm. In practice: $\gamma=2\beta$.

C Position of point *L* with reference to point *T*. *La* lambda; *L′* projection of point *L* on the *TLa* line; $L'\cdot La=2\,T\cdot L'$; $b=2a$.

D Distance between internal cerebral vein and basal vein at the level of line determined by junction of *E* (auditory meatus) and point *L*: average noncorrected value 2.47 (focal distance 100 cm). In practice we employ a value of 2.5

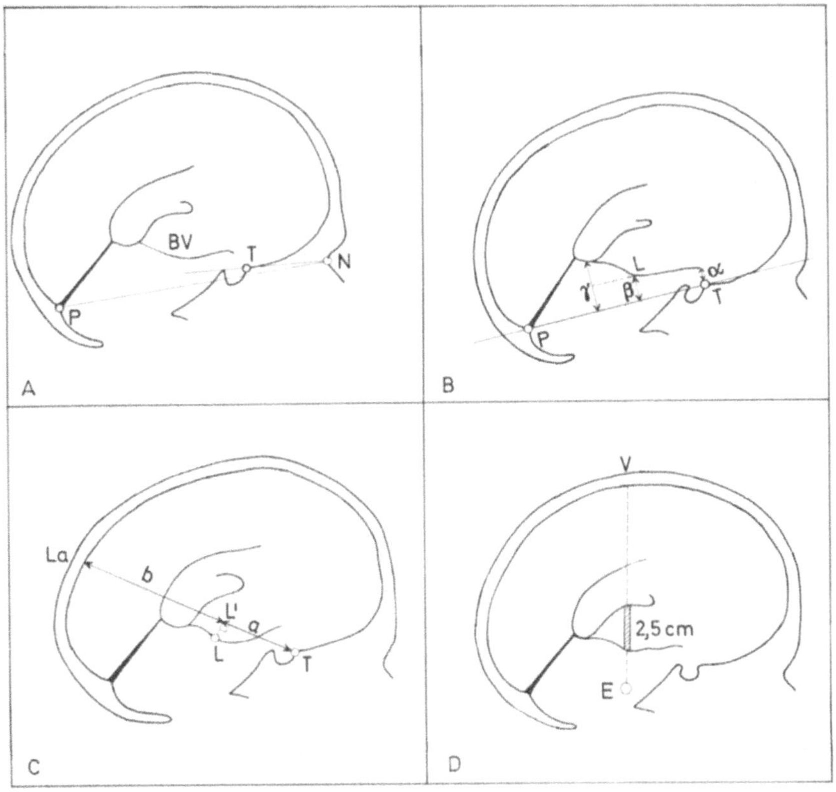

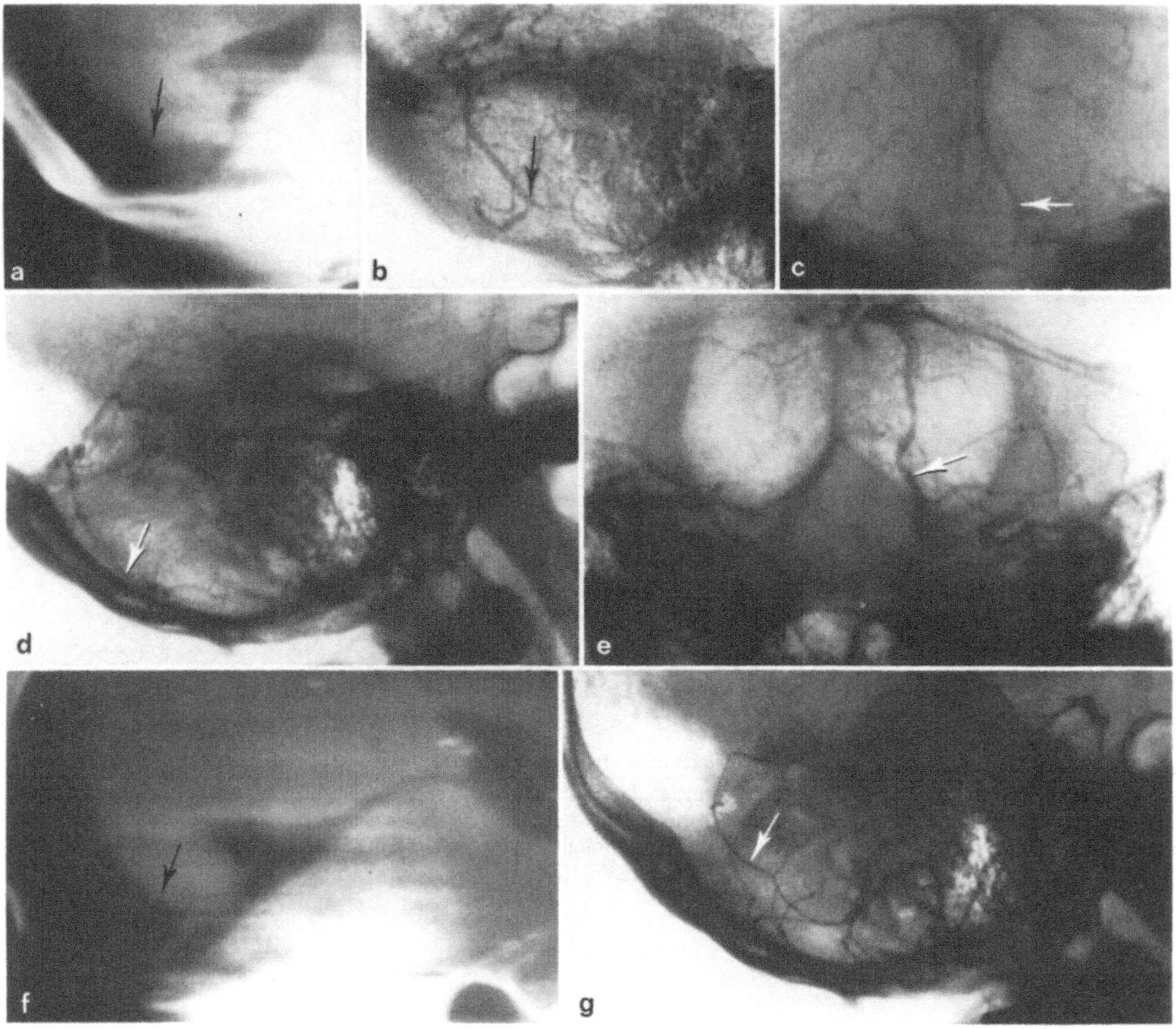

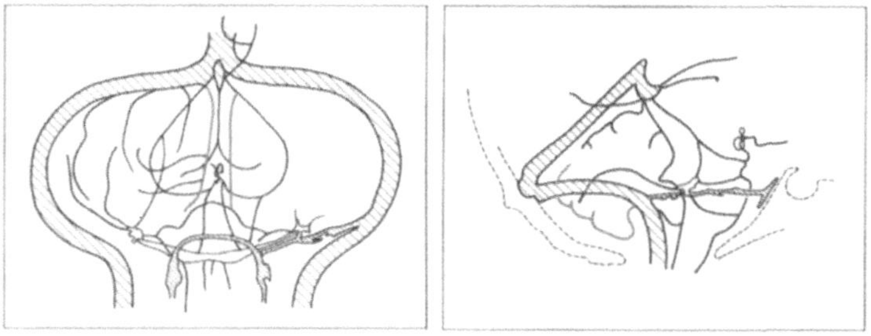

Figure 4 a–g

Inferior vermian vein

a–c When inferior vermian vein is situated on midline, it delimits anterior wall of cisterna magna.

d, e In normal cases when the inferior vermian vein courses more laterally on posterior face of cerebellar hemisphere, it gives, in lateral projection, appearance of compression of great posterior cistern.

f, g Vermian artery of the posteroinferior cerebellar artery can also delimit posterior wall of the great cistern

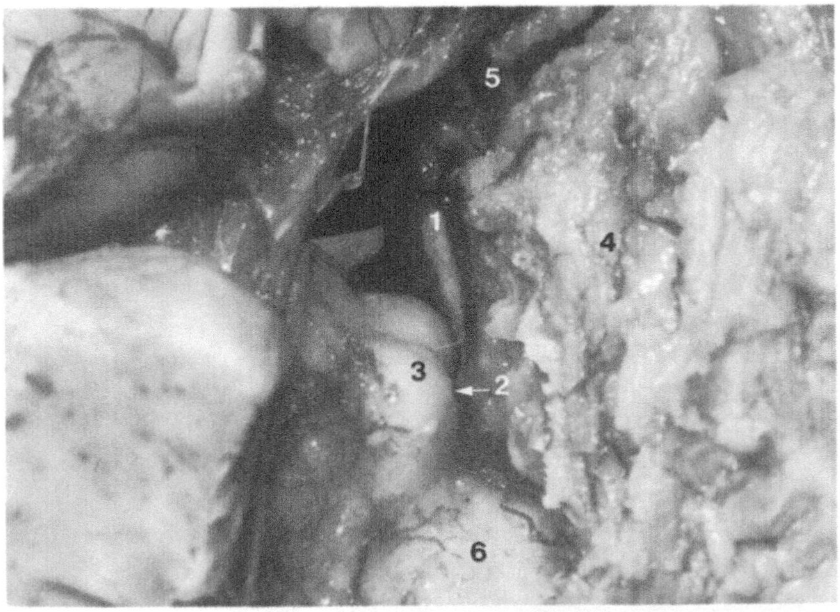

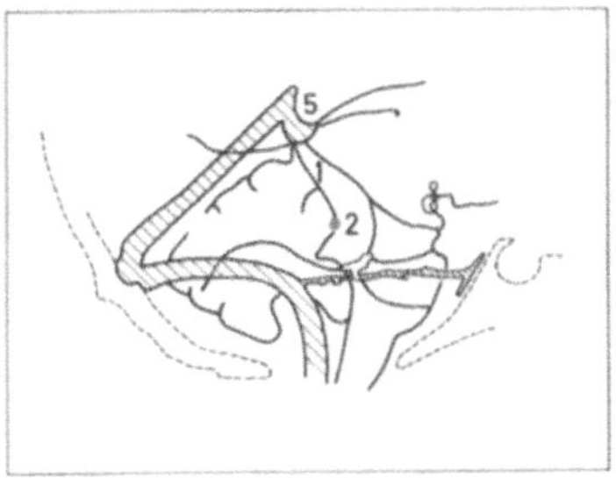

Figure 5

Precentral vein and collicular point (specimen)

1 Precentral vein
2 Collicular point
3 Quadrigeminal plate (posterior tubercle)
4 Cerebellar veins
5 Great vein of Galen
6 Cerebellar peduncle

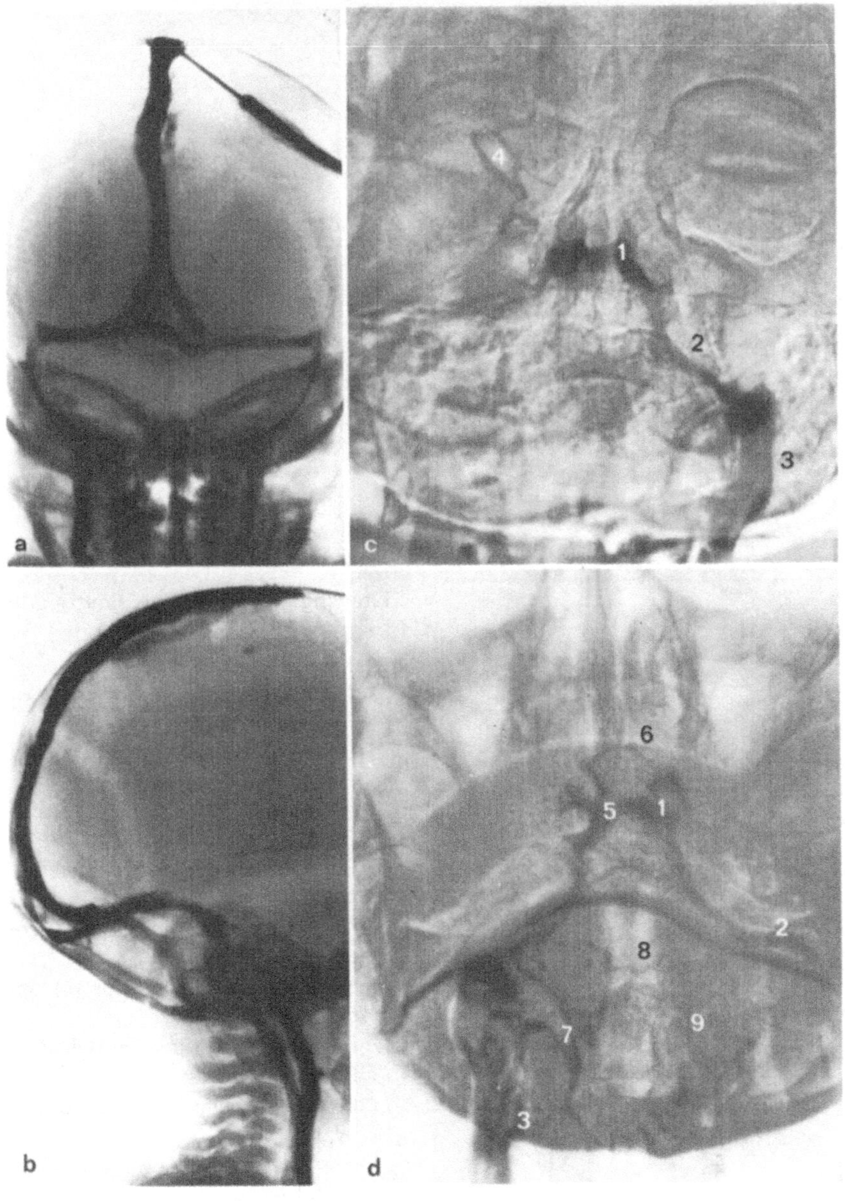

Figure 6 a–d

Sinusography

a, b Puncture of superior longitudinal sinus with anterograde opacification of cranial venous sinuses in child

c, d Retrograde sinusography via jugular passage

1 Cavernous sinus
2 Inferior petrous sinus
3 Jugular vein
4 Ophthalmic vein
5 Internal carotid
6 Coronary sinus
7 Condylar veins
8 Basilar plexus (Virchow)
9 Occipital plexus

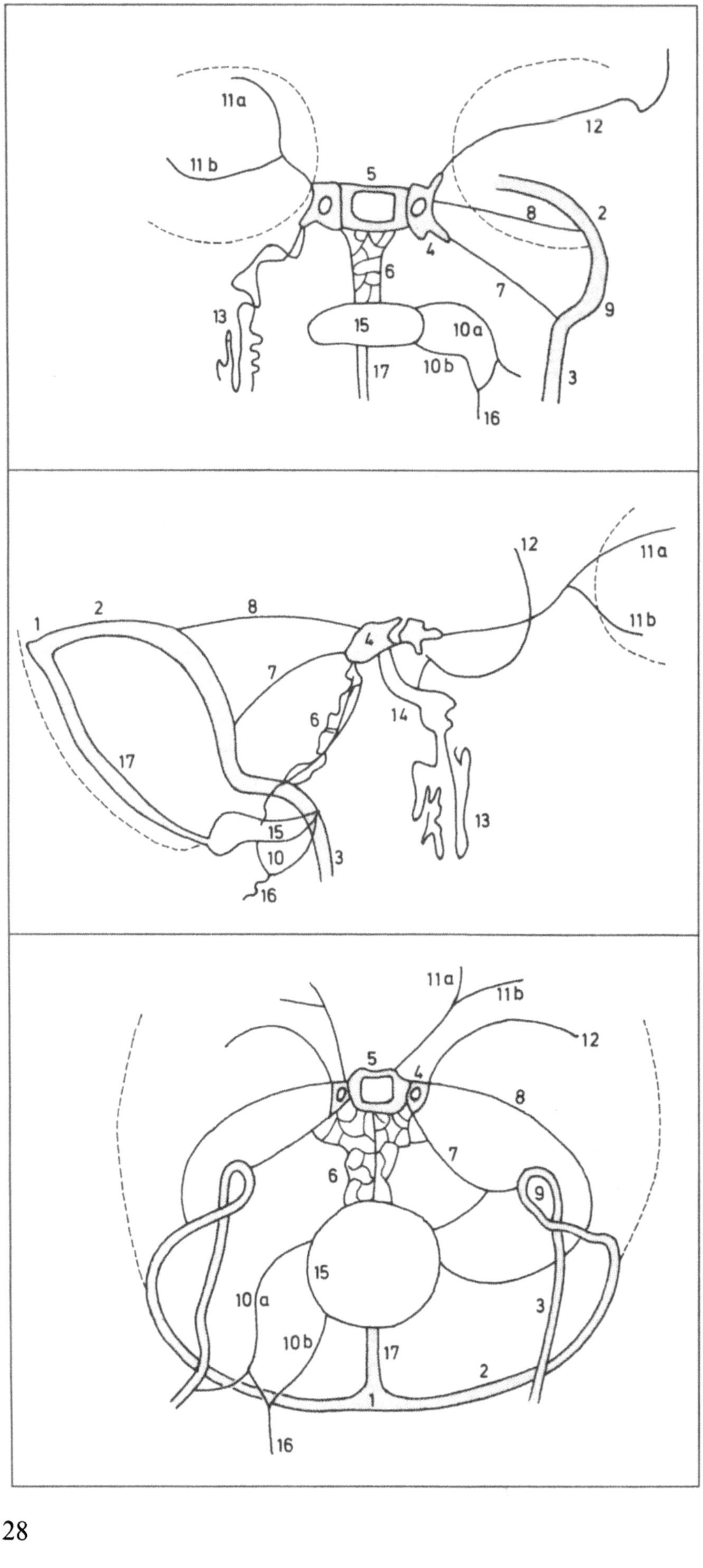

Figure 7

Drawing of main venous sinuses

1 Torcular
2 Lateral sinus
3 Jugular vein
4 Cavernous sinus
5 Coronary sinus
6 Clival plexus or basilar plexus (Virchow)
7 Inferior petrous sinus
8 Superior petrous sinus
9 Jugular groove
10 Anterior (*a*) and posterior (*b*) condylar veins
11 Superior (*a*) and inferior (*b*) ophthalmic veins
12 Frontoparietal sinus (Breschet)
13 Pterygoid plexus
14 Pterygoid veins in the foramen ovale
15 Foramen magnum plexus or periocipital plexus
16 Vertebral veins
17 Occipital sinus (inconstant)

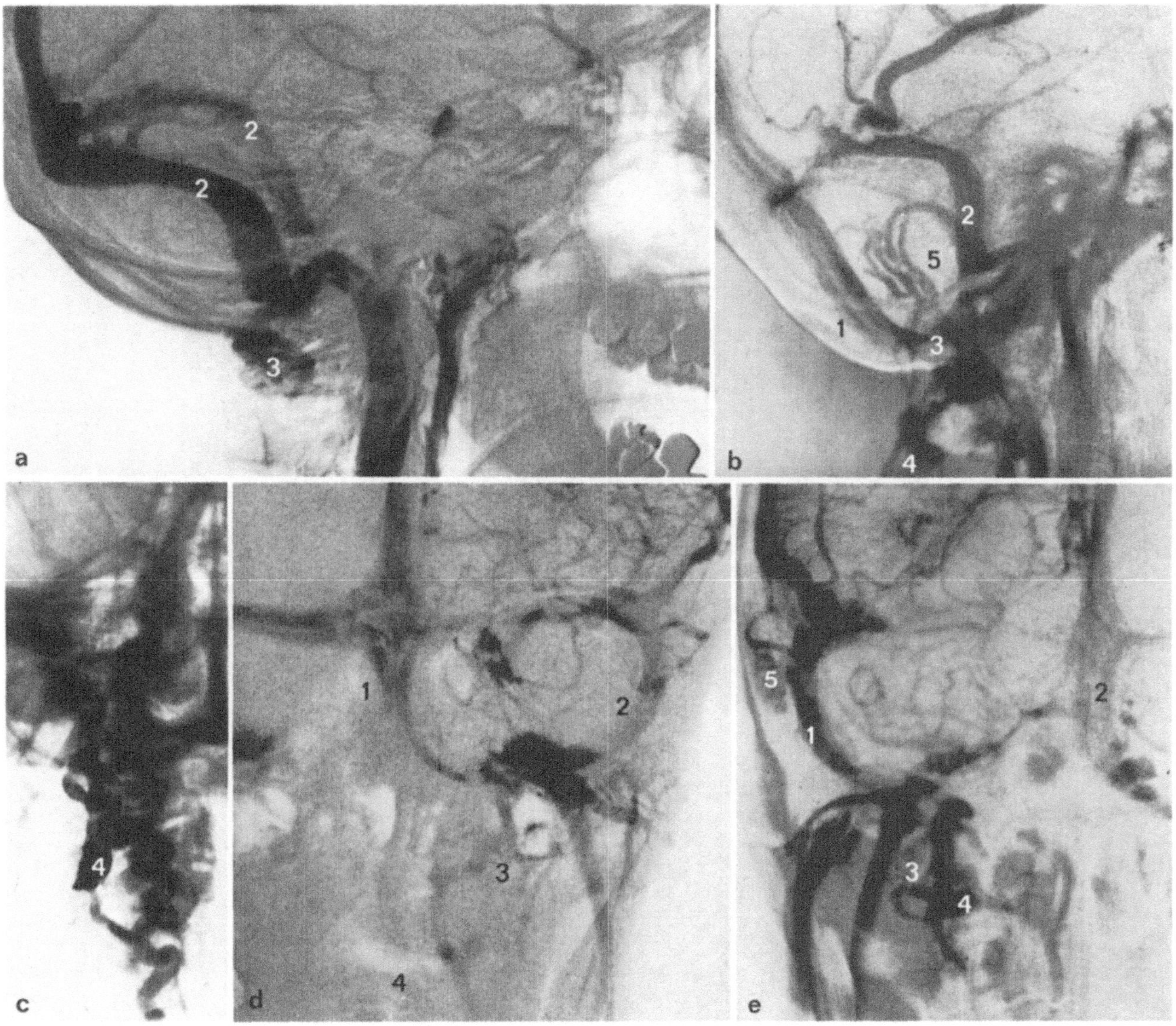

Figure 8 a–e

Venous phase of vertebral angiography

1 Occipital sinus
2 Lateral sinus
3 Condylar veins
4 Vertebral veins
5 Mastoïd emissary vein

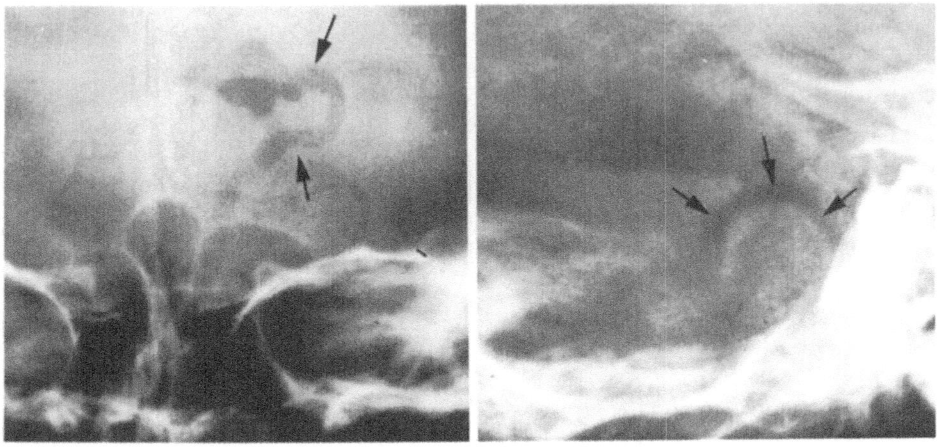

Figure 9

Two types of emissary veins bone canals

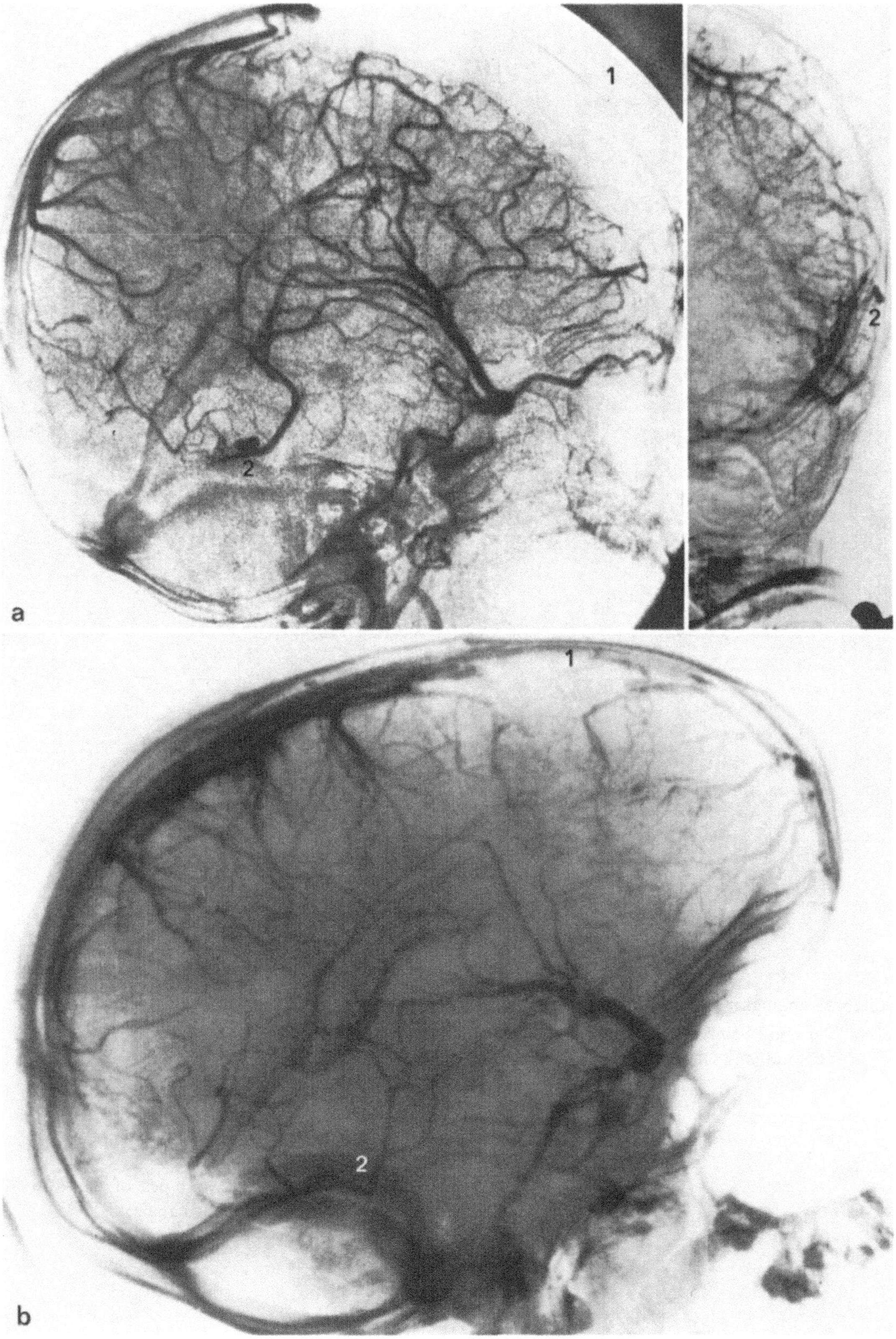

Figure 10a and b

Posterior temporal emissary vein. Dilated posterior temporal emissary vein (*2*) in two cases of an occluded anterior segment of the sagittal sinus (*1*)

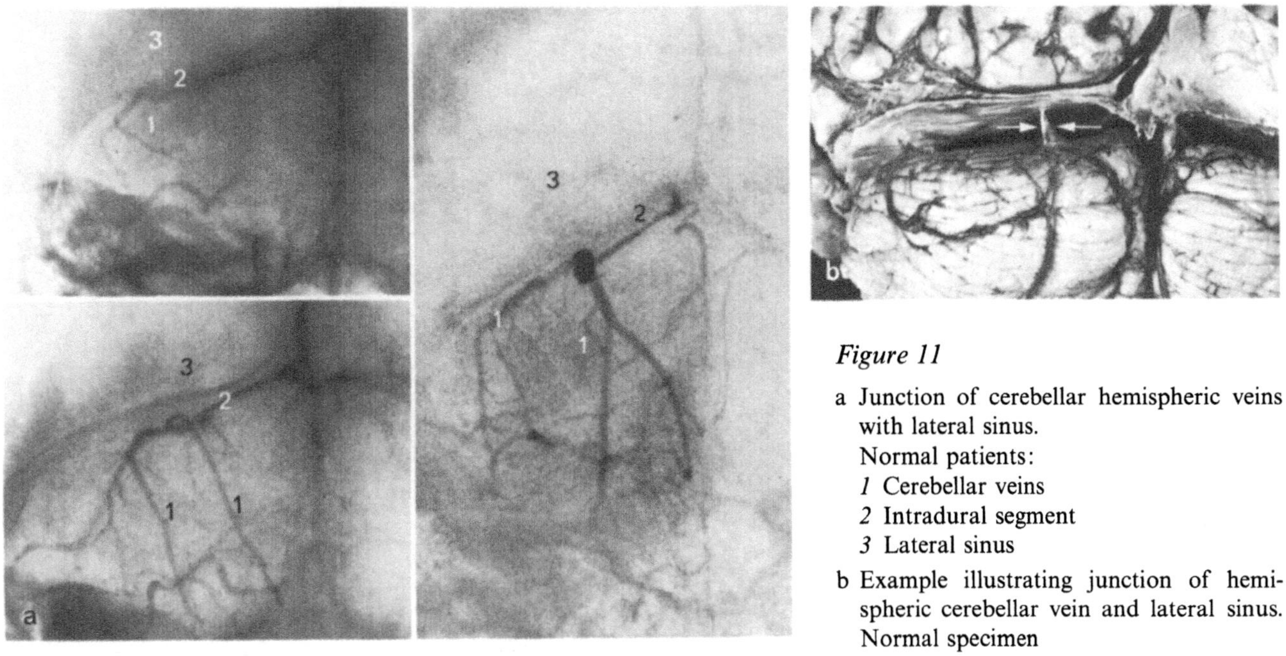

Figure 11

a Junction of cerebellar hemispheric veins with lateral sinus.
Normal patients:
1 Cerebellar veins
2 Intradural segment
3 Lateral sinus

b Example illustrating junction of hemispheric cerebellar vein and lateral sinus. Normal specimen

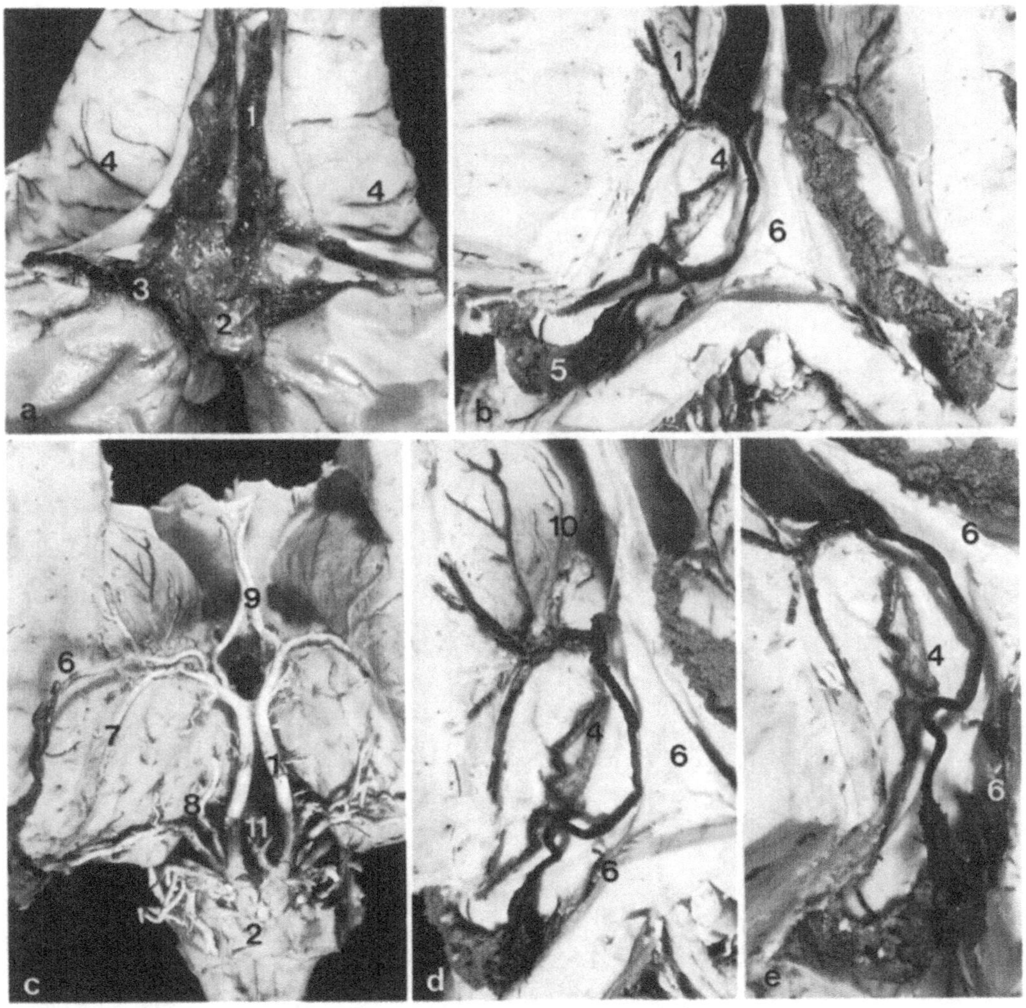

1 Internal cerebral vein
2 Great vein of Galen
3 Basal vein
4 Ventricular vein
5 Choroidal plexus of lateral ventricle
6 Superior choroïdal vein
7 Striothalamic vein
8 Superior thalamic vein
9 Pellucidal vein
10 Vein of caudate nucleus
11 Vein of pineal gland

Figure 12a–e

Specimen demonstrating deep cerebral veins and their connections

Figure 13 a–c

Choroidal veins (X-rays of opacified veins in a progressive anatomical section from a to c)

1 Superior choroidal vein
2 Inferior choroïdal vein
3 Basal vein
4 Thalamic vein
5 Internal cerebral vein
6 Choroidal plexus

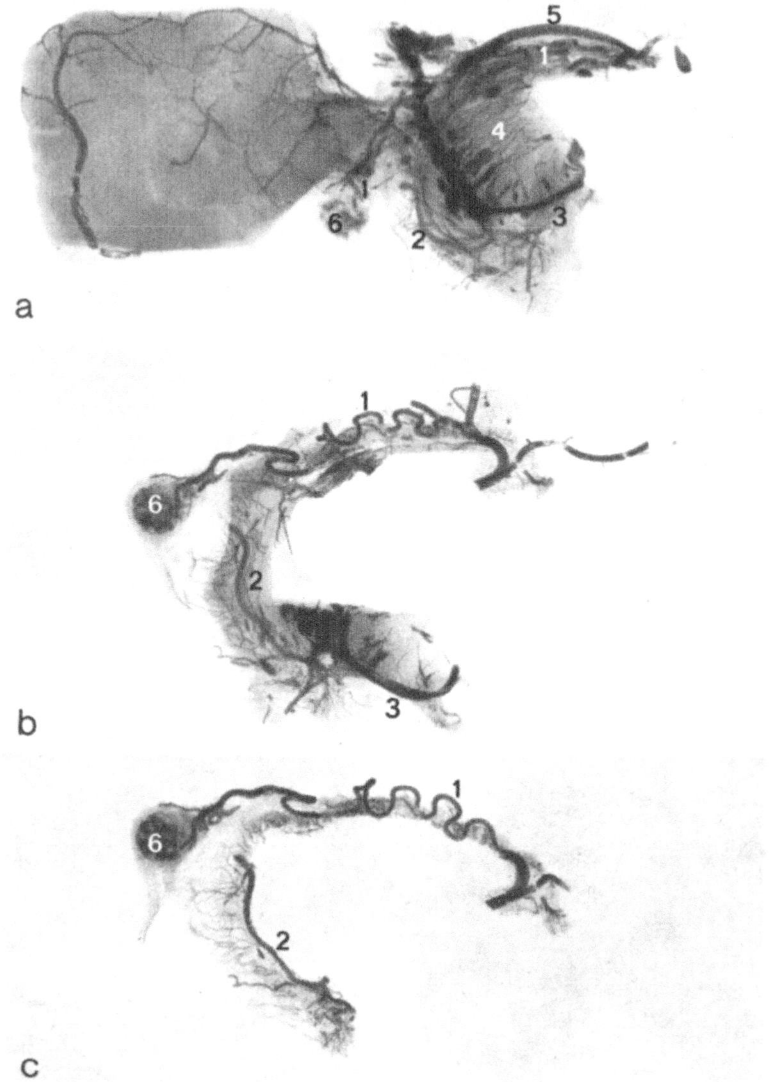

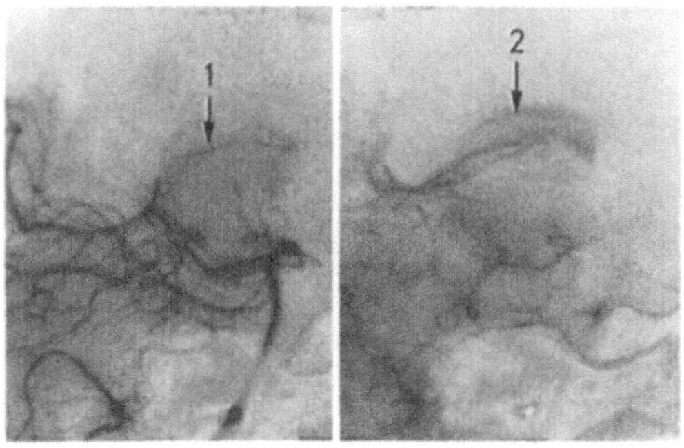

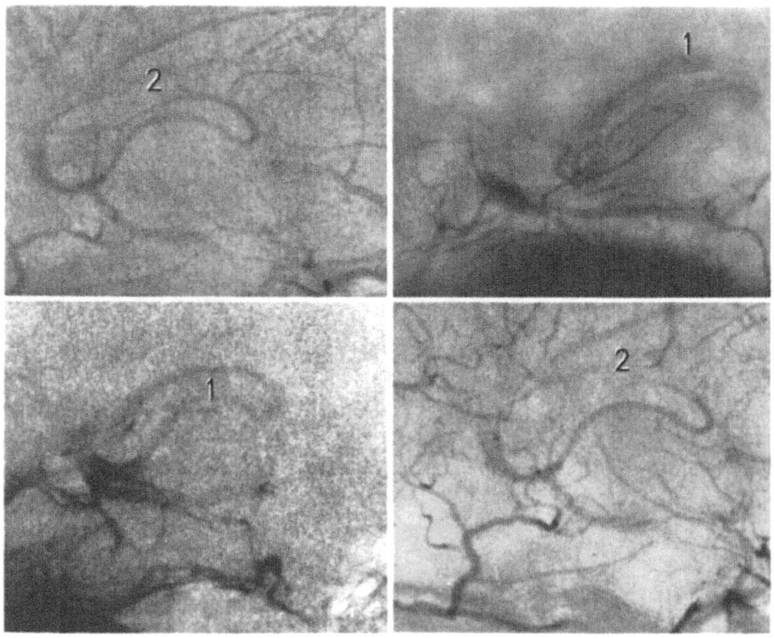

Figure 14

Superior choroidal veins

Above: Posterior choroidal arteries (*1*) and superior choroidal veins (*2*) in same normal patient

Right: Different types of superior choroidal vein in carotid (*2*) and vertebral (*1*) angiography

Figure 15a

Choroidal plexuses of lateral ventricles opacified during vertebral angiography

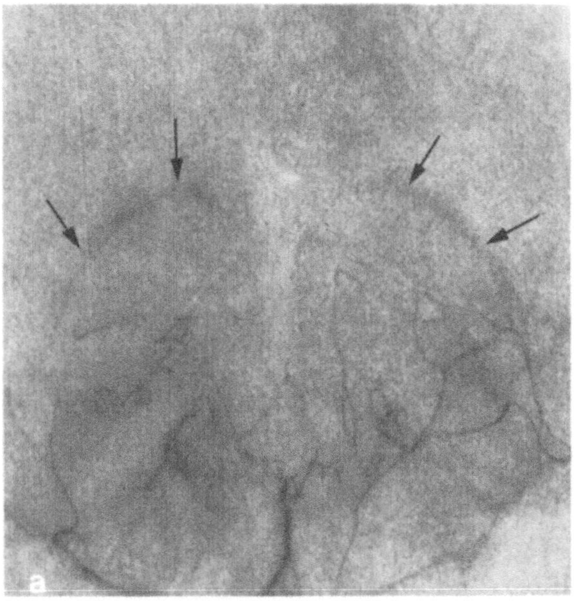

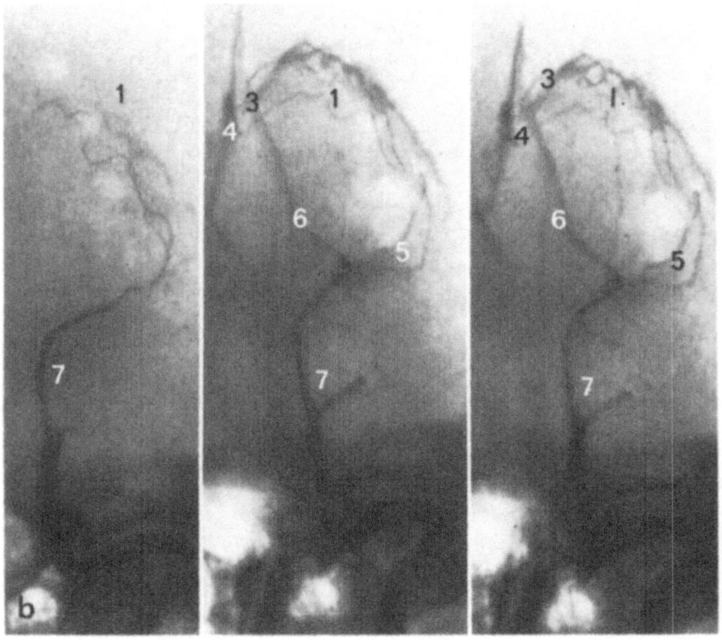

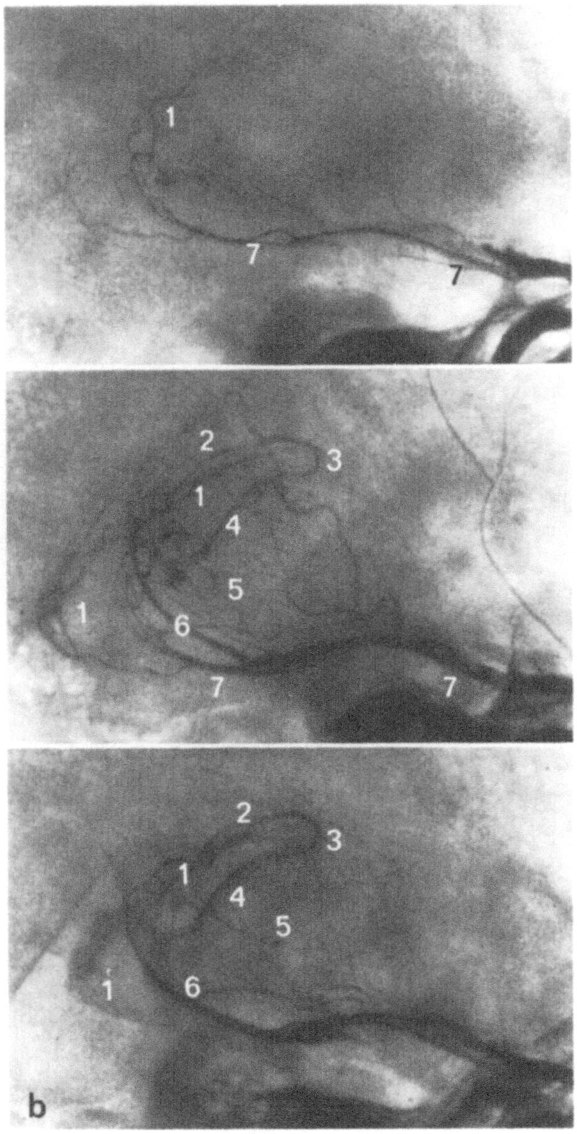

Figure 15b

Thrombosis of siphon of internal carotid artery

1 Choroidal plexus
2 Superior choroidal vein (choroidoventricular vein)
3 Striate vein
4 Internal cerebral vein
5 Inferior part of the choroidoventricular vein
6 Basal vein
7 Anterior choroidal artery

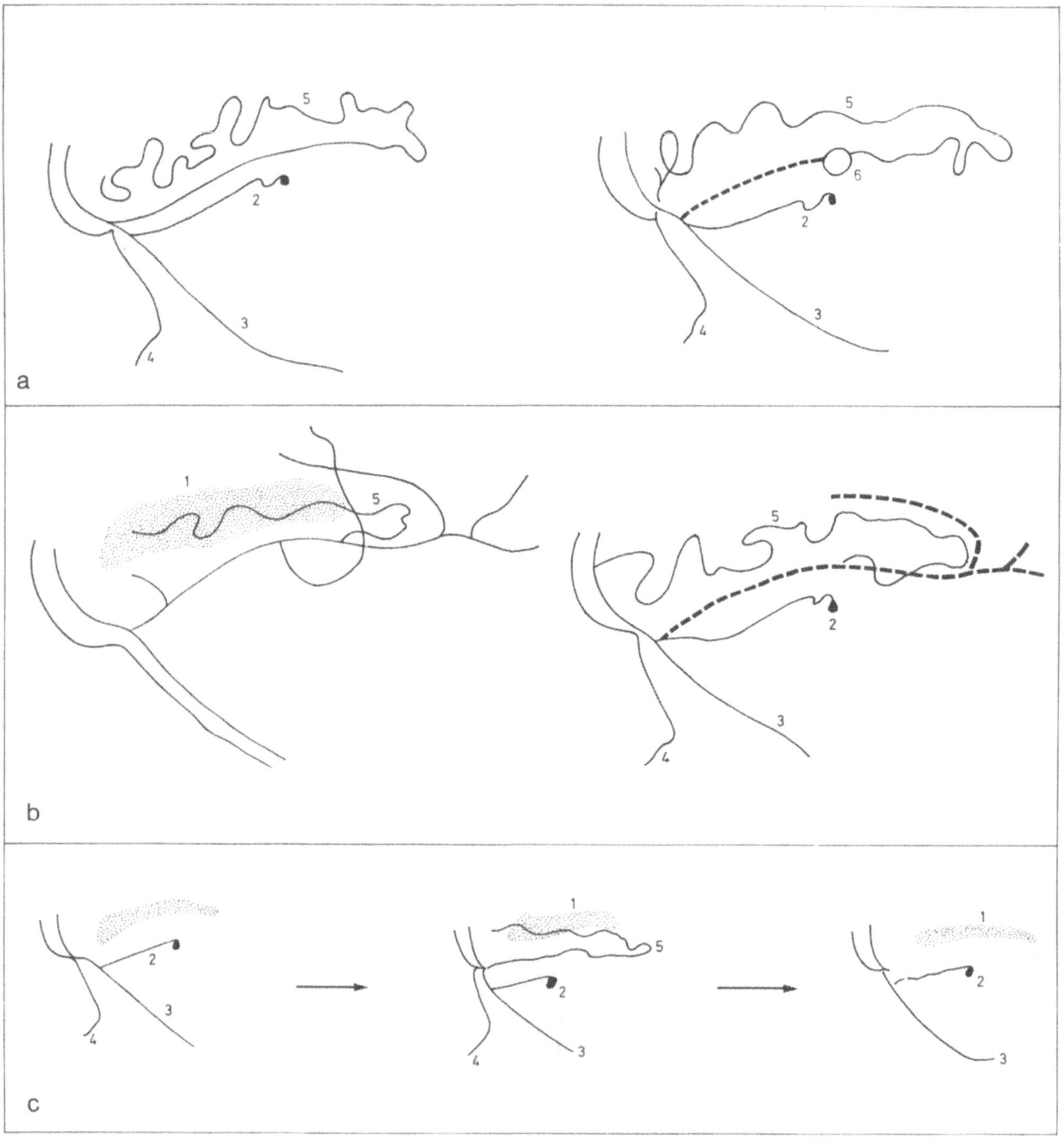

Figure 16

a Superior choroidal vein (*5*) and junction with internal cerebral vein (*6*)
b *Left:* carotid angiography with choroidal plexus (*1*) and superior choroidal vein (*5*). *Right:* vertebral angiography with superior choroidal vein (*5*)
c Chronology of venous opacification in vertebral angiography: choroidal plexus (*1*), superior thalamic vein (*2*), posterior mesencephalic vein (*3*), precentral vein (*4*)

Figure 17

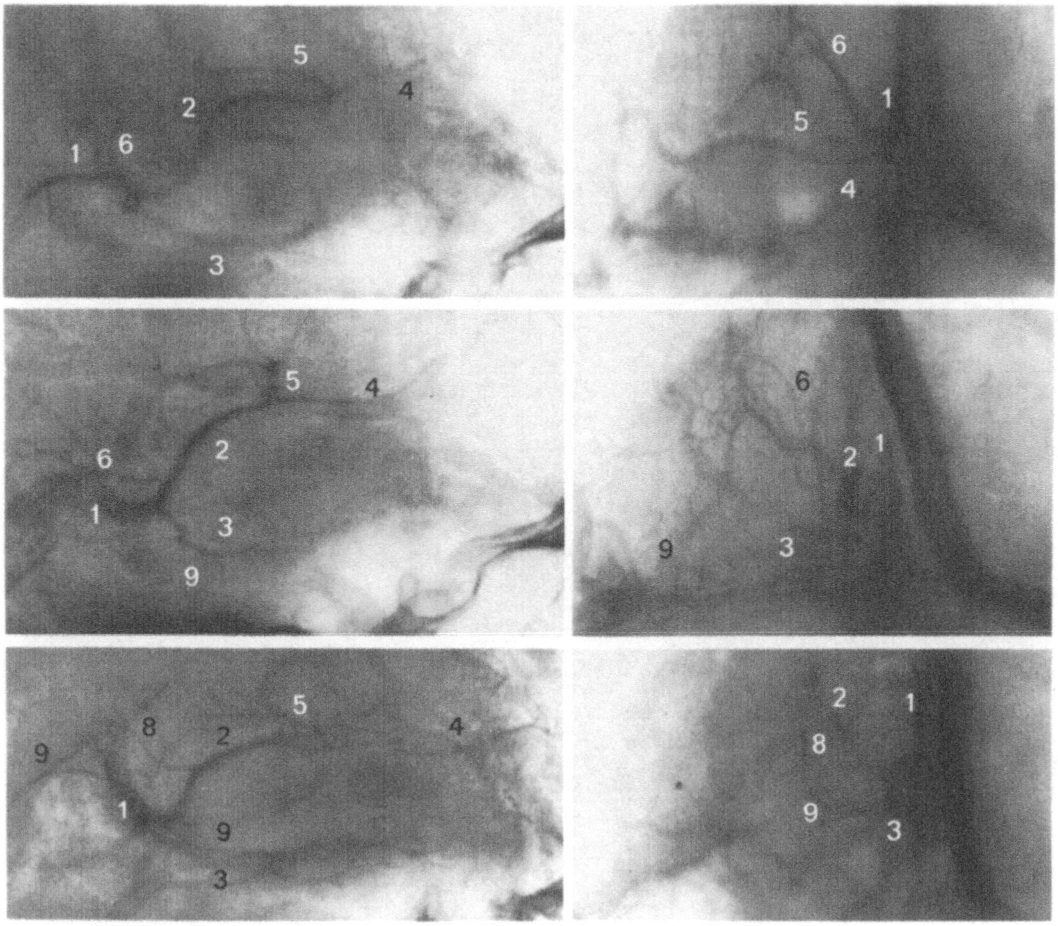

◁ *Figure 18*

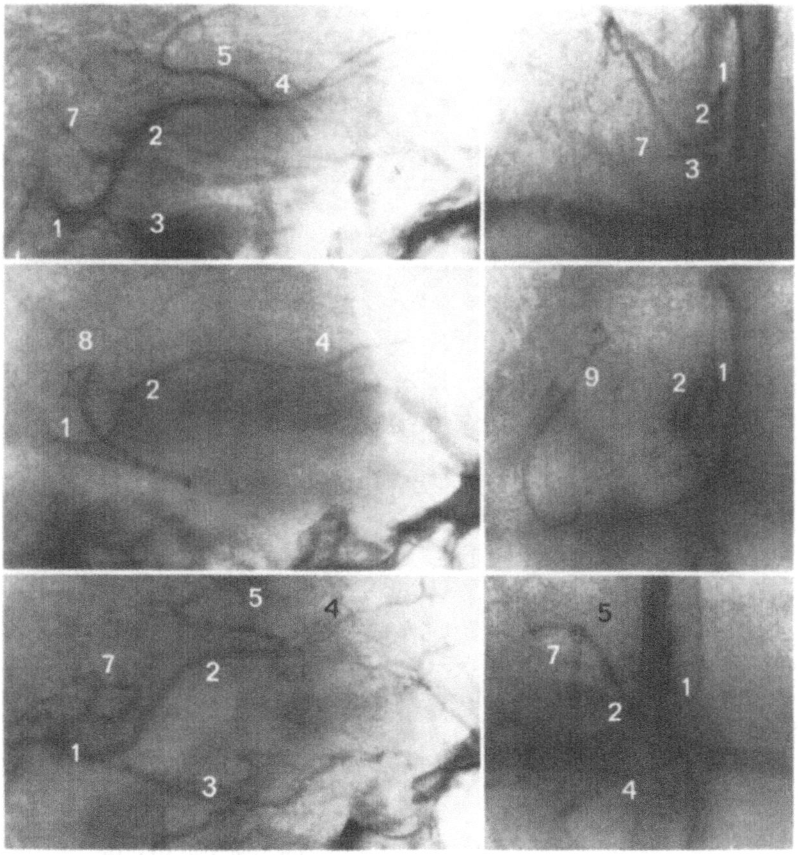

Figures 17 and 18

Ventricular or subependymal veins (according to Billewicz and Ben Amor)

1 Great vein of Galen
2 Internal cerebral vein
3 Basal vein
4 Pellucidal vein
5 Striothalamic vein
6 Medial atrial vein (vein of internal wall of trigonum)
7 Direct lateral vein (vein of external wall of body of lateral ventricle)
8 Direct medial vein (vein of internal wall of body of lateral ventricle)
9 Lateral atrial vein (vein of external wall of trigonum)

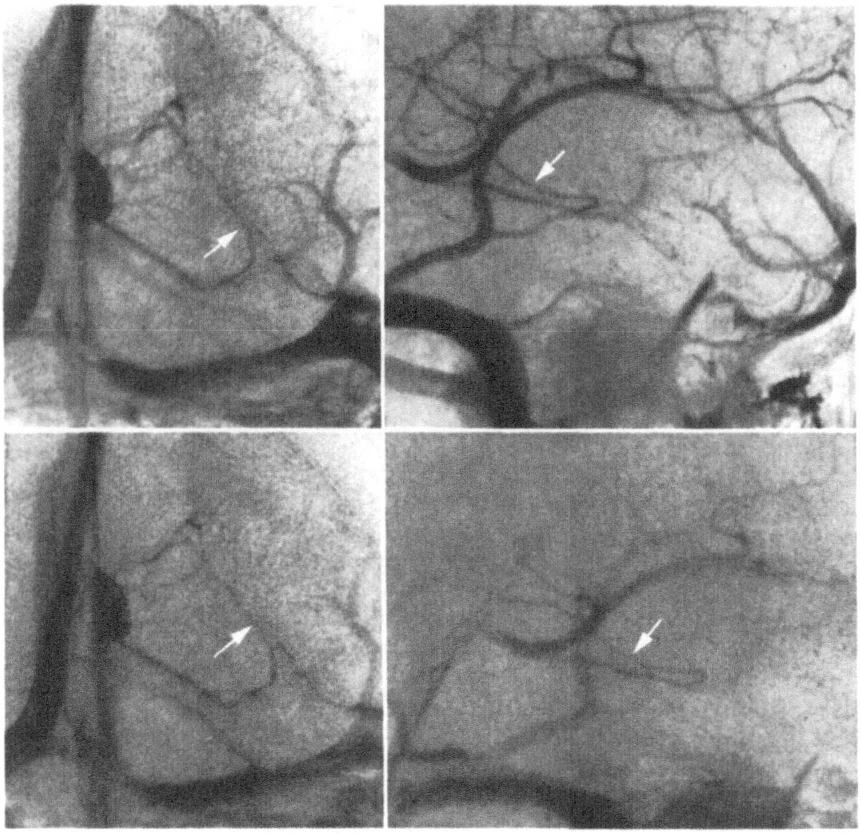

Figure 19

Inferior ventricular vein visualized in carotid angiography (choroidoventricular vein)

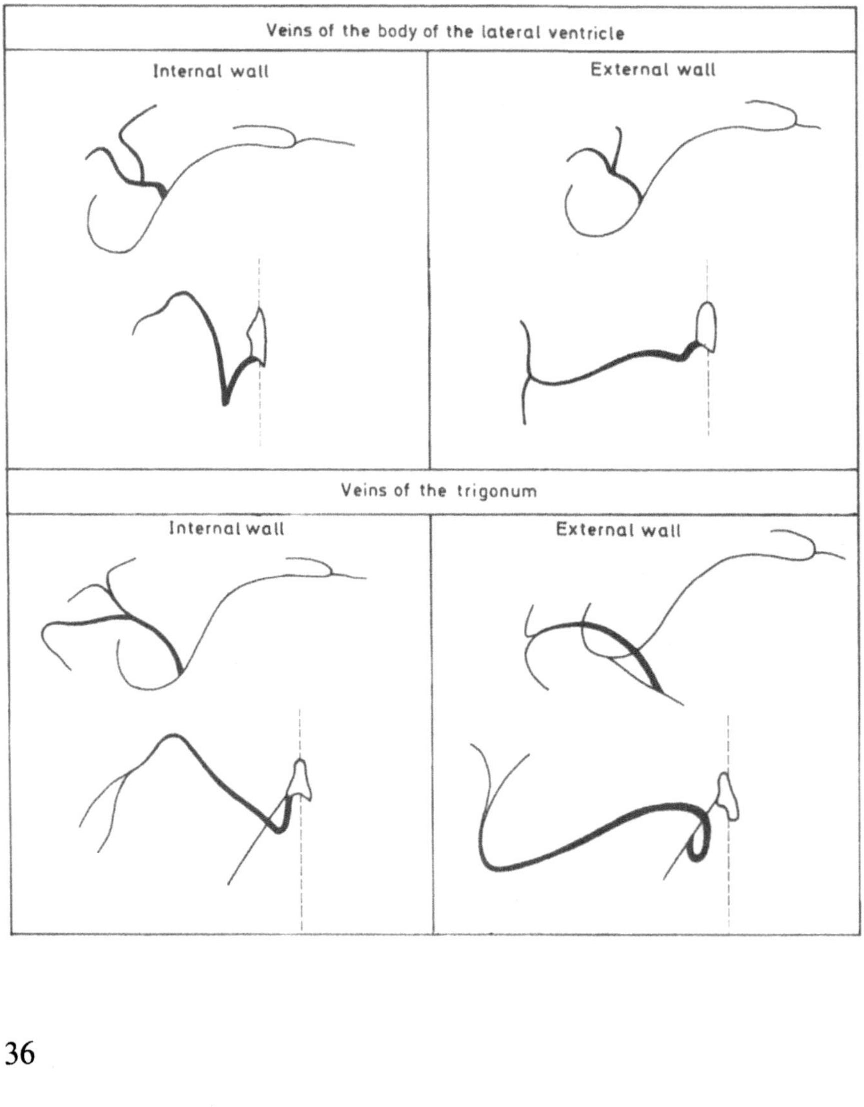

Figure 20

Drawing of ventricular or subependymal veins (see also Fig. 19)

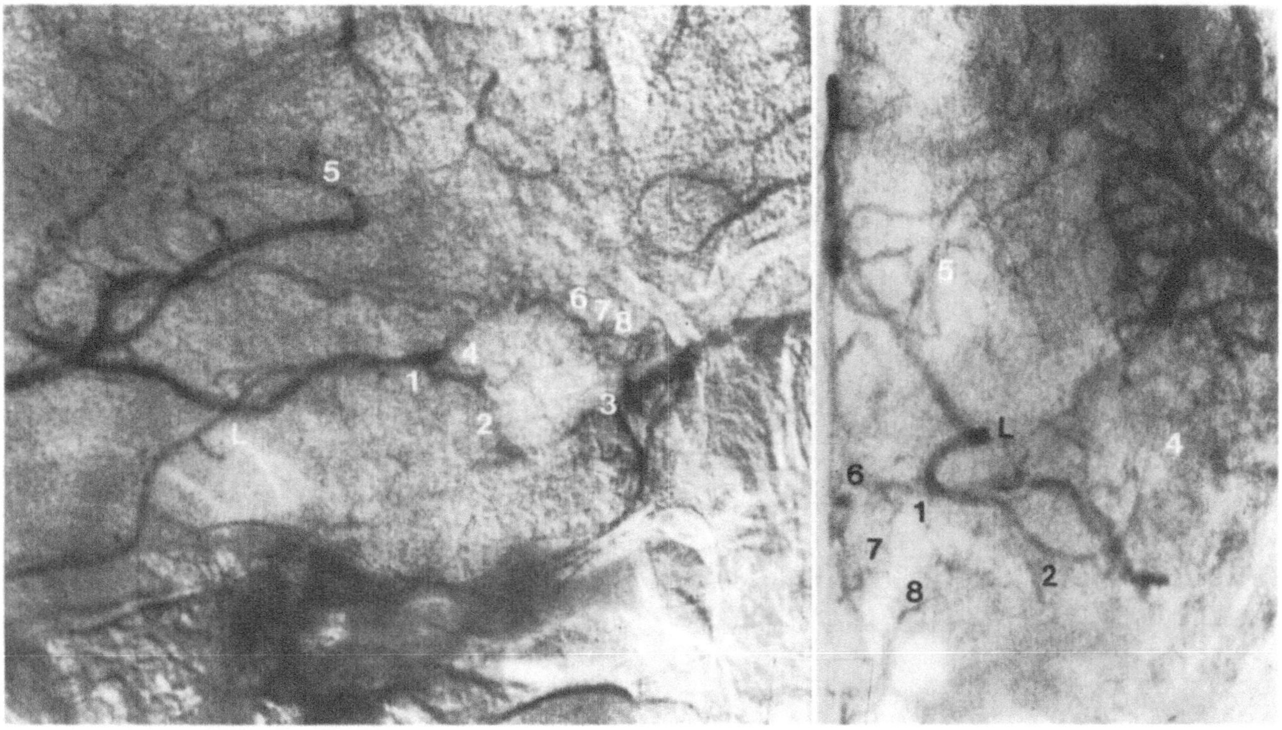

△
Figure 21

Basal vein

1 Origin of the basal vein
2 Deep sylvian vein
3 Superficial sylvian vein
4 Insular vein
5 Thalamostriate vein
6 Anterior cerebral vein
7 Internal frontobasal (olfactive) vein
8 External frontobasal vein (orbitary)
L Most external point of peripeduncular segment of basal vein

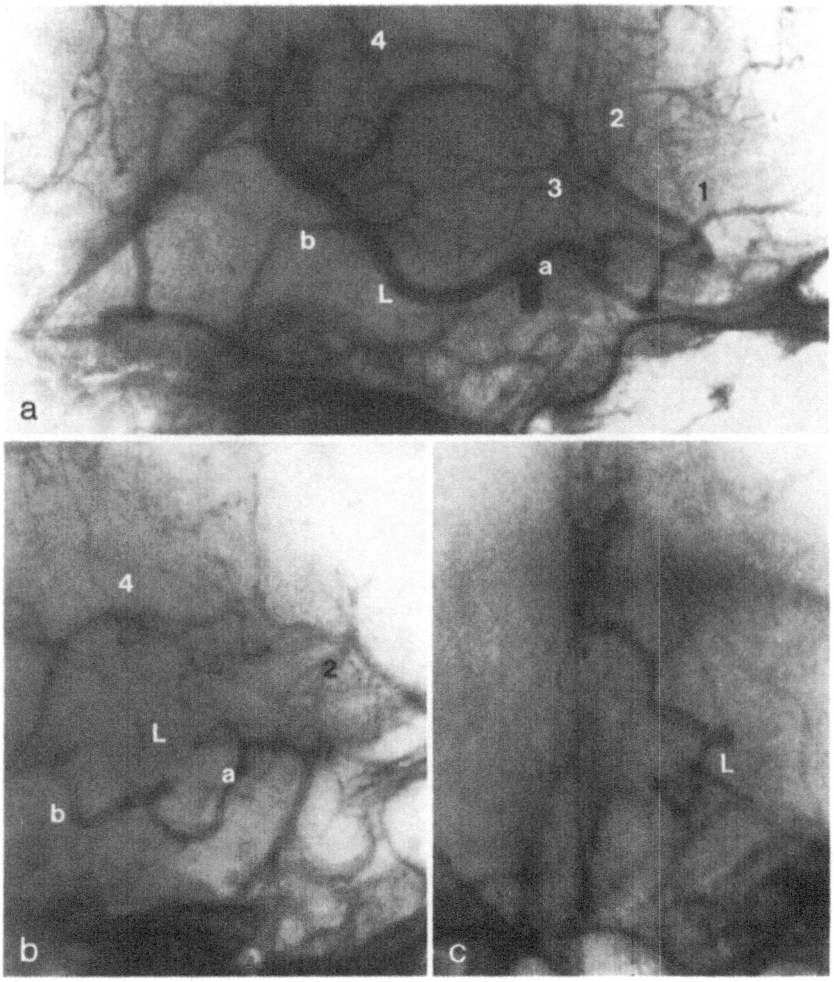

Figure 22 a–c

Mega-dolicho basal vein

a Anterior segment
b Posterior segment
c Point *L* (most external point of peripeduncular segment of basal vein)
1 Olfactive vein
2 Superficial sylvian vein
3 Deep sylvian vein
4 Internal cerebral vein

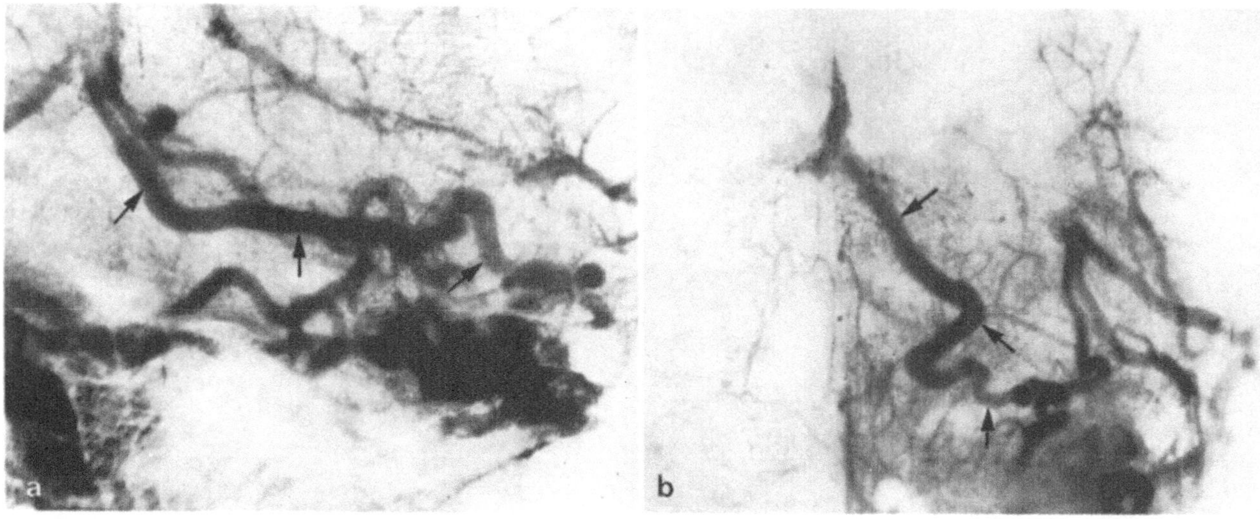

Figure 23a and b

Cerebral varicose veins which permit a better identification of basal vein tributaries

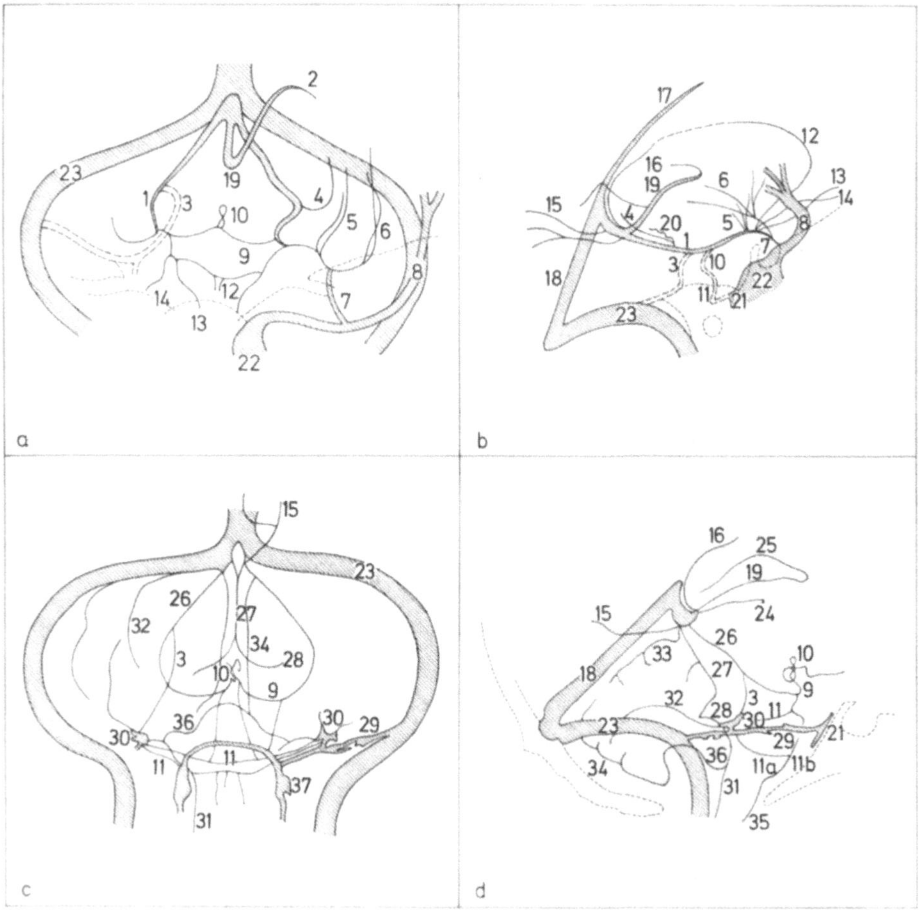

Figure 24a–d

Frontal (a) and lateral (b) drawing of system of basal vein and of veins of posterior fossa (c and d)

1 Basal vein and point *L*
2 Striothalamic vein
3 Lateral mesencephalic vein
4 Inferior ventricular vein (choroidoventricular vein)
5 Inferior striate veins
6 Insular veins
7 Deep sylvian veins
8 Superficial sylvian vein
9 Posterior communicating vein
10 Optopeduncular veins
11 Pontine veins: transversal (*a*) and longitudinal (*b*)
12 Anterior pericallosal vein
13 Olfactory vein (internal frontobasal vein)
14 Orbital vein (external frontobasal vein)
15 Internal occipital veins
16 Posterior pericallosal vein
17 Inferior longitudinal sinus
18 Straight sinus
19 Internal cerebral vein
20 Inferior choroidal vein
21 Clival plexus
22 Cavernous sinus
23 Lateral sinus
24 Superior thalamic vein
25 Superior choroidal vein
26 Posterior mesencephalic vein
27 Precentral vein
28 Brachial vein
29 Superior petrous sinus
30 Superior petrous vein
31 Lateral medullopontine vein
32 Cerebellar hemispheric veins
33 Superior vermian vein
34 Inferior vermian vein
35 Anterior bulbar vein
36 Vein of lateral recess of fourth ventricle
37 Perioccipital plexus

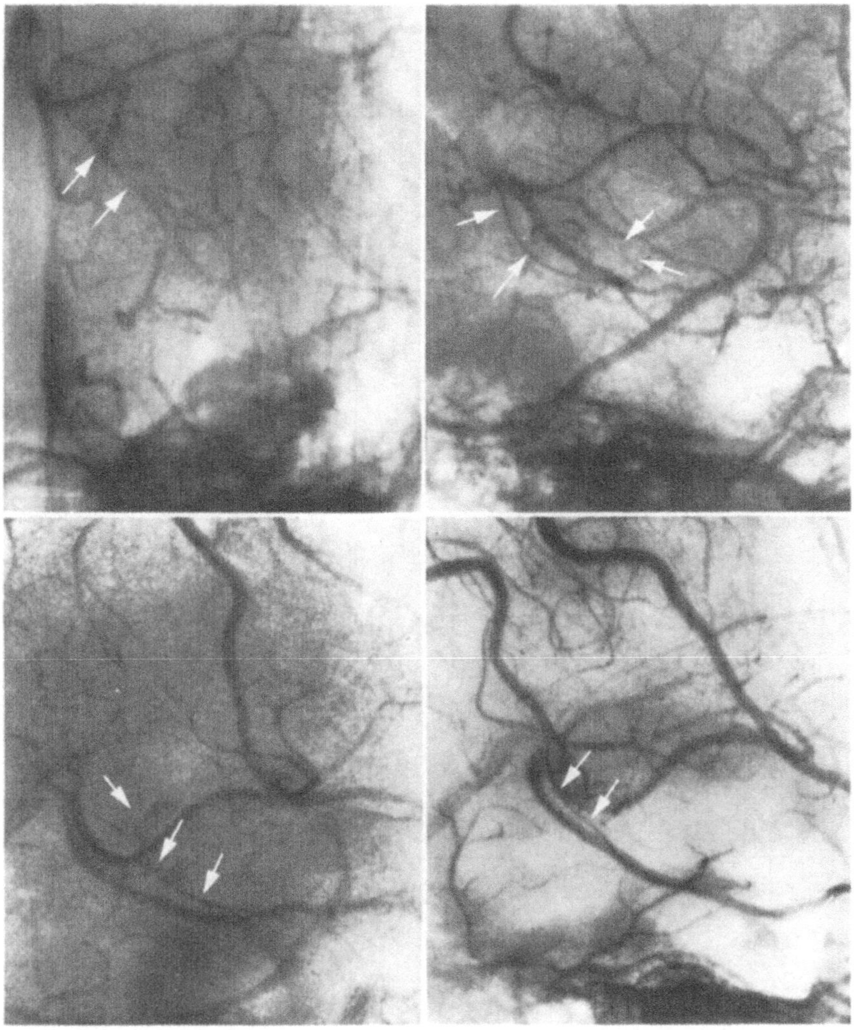

Figure 25

Inferior ventricular vein

Inferior ventricular vein or choroidoventricular vein (arrows) drains choroidal plexus of lateral ventricle in basal vein near point *L*. It can nevertheless have a recurrent course toward great vein of Galen, parallel to basal vein. In this case, recurrent segment is called posterior mesencephalic vein. See also Figure 19

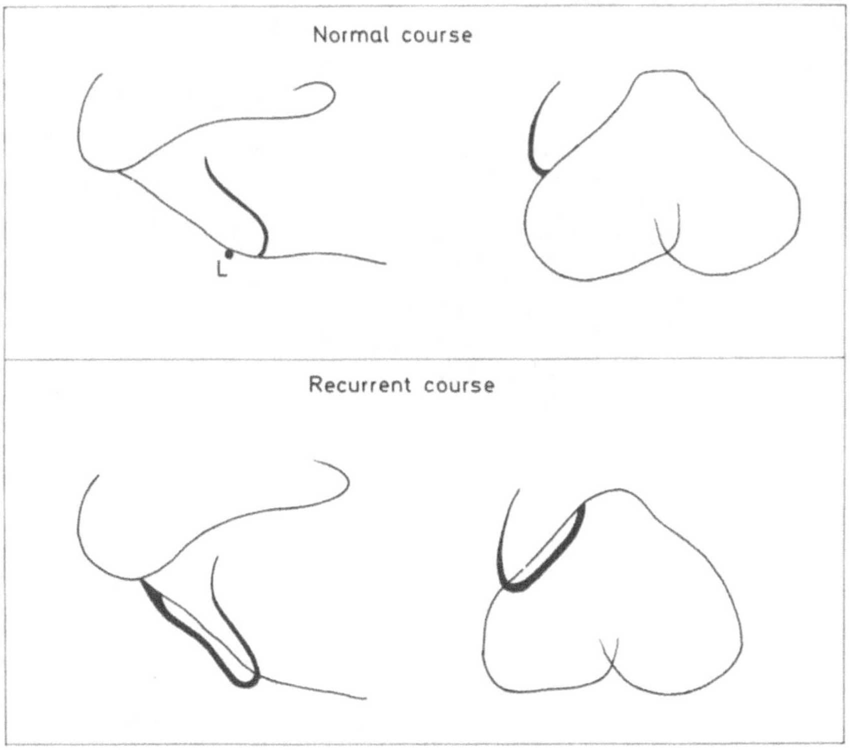

Figure 26

Drawing of inferior ventricular vein or choroidoventricular vein (normal and recurrent course). See also Figures 19 and 25

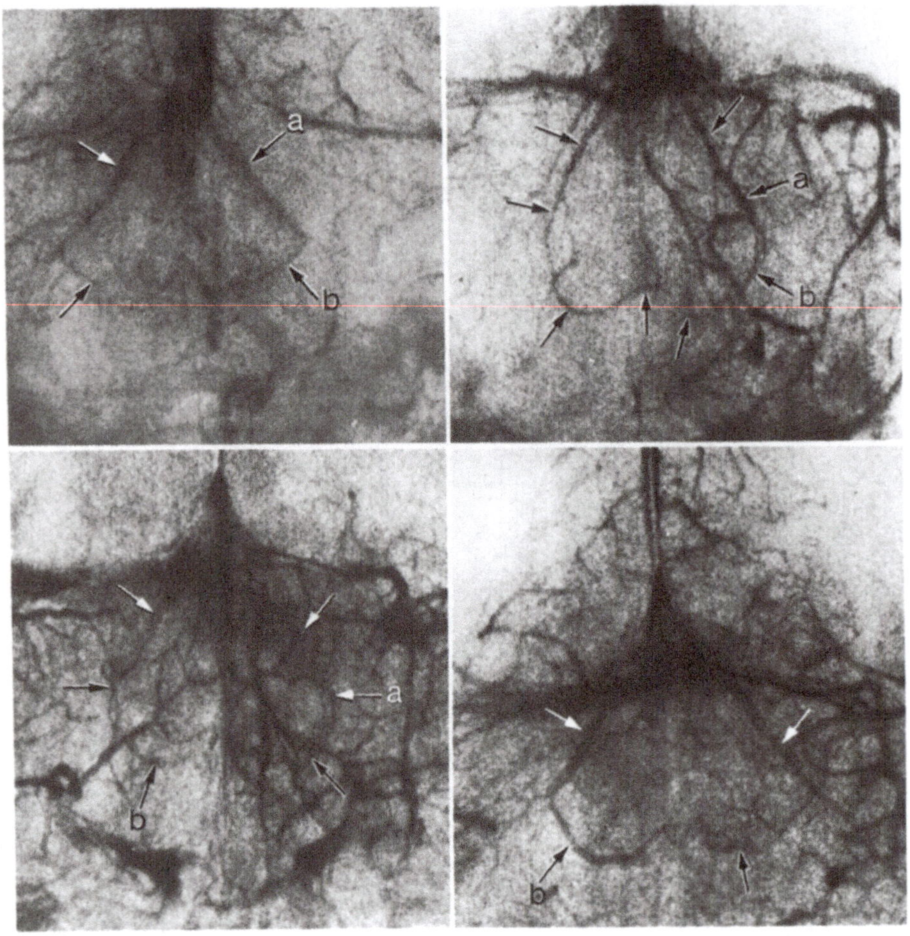

Figure 27

Brain stem bounded by following veins

a Posterior segment of basal vein
b Inferior and lateral peduncular veins (peripeduncular vein)

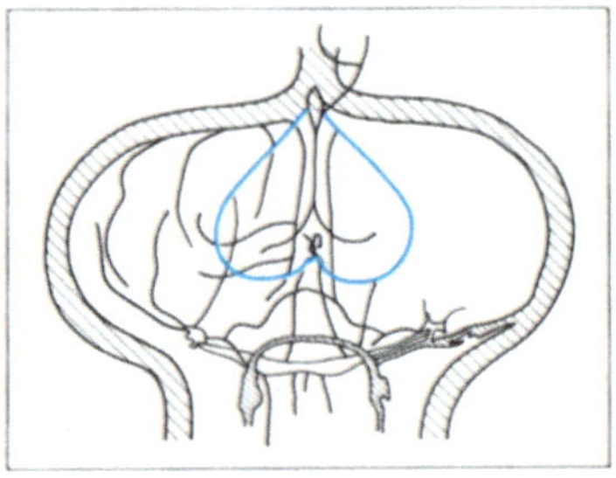

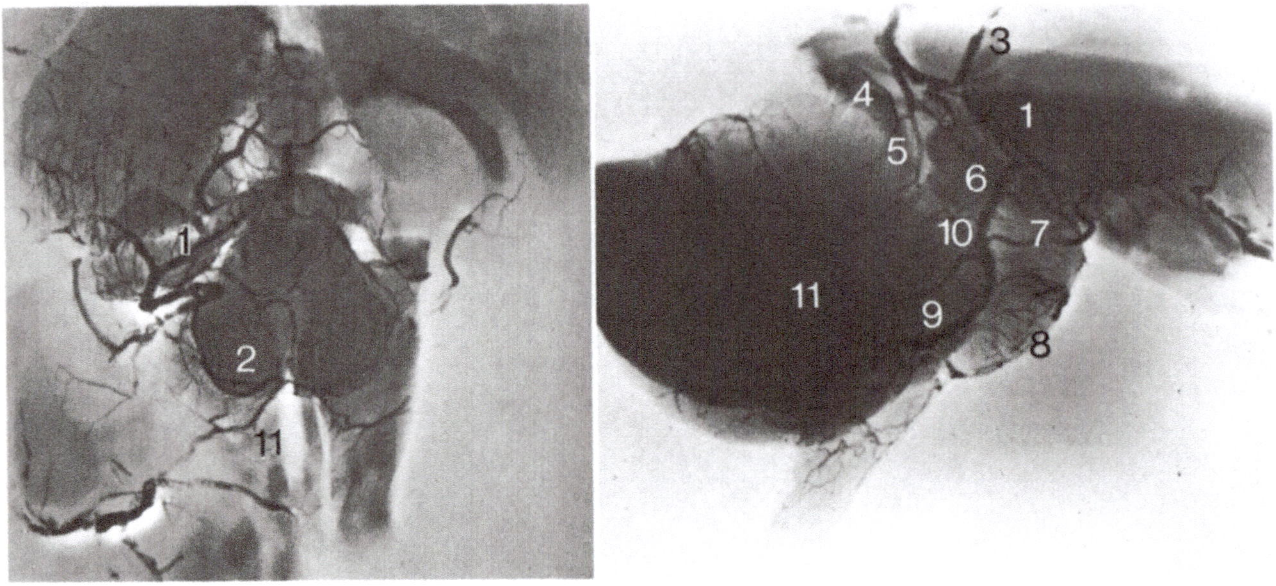

Figure 28

X-ray of specimen with opacified veins

1 Basal vein
2 Inferior peduncular vein (posterior communicating vein)
3 Internal cerebral vein
4 Superior vermian vein
5 Precentral vein
6 Lateral mesencephalic vein
7 Transversal pontine vein
8 Anterior longitudinal pontine vein
9 Lateral medullopontine vein
10 Tributary of superior petrosal vein
11 Vein of lateral recess of fourth ventricle

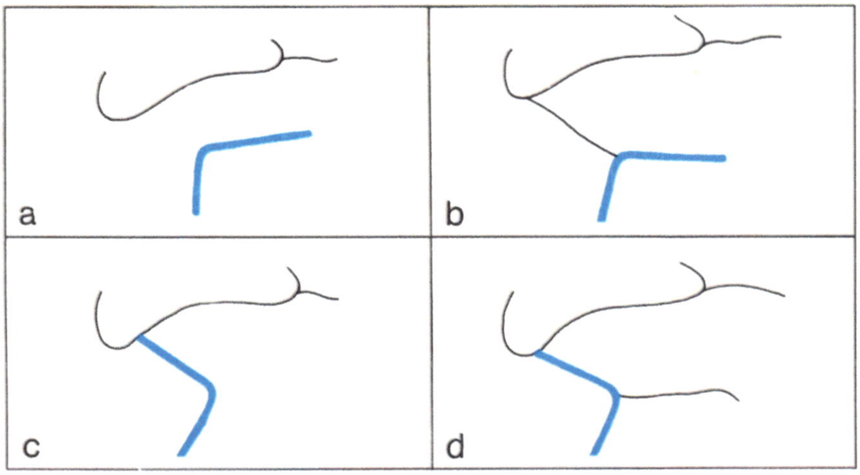

Figure 29a–d

Variation in drainage of basal vein

a Lateral anastomotic mesencephalic vein
b Incomplete form of above (a) with a posterior hypoplastic segment of basal vein
c Aplasia of anterior segment of basal vein
d Incomplete form of (c) with anterior hypoplastic segment of basal vein

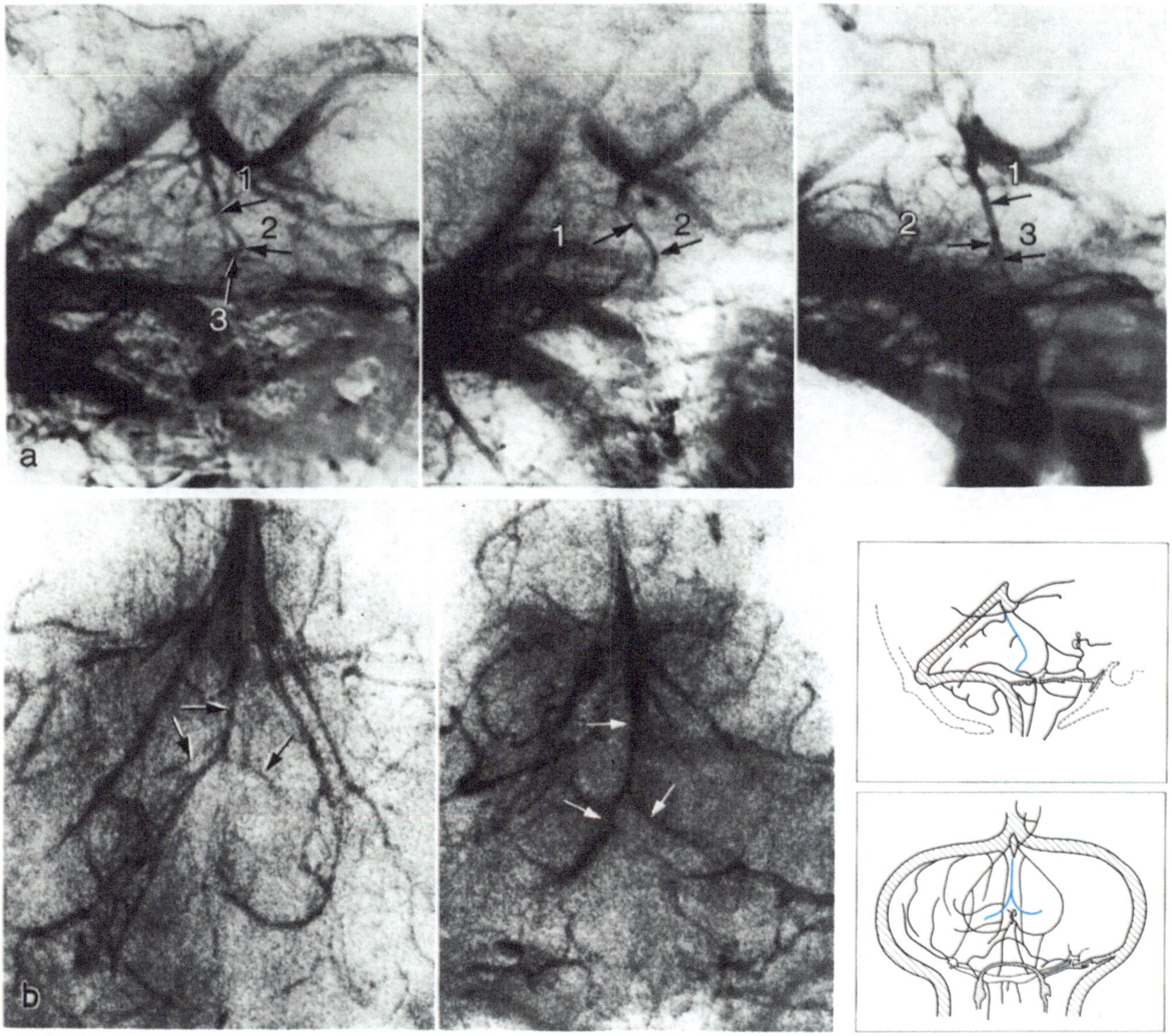

Figure 30

a Precentral vein in lateral projection. *1* Vermian segment, *2* Collicular point, *3* Vein of brachium conjonctivum (superior cerebellar peduncular vein or brachial vein)
b Precentral vein, frontal projection (arrows)

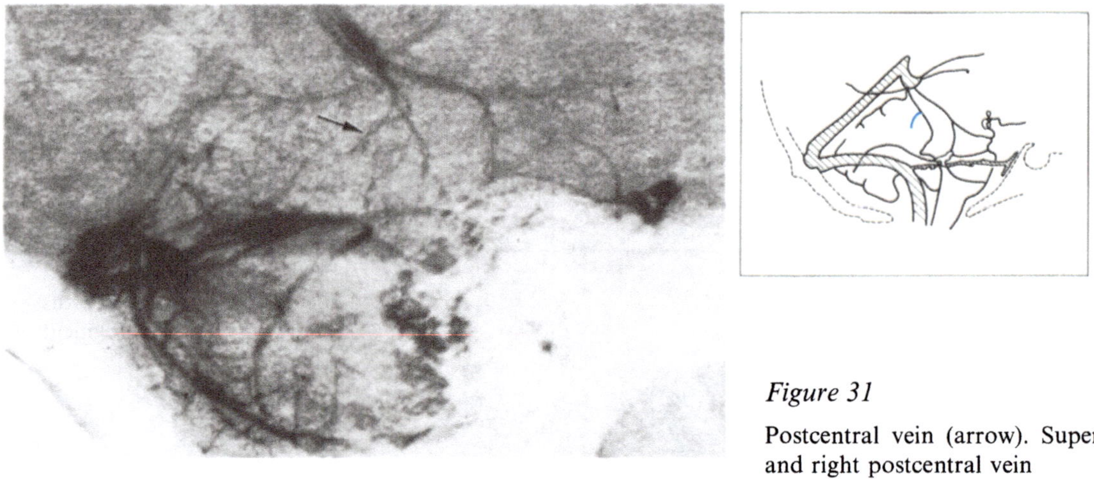

Figure 31

Postcentral vein (arrow). Superimposition of left and right postcentral vein

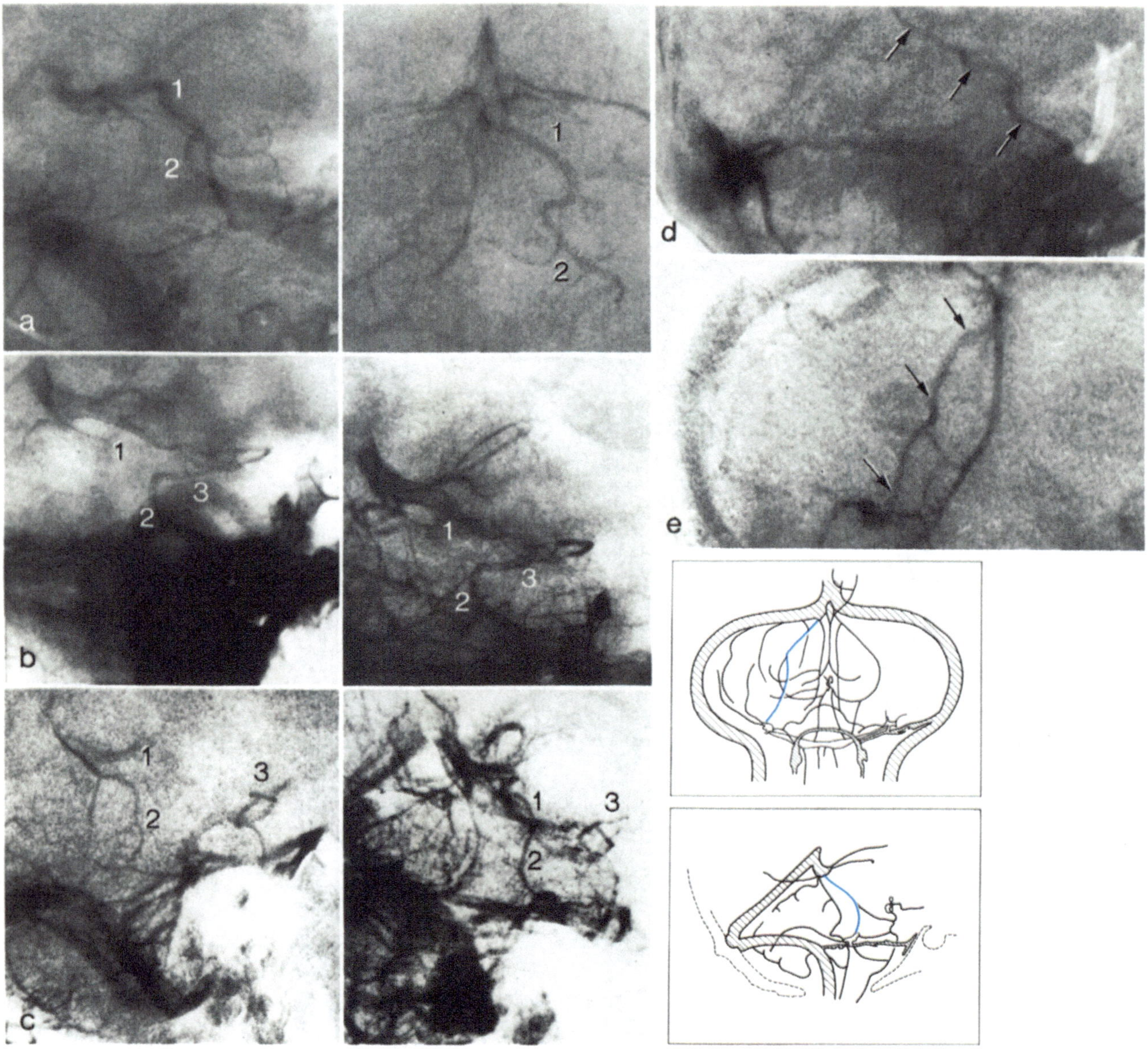

Figure 32

a–c Different types of drainage:
1 Posterior segment of basal vein, *2* Lateral mesencephalic vein, *3* Optopeduncular veins

d, e Superior internal cerebellar hemispheric vein

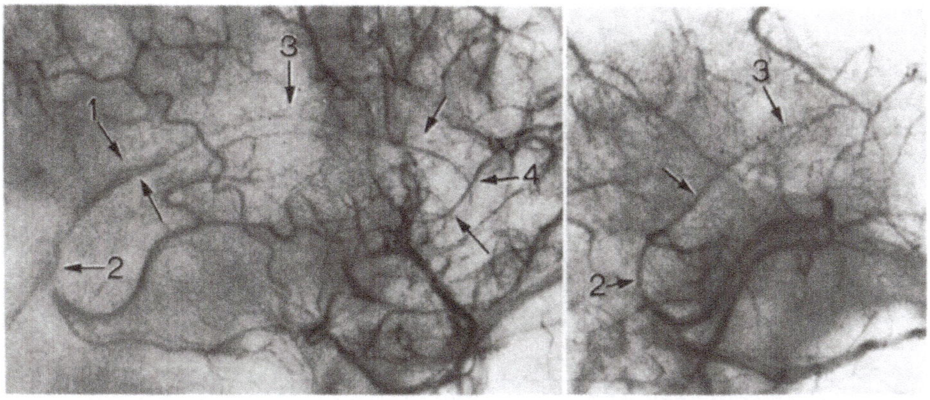

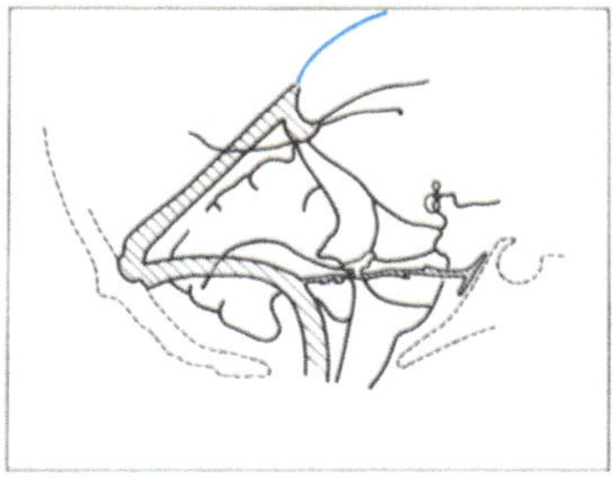

Figure 33 Posterior cerebral vein

1 Anteroposterior segment split in two branches
2 Terminal segment around splenium
3 Anastomosis with anterior pericallosal vein
4 Anterior pericallosal vein

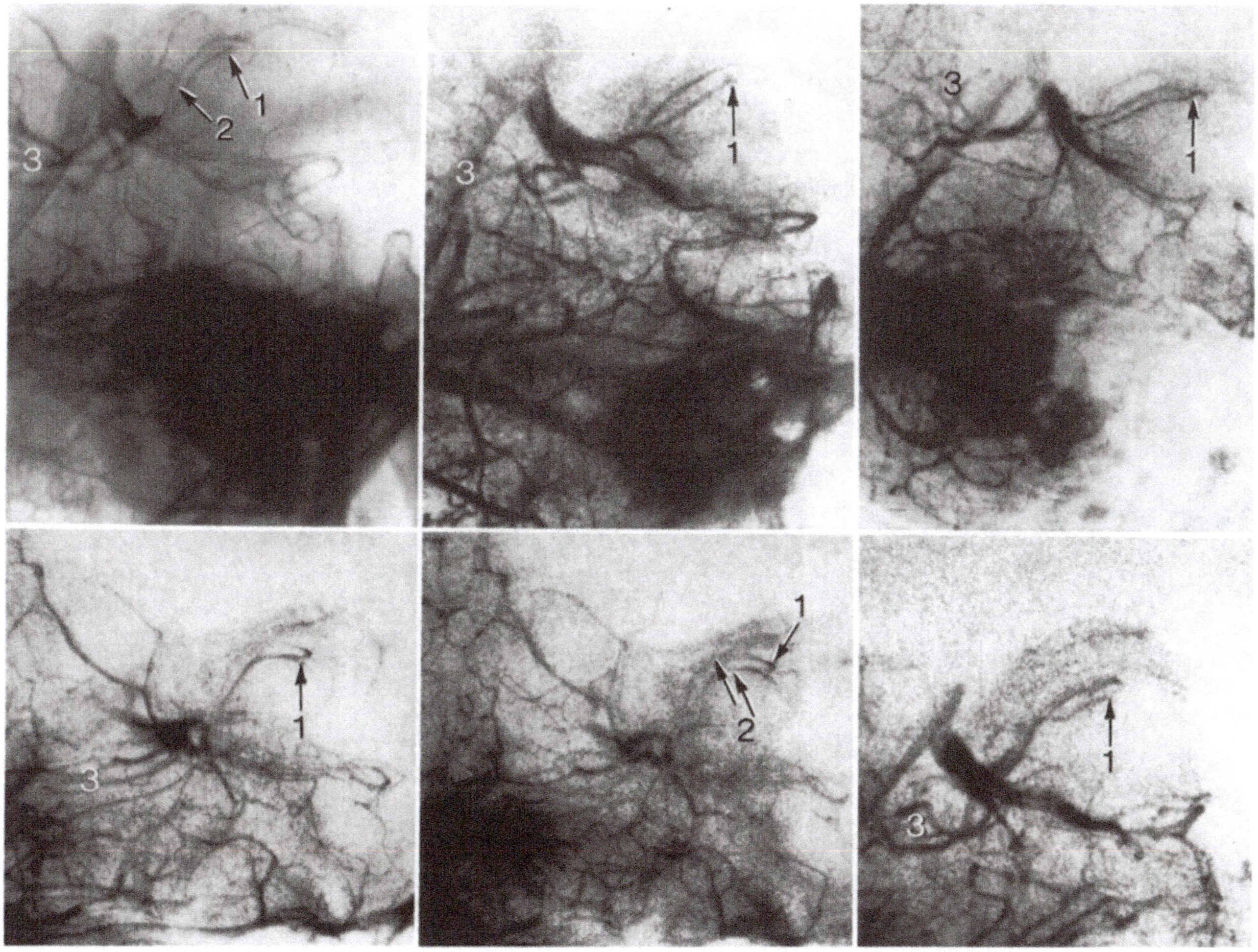

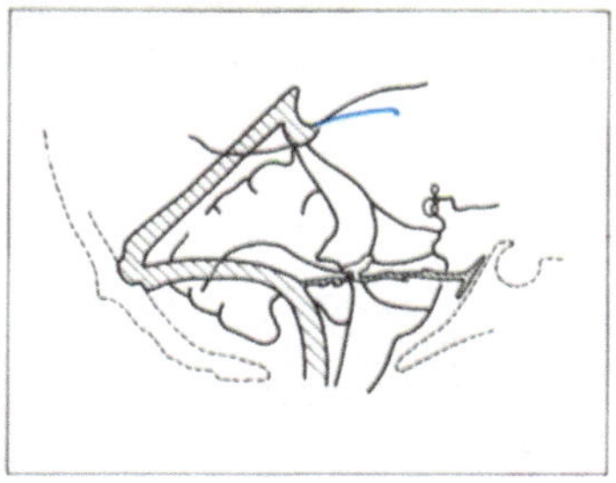

Figure 34

Superior thalamic vein

1 Origin of superior thalamic vein
2 Curvilinear course of superior thalamic veins
3 Internal occipital veins

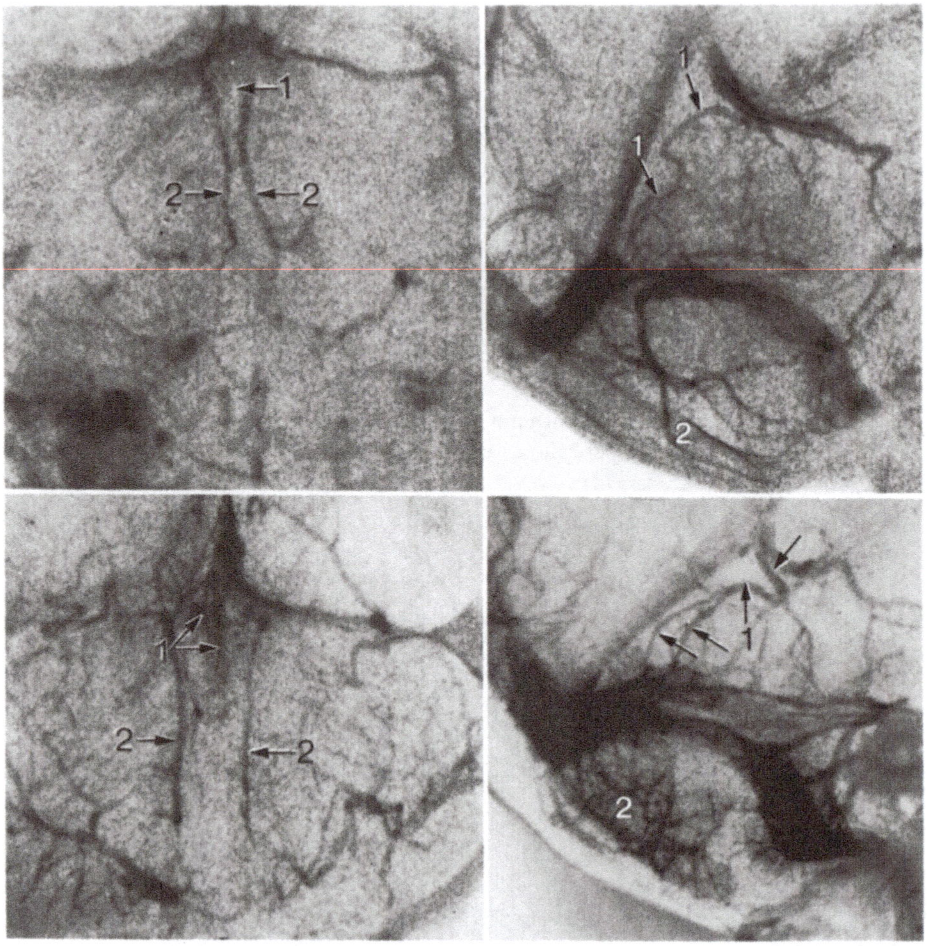

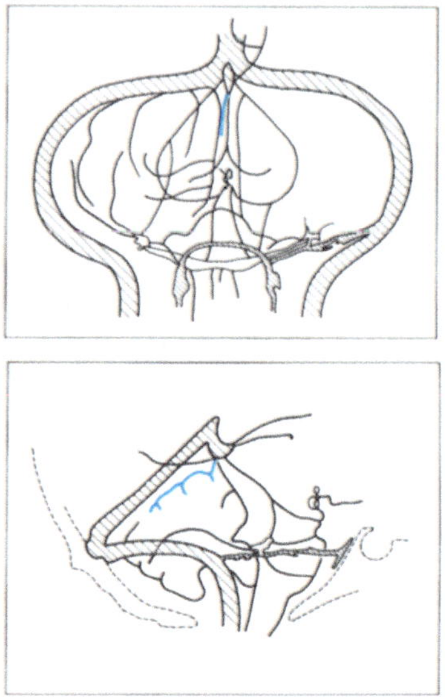

Figure 35

Superior vermian vein in two cases
Frontal and lateral projection:

1 Superior vermian vein
2 Inferior vermian vein (paravermian vein)

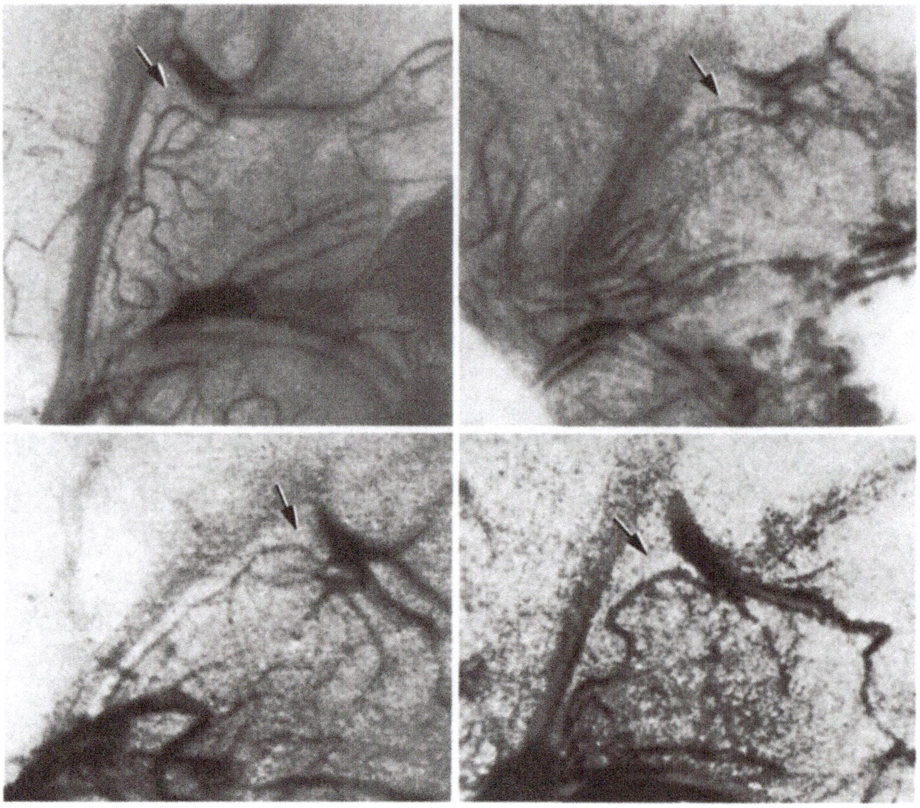

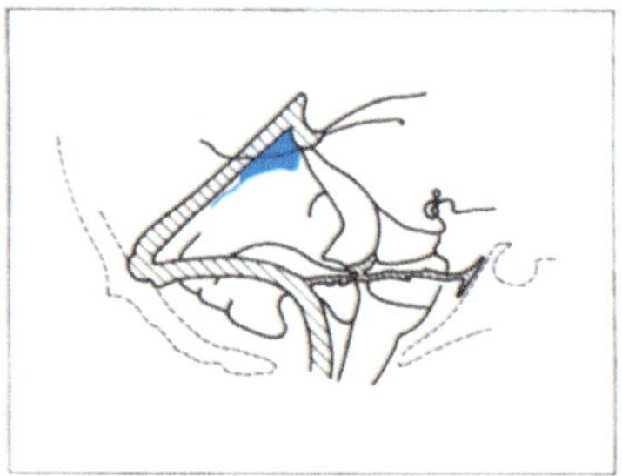

Figure 36

Superior vermian vein delineates a triangular space which is limited by straight sinus behind and great vein of Galen in front. This space corresponds to the subdural space above superior cerebellar vermis

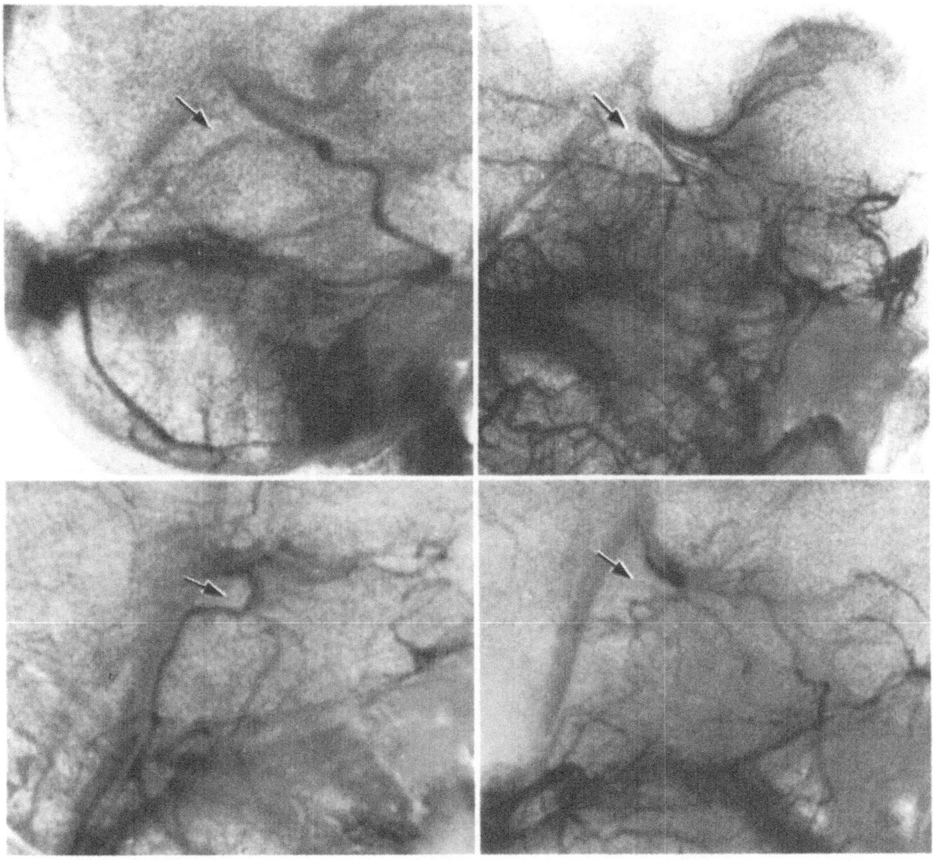

Figure 37

Four other examples of subdural space delineated by veins

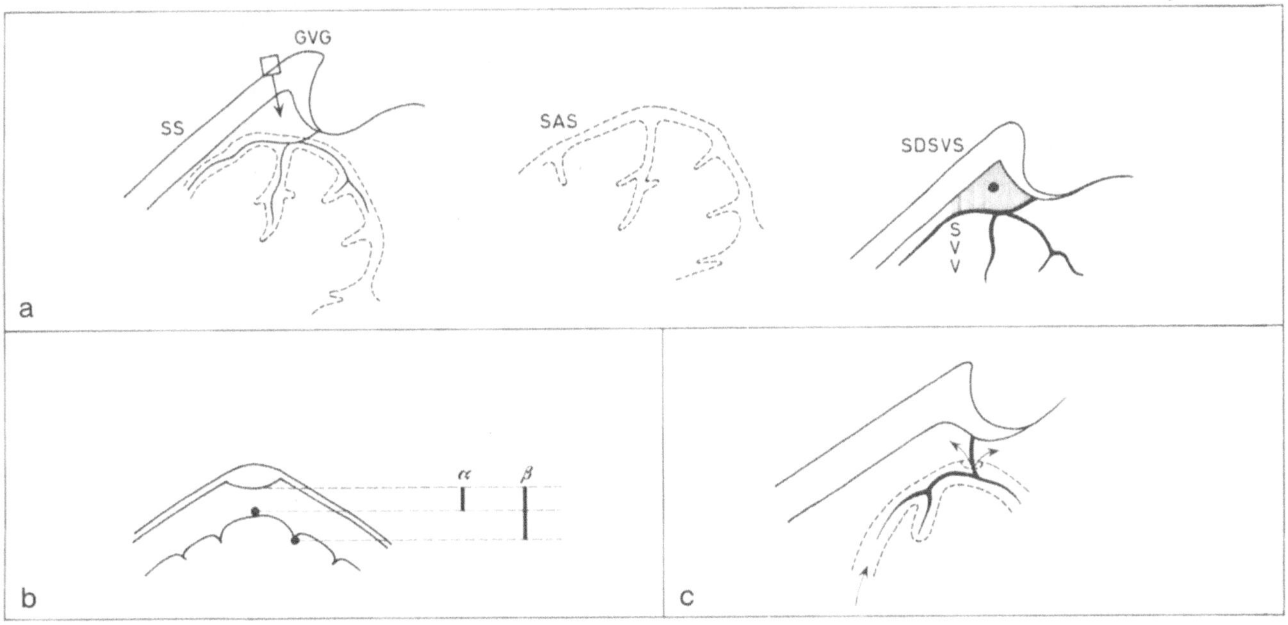

Figure 38

a Venous limits of subdural space above superior cerebellar vermis

SS	Straight sinus
GVG	Great vein of Galen
SVV	Superior vermian vein
SAS	Subarachnoïd space
SDSVS	Subdural supravermian space

b Height of triangular projection surface corresponding at subdural space above superior cerebellar vermis depends on more or less lateral course of superior vermian vein. $\alpha=$ normal height in cases of a strictly median course of superior vermian vein, $\beta=$increased height in cases of a paramedian or lateral course of vein

c During encephalography, gas in subarachnoïd space can progress along superior vermian vein and fill subdural space

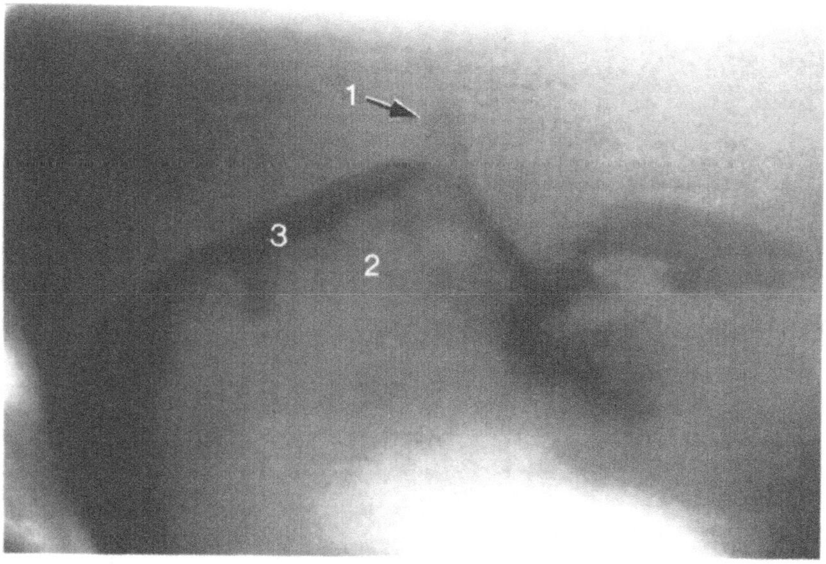

Figure 39

Air (*1*) penetrating in subarachnoïdal space (*3*) above cerebellar vermis (*2*). Air courses very probably along superior vermian vein

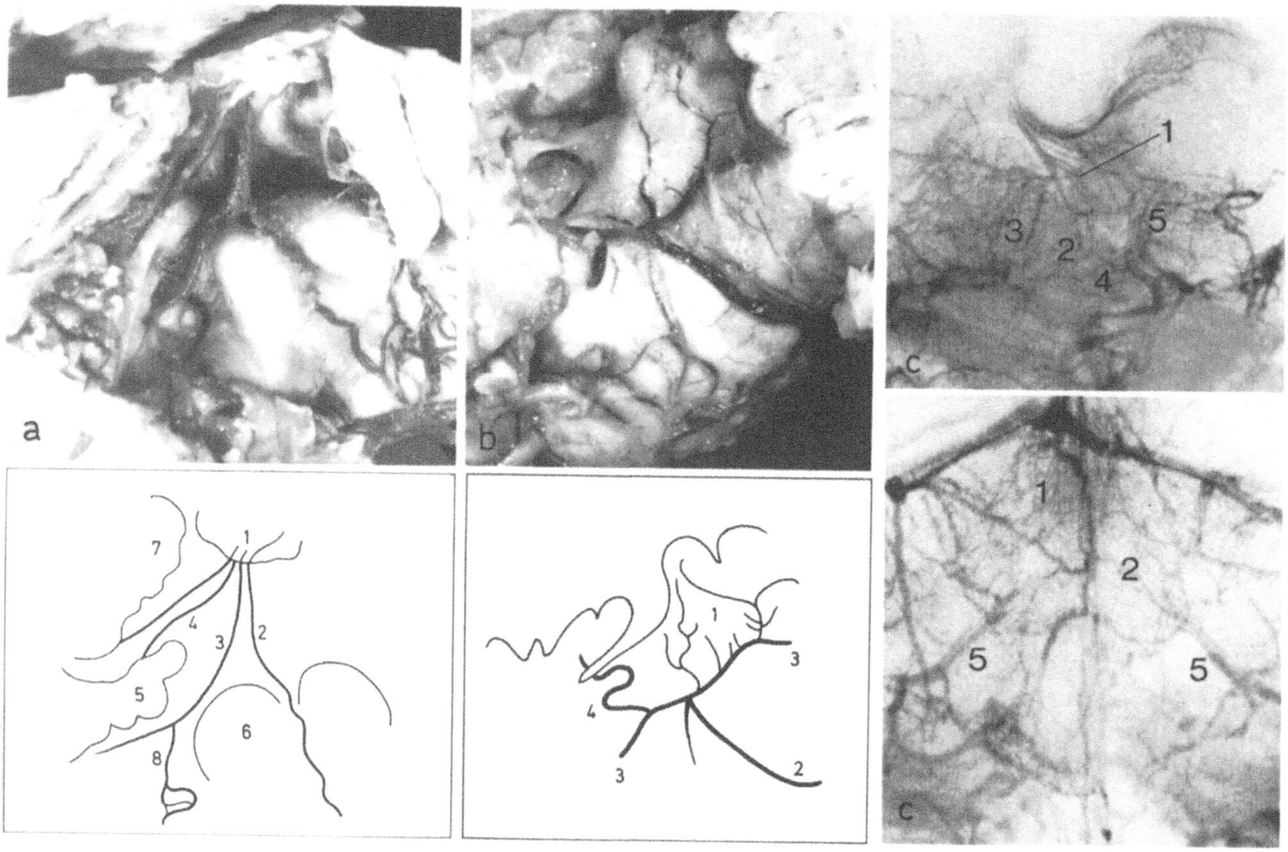

Figure 40

a Specimen of veins in mesencephalic area
- *1* Great vein of Galen
- *2* Posterior mesencephalic vein
- *3* Precentral vein
- *4* Postcentral vein
- *5* Lobulus centralis
- *6* Quadrigeminal plate
- *7* Culmen
- *8* Brachial vein joins precentral vein and superior petrous vein (brachial or superior cerebellar peduncular vein)

b Veins of quadrigeminal plate
- *1* Veins of colliculi quadrigemini
- *2* Vein of pontopeduncular sulcus (superior transversal pontine vein)
- *3* Lateral mesencephalic vein
- *4* Brachial vein (superior cerebellar peduncular vein)

c Venous phase of normal vertebral angiography
- *1* Veins of quadrigeminal plate
- *2* Precentral vein
- *3* Postcentral vein
- *4* Brachial vein
- *5* Lateral mesencephalic veins

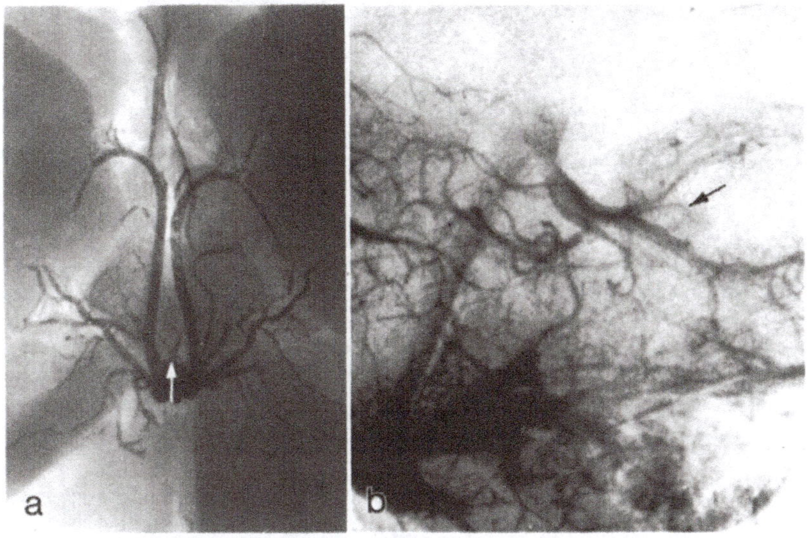

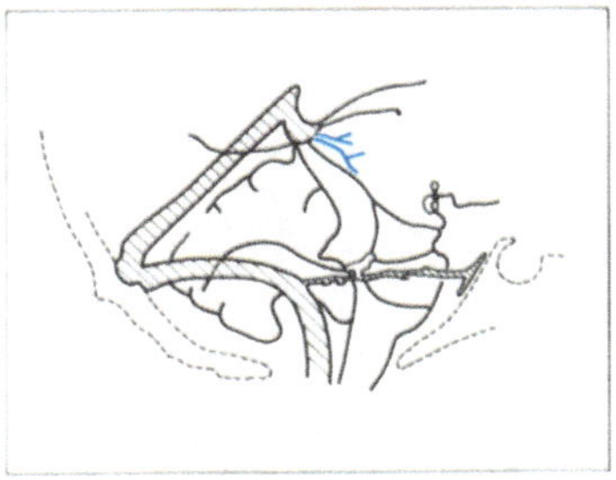

Figure 41 a and b

Veins of pineal gland (posterior thalamic veins). See also Figure 52 d

a Specimen with opacified veins (axial view)

b Venous phase of a normal vertebral angiography (lateral view)

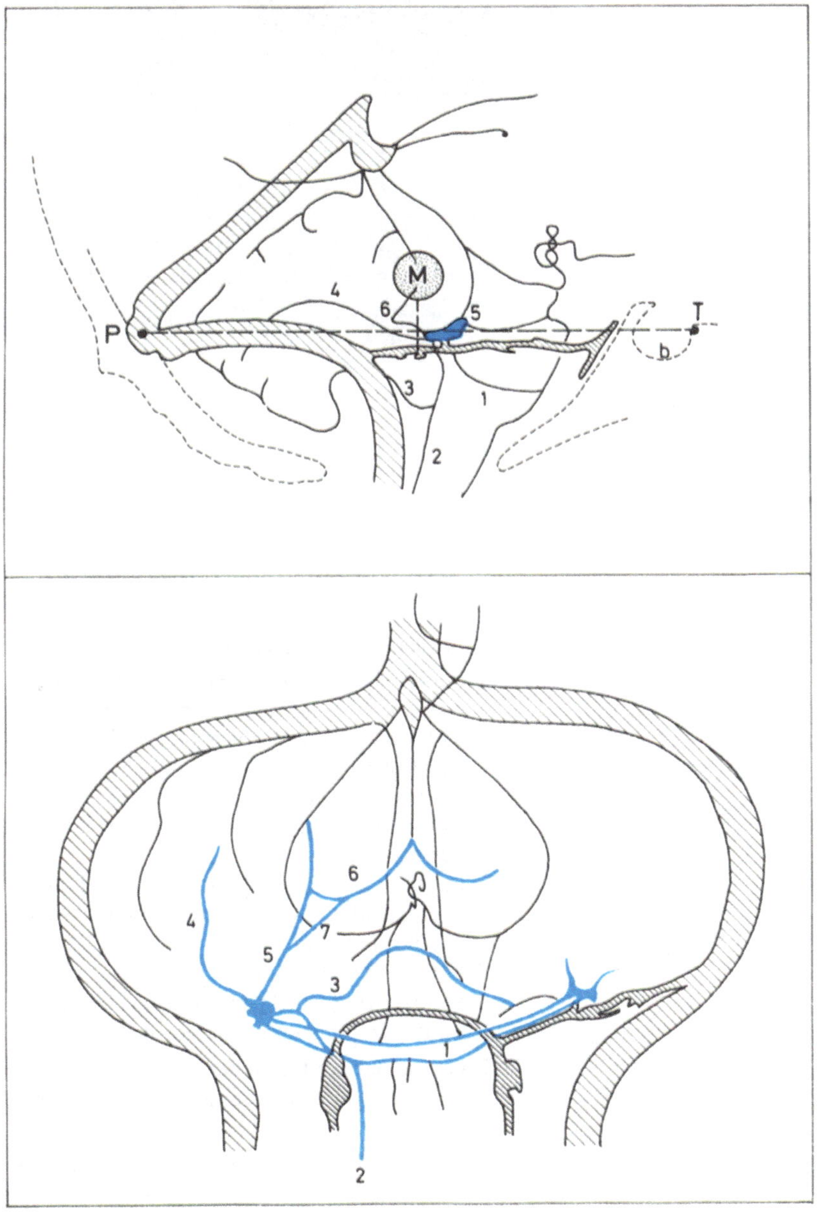

Figure 42

Drawing of superior petrous vein and its affluents in frontal and in lateral projection

a Internal auditory canal
b Sella turcica
T Tuberculum sellae
M Midpoint of TP (Twining's line)
P Internal occipital protuberance
1 Transversal pontine veins
2 Lateral medullopontine vein
3 Vein of lateral recess of fourth ventricle
4 Cerebellar hemispheric vein
5 Lateral mesencephalic vein
6 Brachial vein
7 Venous anastomoses between precentral and lateral mesencephalic vein

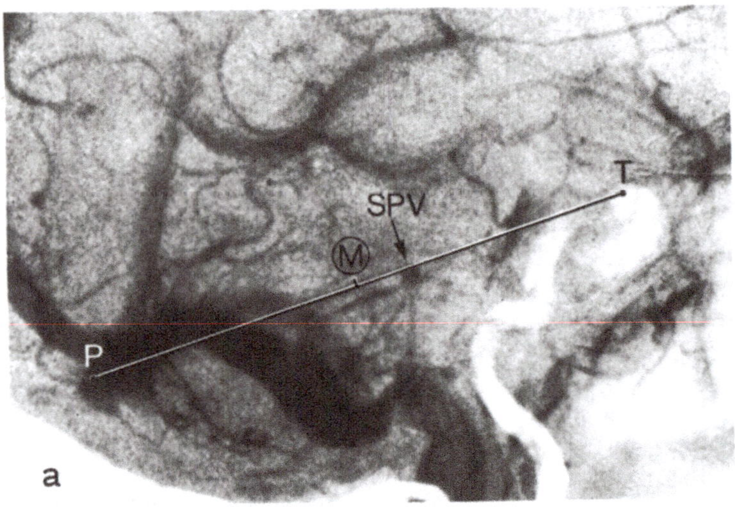

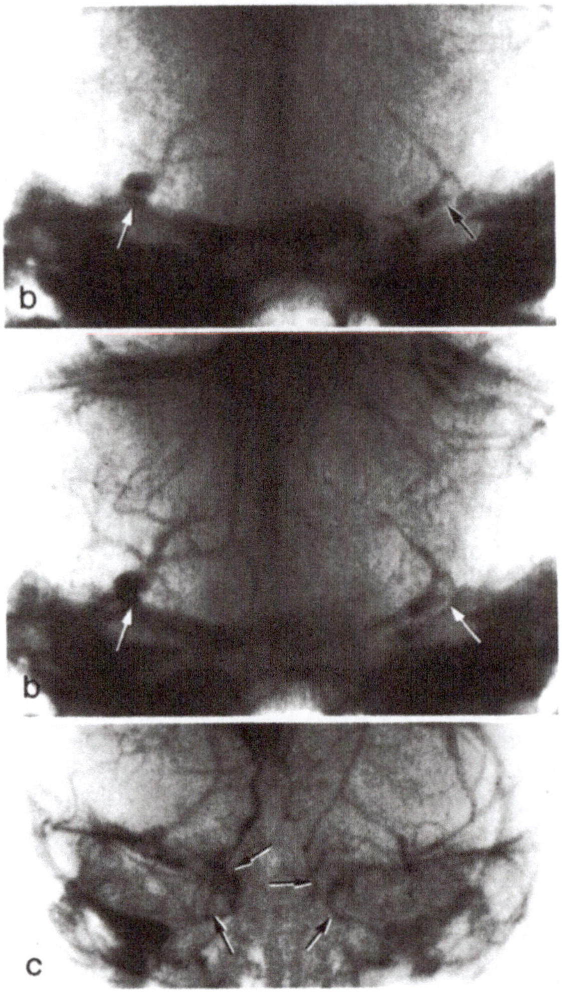

Figure 43

a The superior petrous vein (SPV) lies in front of midpoint (*M*) of *TP* (Twining's line)
b Superior petrous veins in two adults
c Inferior petrous vein in child (IPV)

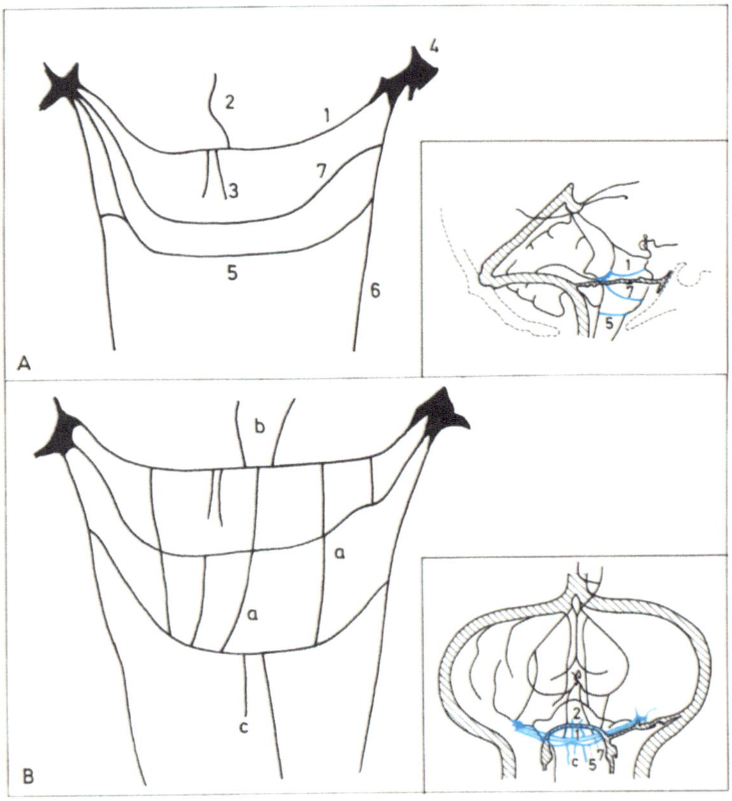

Figure 44

A Transversal pontine veins
- *1* Superior transversal pontine vein running in pontopeduncular sulcus and communicating with optopeduncular veins (*2*) and with the clival plexus (*3*)
- *2* Optopeduncular veins
- *3* Communication with the clival plexus
- *4* Superior petrous vein
- *5* Inferior transversal pontine vein joining lateral medullopontine vein (*6*) and/or superior petrous vein
- *6* Lateral medullopontine vein
- *7* Median transversal pontine vein. This inconstant vein joins both superior petrous veins

B Longitudinal pontine veins
- *a* Longitudinal veins connecting transversal veins
- *b* Longitudinal veins connecting superior transversal veins with optopeduncular veins
- *c* Longitudinal veins connecting inferior transversal veins with prebulbar veins

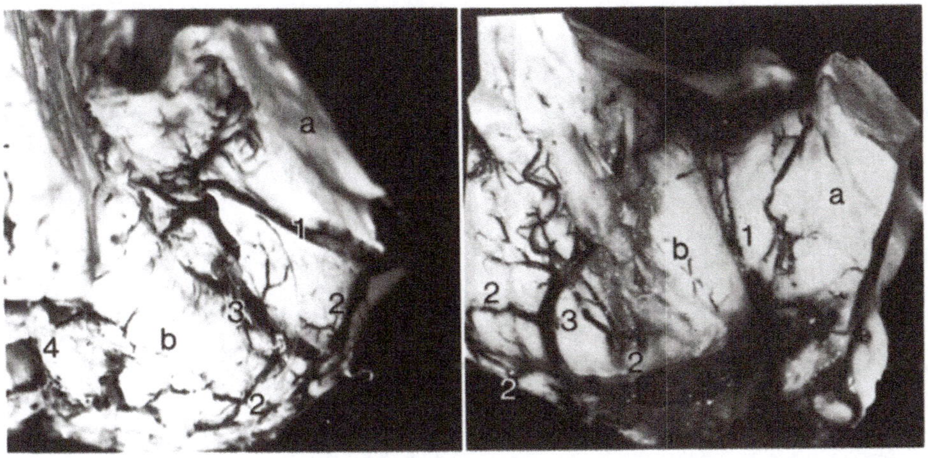

Figure 45

Specimen

a Cerebral peduncle
b Pons
1 Superior transversal pontine vein (pontopeduncular sulcus)
2 Longitudinal pontine vein
3 Median transversal pontine vein
4 Inferior transversal pontine vein (pontobulbar sulcus)

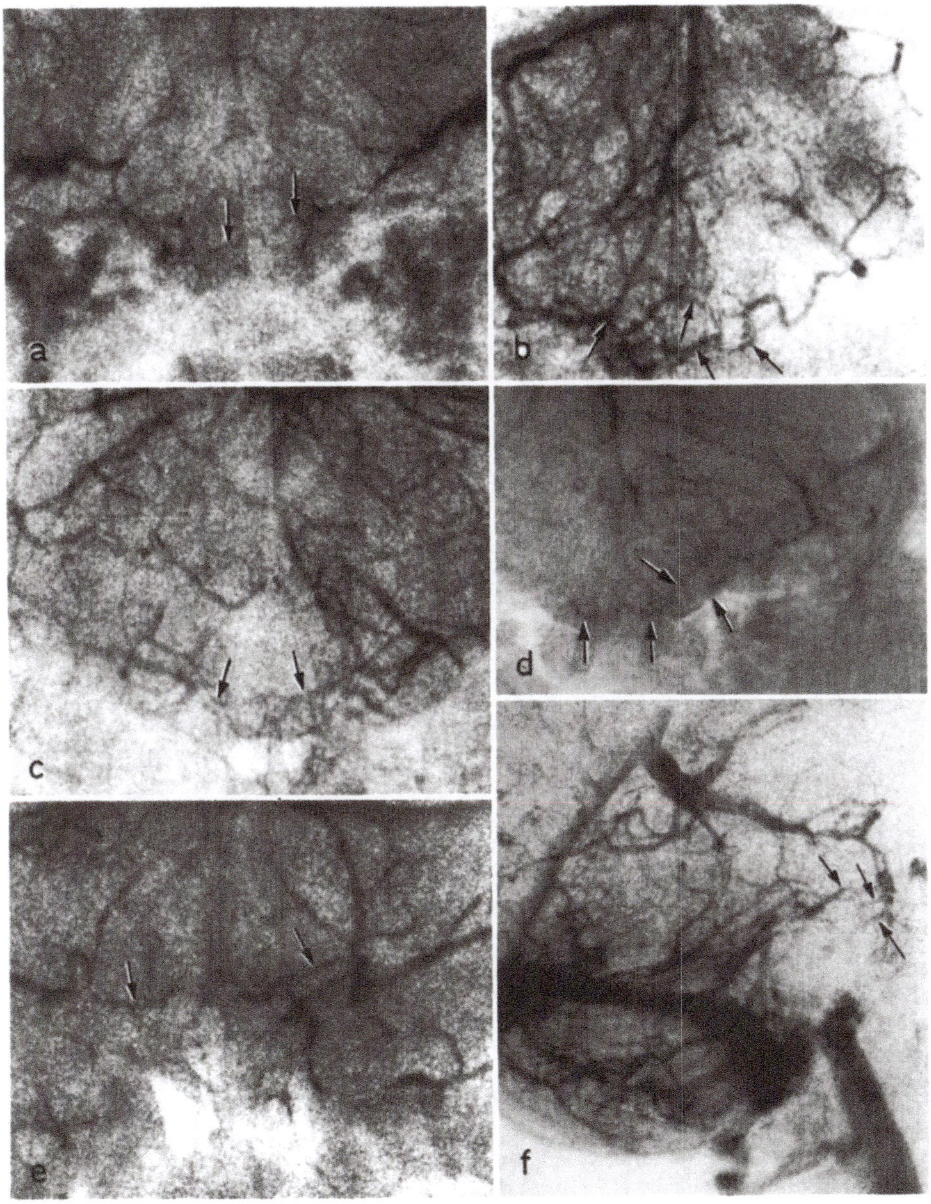

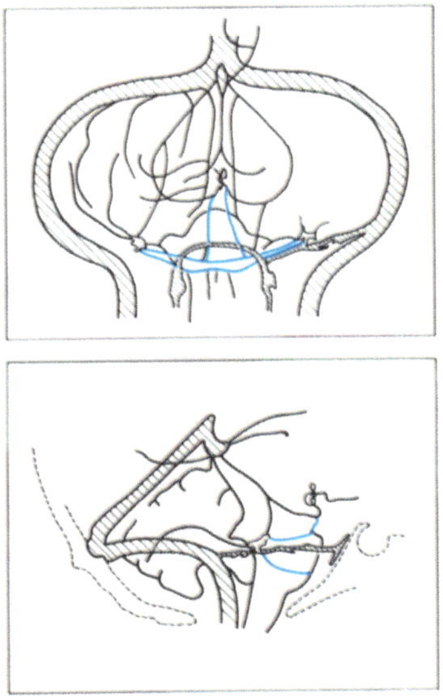

Figure 46a–f

Transversal pontine veins

a–e Frontal projection of these veins demonstrates one or two main trunks. In some cases there is a complex network of transversal pontine veins

f In a lateral projection, left and right veins are superimposed and occupy area between longitudinal pontine vein and superior petrous vein

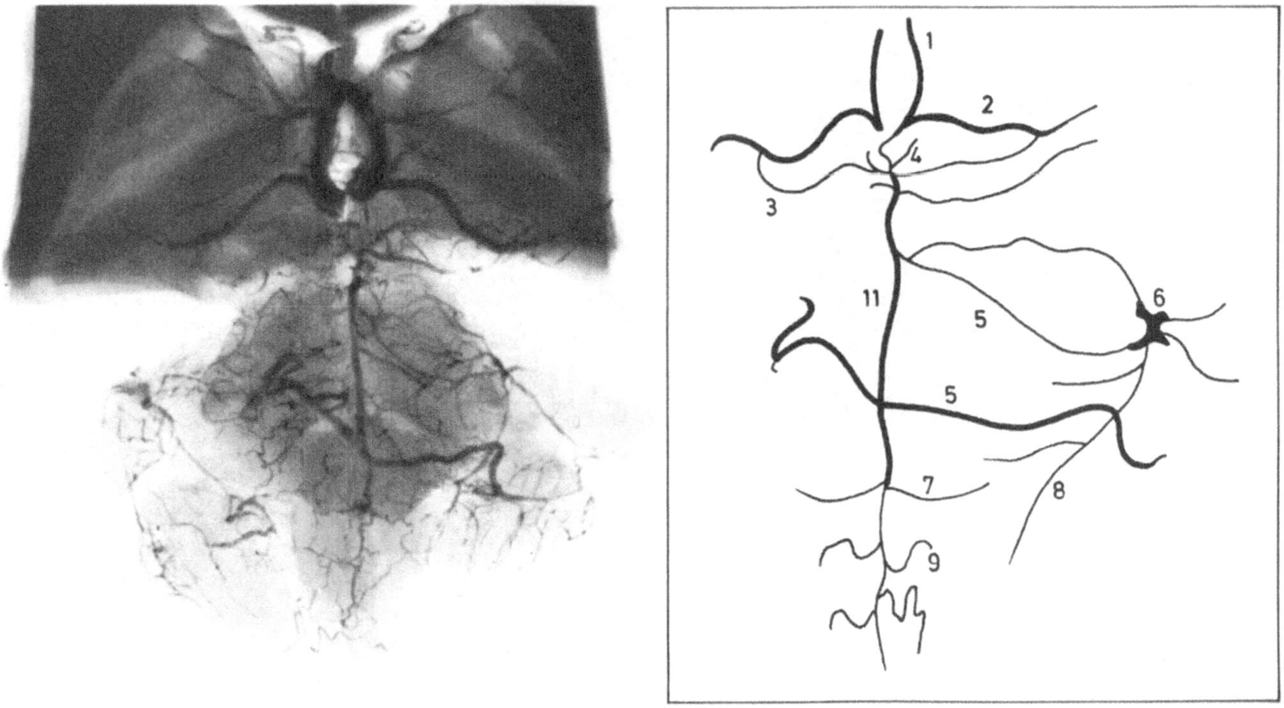

Figure 47

Plain radiography of a specimen with opacified veins (axial view). See legend of Figure 48

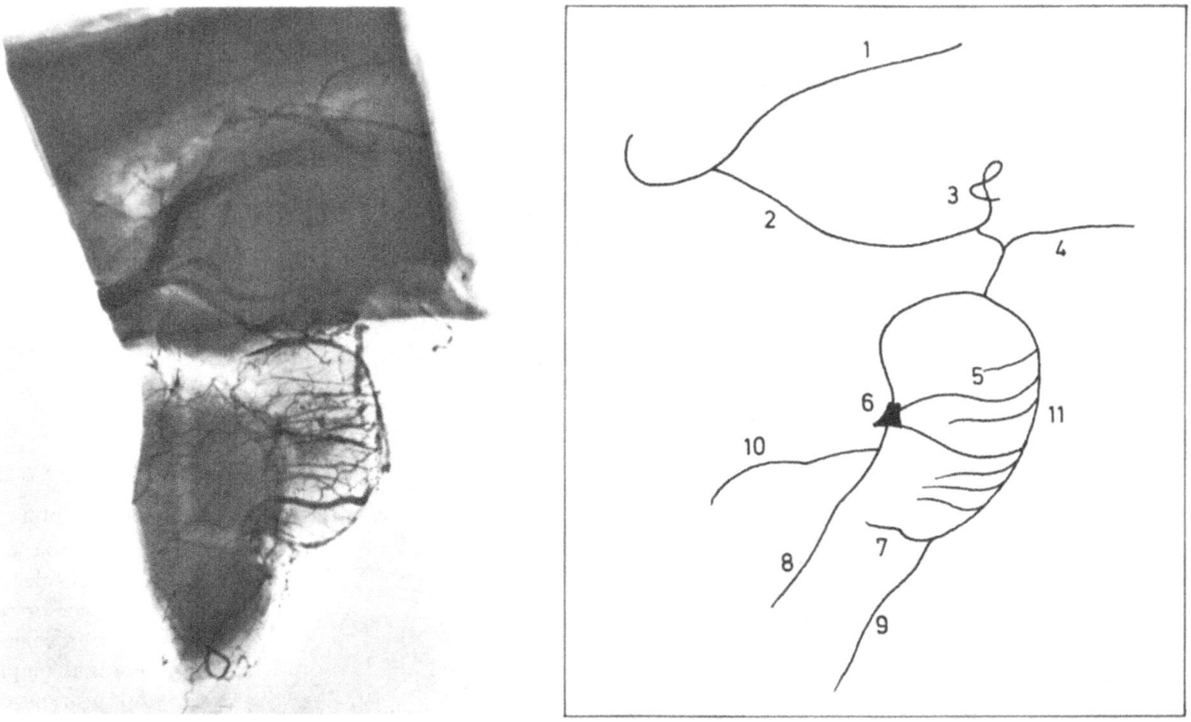

Figure 48

Plain radiography of a specimen with opacified veins (lateral view)

1 Internal cerebral vein
2 Basal vein
3 Posterior communicating vein
4 Optopeduncular veins
5 Transversal pontine veins
6 Superior petrous vein
7 Inferior transversal pontine vein (pontop duncular sulcus)
8 Lateral medullopontine vein
9 Bulbar veins
10 Vein of lateral recess of fourth ventricle
11 Longitudinal pontine vein

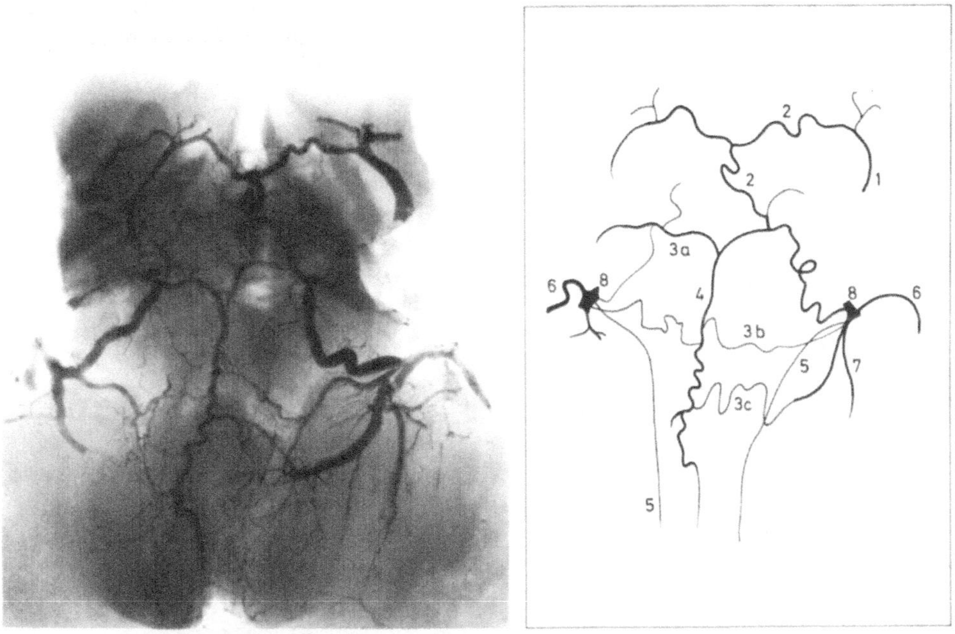

Figure 49

Plain radiography of a specimen with opacified veins

1 Basal vein
2 Optopeduncular veins
3a Superior transversal pontine vein
3b Median transversal pontine vein
3c Inferior transversal pontine vein
4 Longitudinal pontine vein
5 Lateral medullopontine vein
6 Vicq d'Azyr's vein
7 Cerebellar hemispheric vein
8 Superior petrous vein

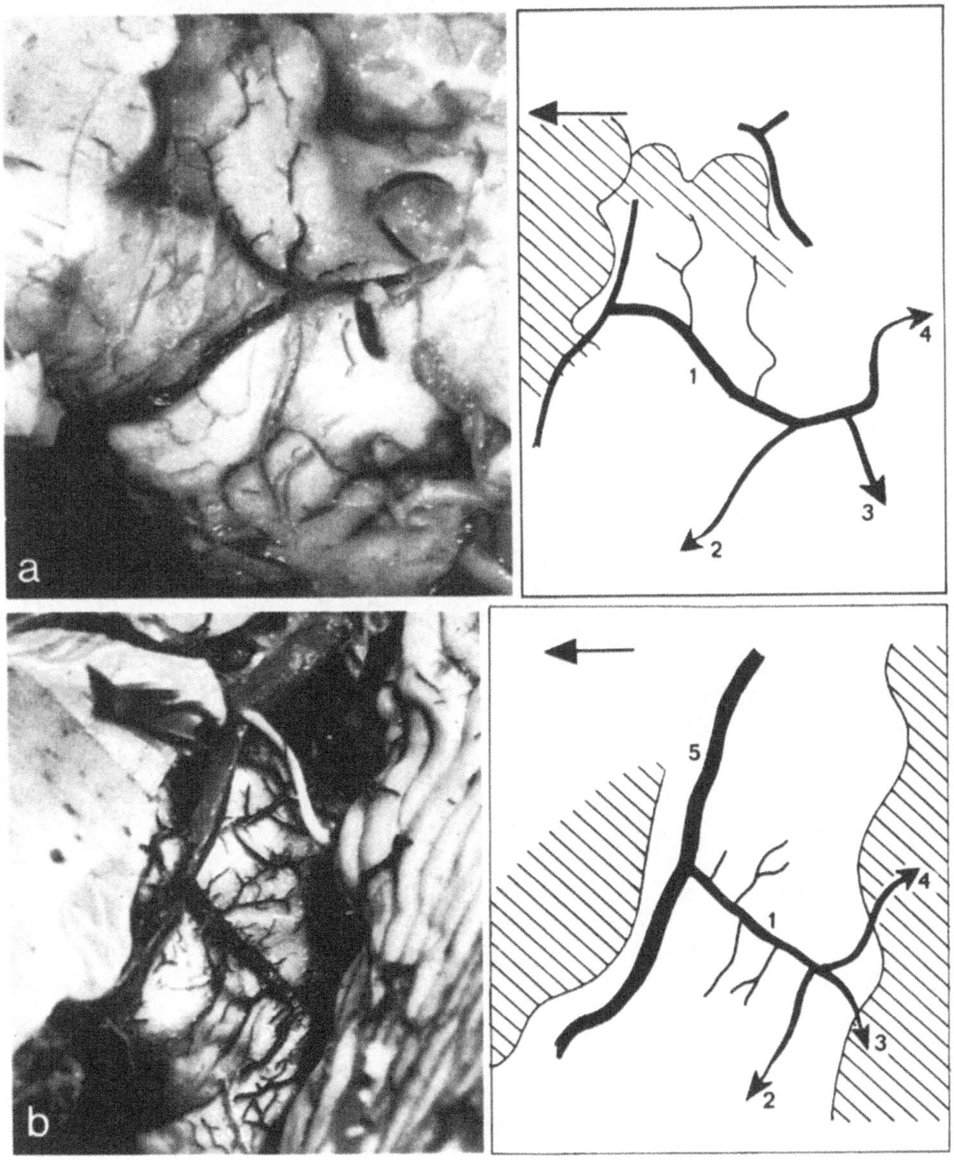

Figure 50a and b

Lateral mesencephalic vein in two specimens

1 Lateral mesencephalic vein
2 Junction with pontine veins
3 Junction with superior petrous vein
4 Junction with cerebellar hemispheric veins
5 Basal vein

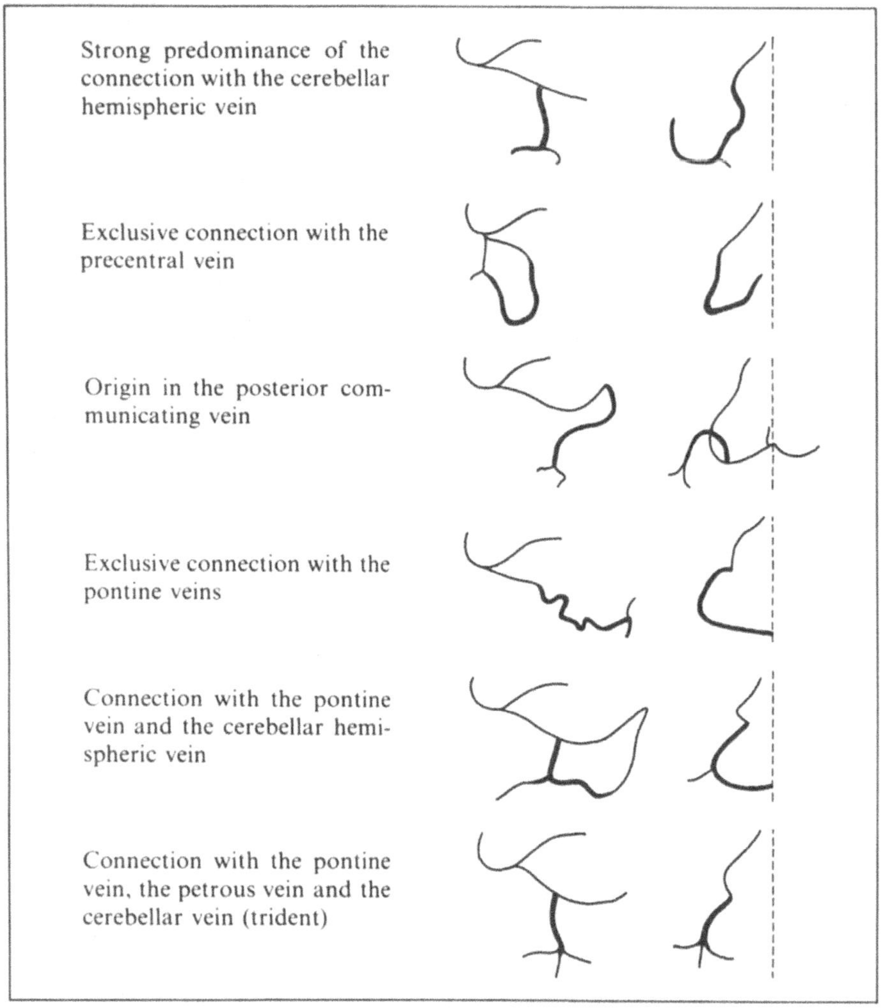

Figure 50 c

Drawing of main variations in drainage of lateral mesencephalic vein

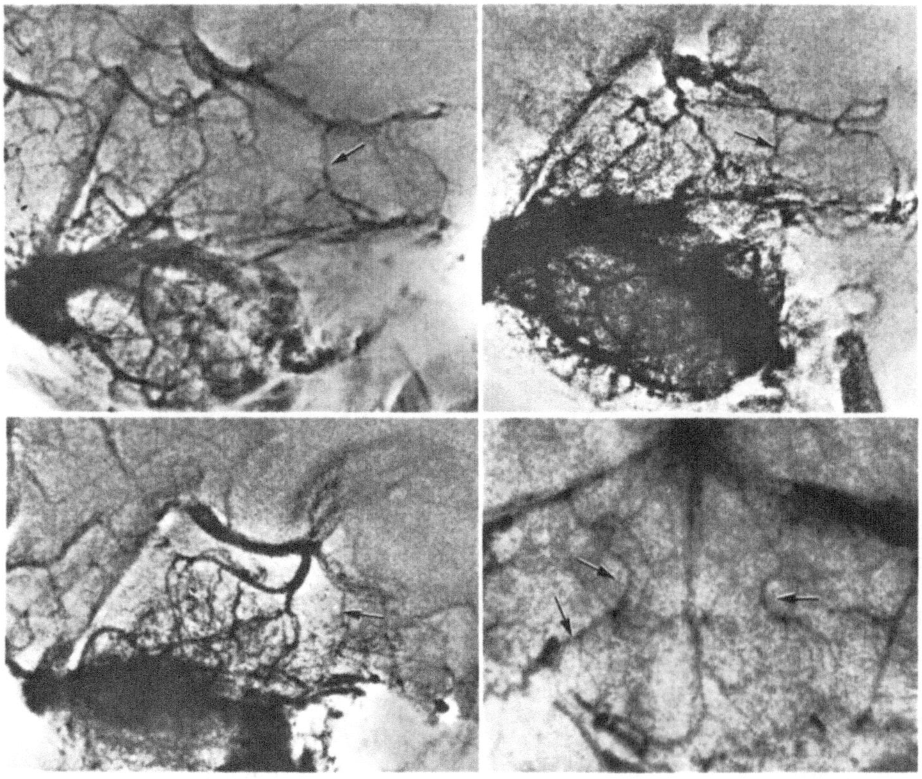

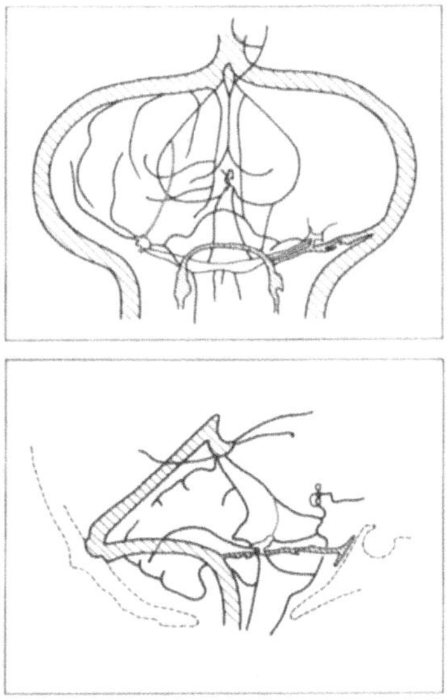

Figure 51

Lateral mesencephalic vein. Normal course of the lateral mesencephalic vein

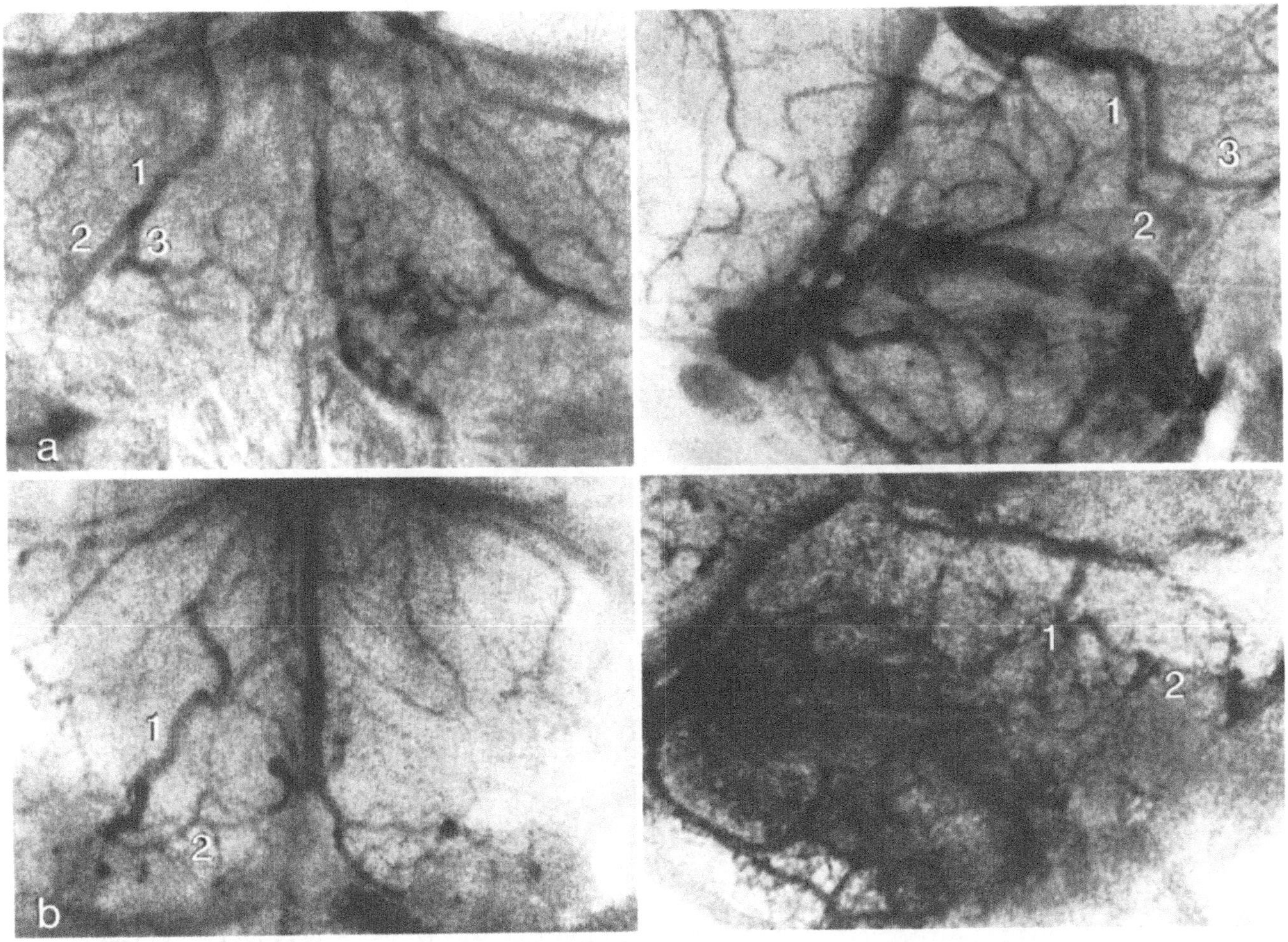

Figure 52a and b Variations of the lateral mesencephalic vein

a *1* Hypertrophic lateral mesencephalic vein
2 Anastomosis with superior petrous vein
3 Anastomosis with pontine veins

b *1* Hypertrophic and undulated lateral mesencephalic vein
2 Anastomosis with pontine veins

Figure 53

Specimen with opacified veins demonstrating anterior bulbar network

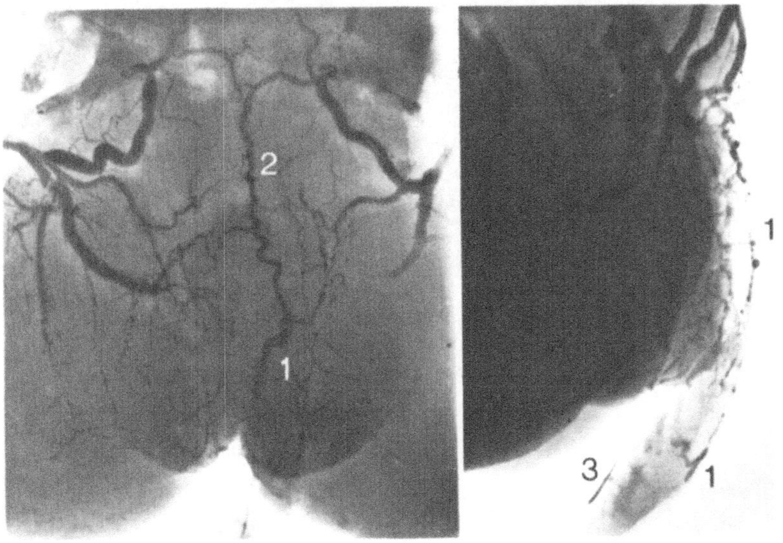

Figure 54

Plain radiograph of a specimen with opacified veins of medulla oblongata. *1* anterior medial vein of medulla oblongata; *2* anterior medial vein of pons; *3* posterior medial vein of medulla oblongata

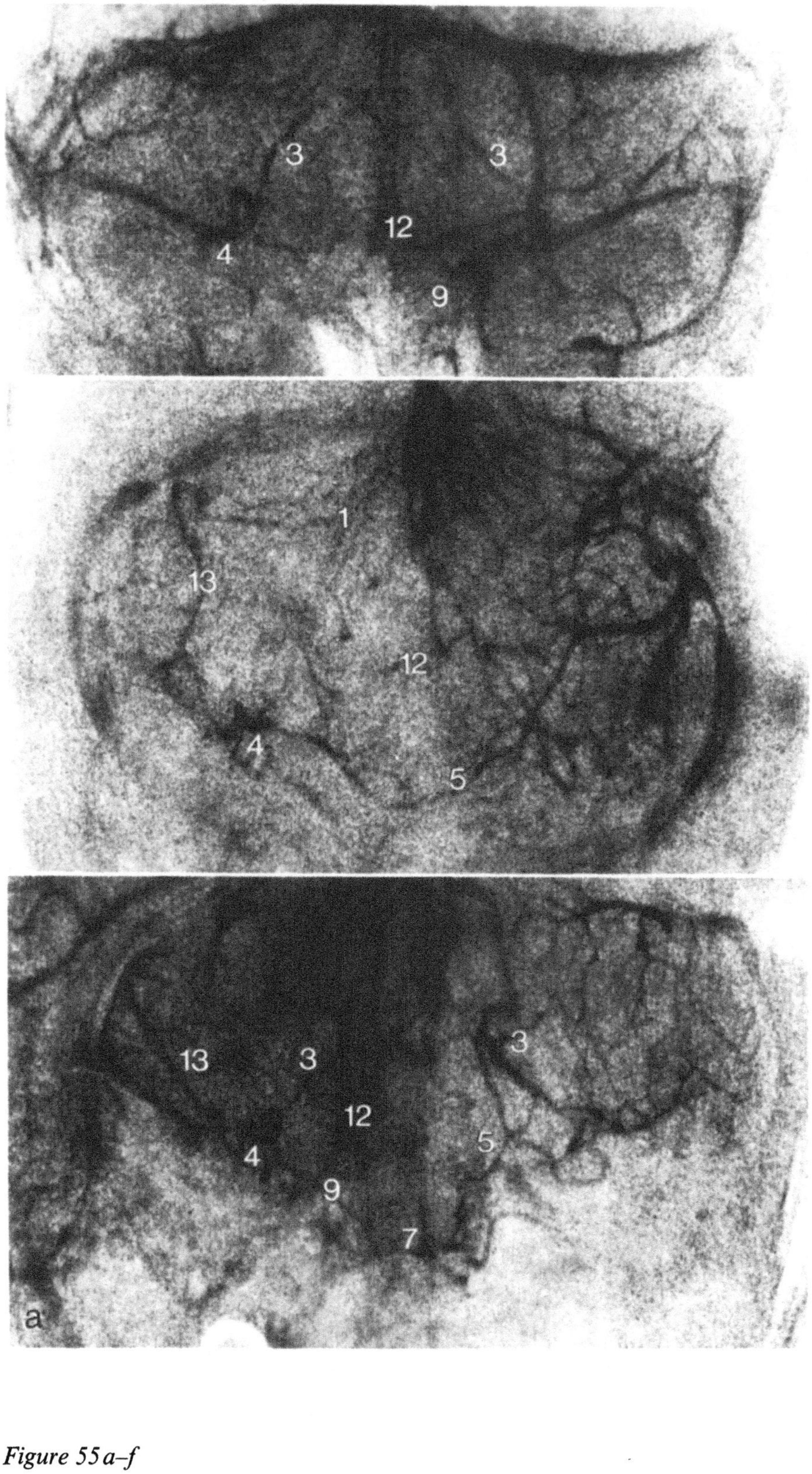

1 Basal vein
2 Optopeduncular veins
3 Lateral mesencephalic vein
4 Superior petrous vein
5 Transversal pontine veins
6 Superior transversal pontine vein (transversal ponto-peduncular vein)
7 Inferior transversal pontine vein (transversal pontobulbar vein)
8 Longitudinal medial pontine vein
9 Lateral medullopontine vein
10 Anterior medial bulbar vein
11 Posterior medial bulbar vein
12 Vein of lateral recess of fourth ventricle
13 Vein of great horizontal fissure of Vicq d'Azyr
14 Veins of olive
15 Lateral vein of floor of fourth ventricle

Figure 55 a–f

Lateral medullopontine vein

a Frontal views
b Lateral views
c Frontal drawing
d Lateral drawing
e Localization of junction point between vein of lateral recess of fourth ventricle (*3*) and lateral medullopontine vein (*4*) in lateral view. This point is located on vertical line (*2*) dropped on midpoint *M* of Twining's line (*1*)
f Drawing of lateral medullopontine vein as seen in phlebographic frontal projection on right side. Lateral projection reported in Figure 63

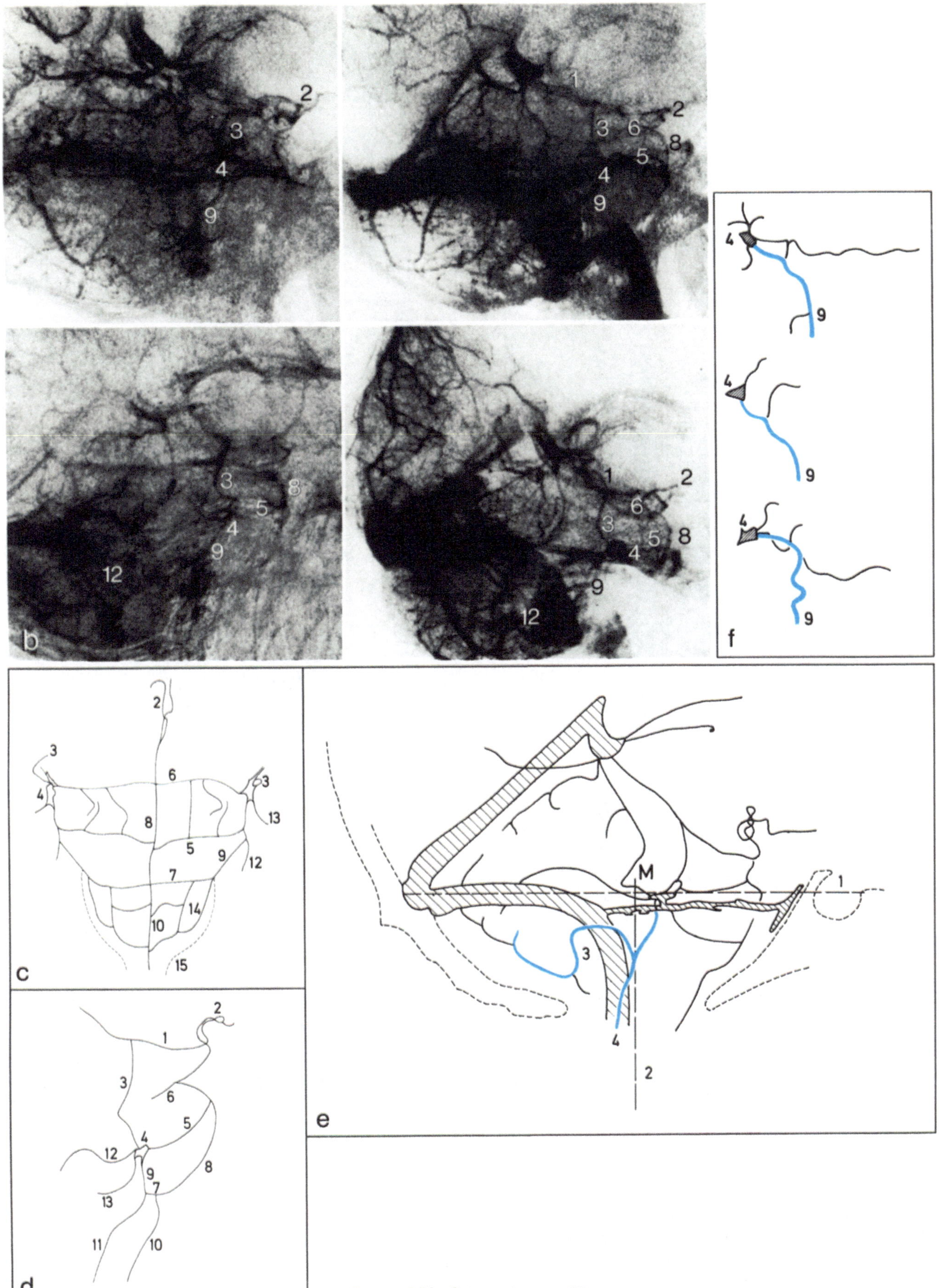

Figure 55b–f (legend see p. 54)

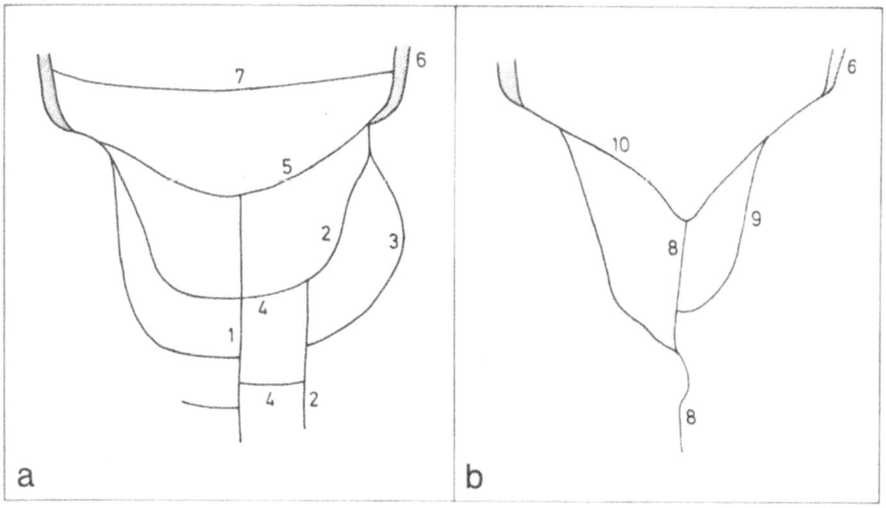

Figure 56

Anterior (a) and posterolateral (b) veins of medulla oblongata

1 Anterior medial bulbar vein
2 Preolivary vein
3 Retroolivary vein
4 Anterior transverse bulbar vein
5 Anterior medullopontine vein
6 Lateral medullopontine vein
7 Inferior transverse pontine vein (transverse medullopontine vein)
8 Posterior medial bulbar vein
9 Restiform body vein
10 Bordering veins of floor of fourth ventricle

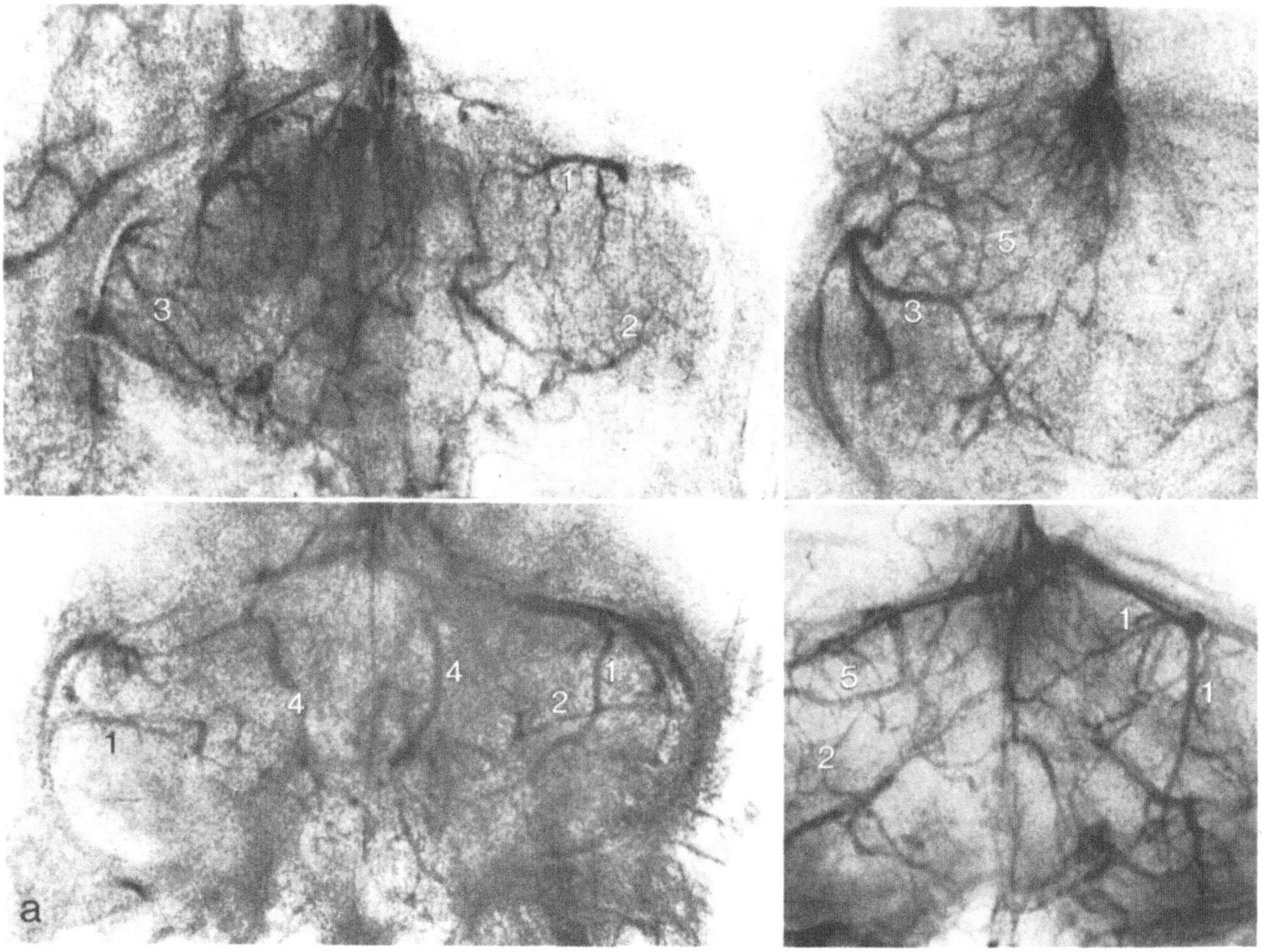

Figure 57 a–d

Veins of cerebellar hemispheres

a Frontal views
b Lateral view
c Frontal drawing
d Lateral drawing

1 Posterior cerebellar veins draining into lateral sinus
2 Anterior cerebellar veins draining into superior petrous vein
3 Vein of great horizontal fissure of Vicq d'Azyr
4 Paravermian vein
5 Transverse cerebellar vein joining paravermian vein

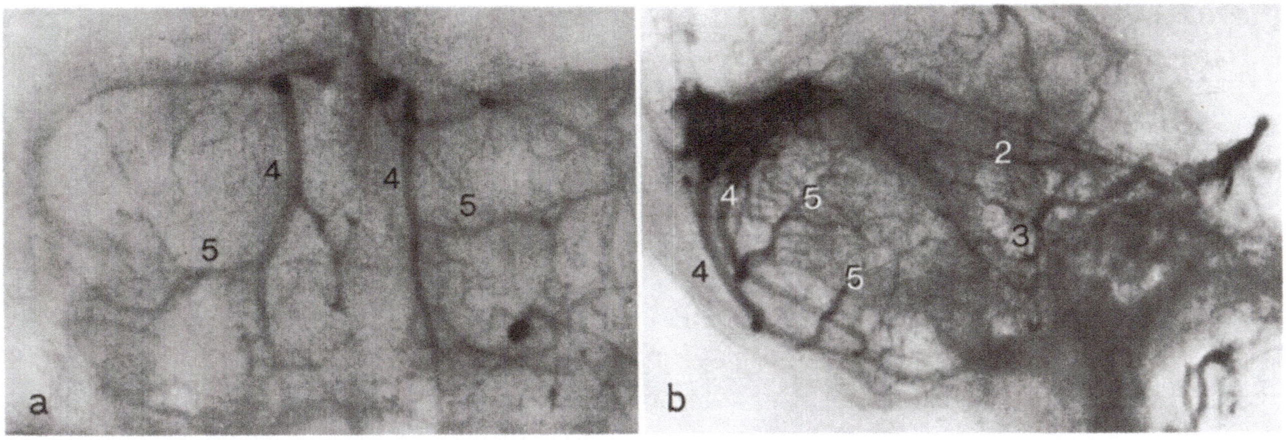

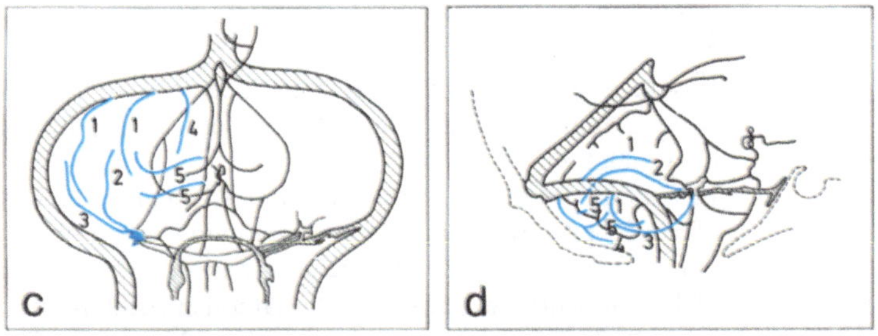

Figure 57 a–d (legend see p. 56)

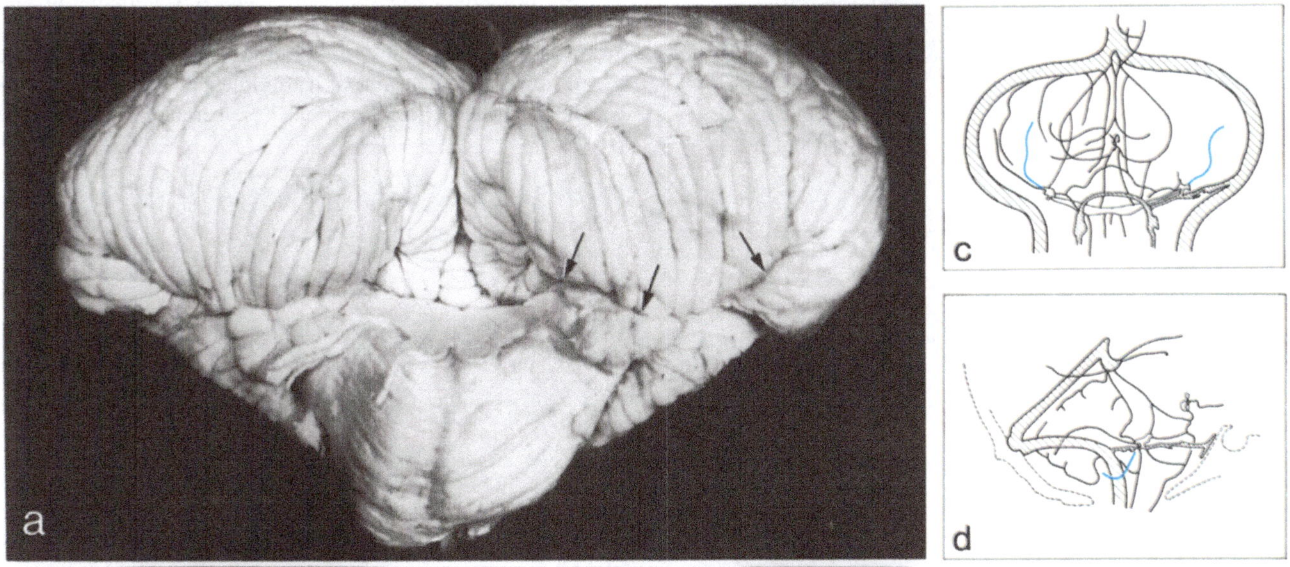

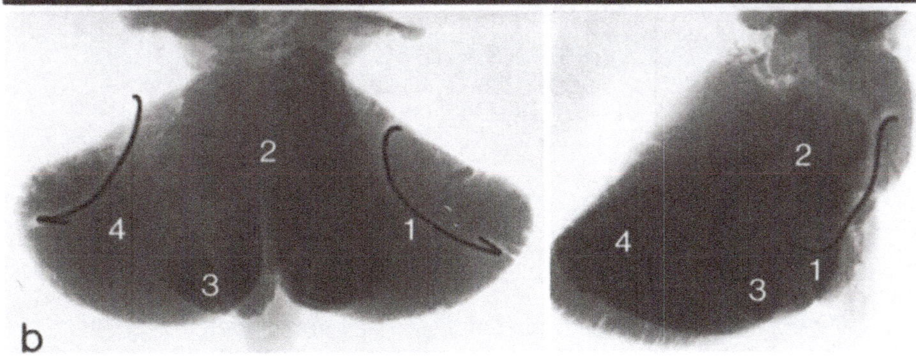

1 Great horizontal fissure of Vicq d'Azyr
2 Fourth ventricle
3 Tonsil
4 Cerebellar hemisphere

Figure 58 a–d

Specimen of cerebellar hemispheres with localization of great horizontal fissure of cerebellum (Vicq d'Azyr) in frontal and in lateral projection. a Photography of specimen; b Radiography of specimen with metal wire in great horizontal fissure; c and d Drawings of radiographic frontal and lateral projection

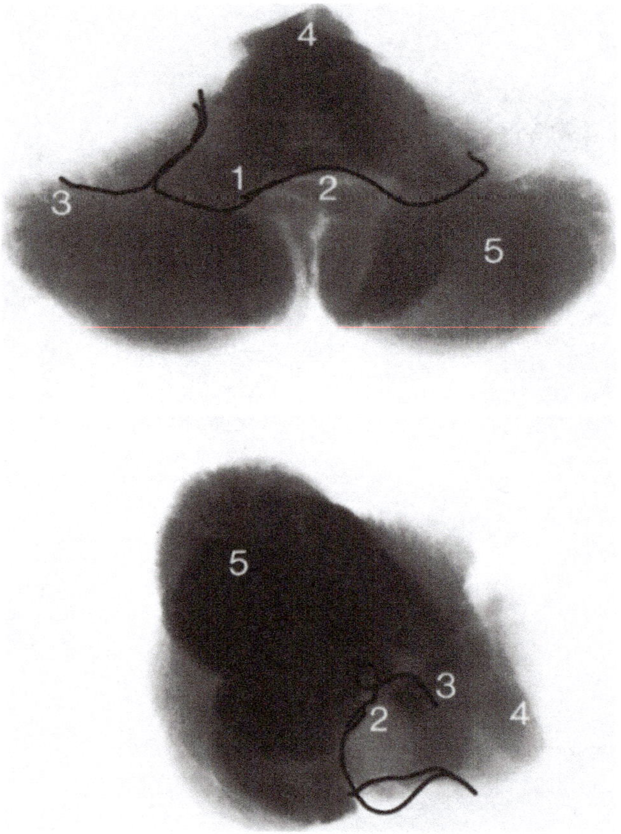

Figure 59

Frontal and lateral film of specimen of cerebellum demonstrating course of vein of lateral recess of fourth ventricle by a metal wire (*1*).

In frontal projection: demonstration of medial segment of roof of fourth ventricle corresponding to cerebellar part of vein (*2*). Pontine segment of vein of lateral recess of fourth ventricle flows into the lateral medullopontine vein (*3*). This point is easily localized, as depicted in Figure 55.

In lateral projection: topographic demonstration of roof of fourth ventricle and of its middle part by vein of recess of fourth ventricle (*2*), pons (*4*), cerebellar hemisphere (*5*)

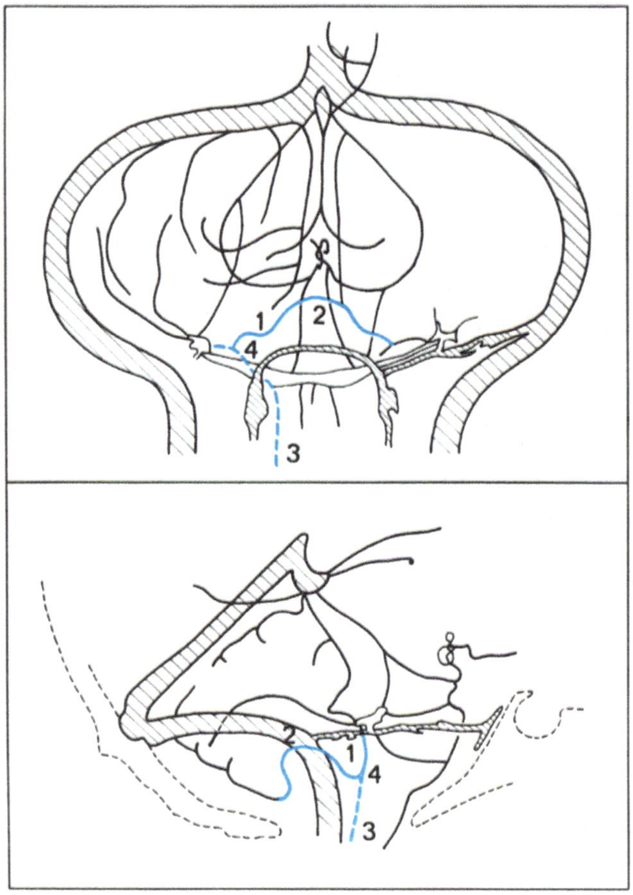

Figure 60

Drawing of vein of lateral recess of fourth ventricle

1 Lateral recess
2 Roof of fourth ventricle
3 Lateral medullopontine vein
4 Junction between the two veins (see Figure 55)

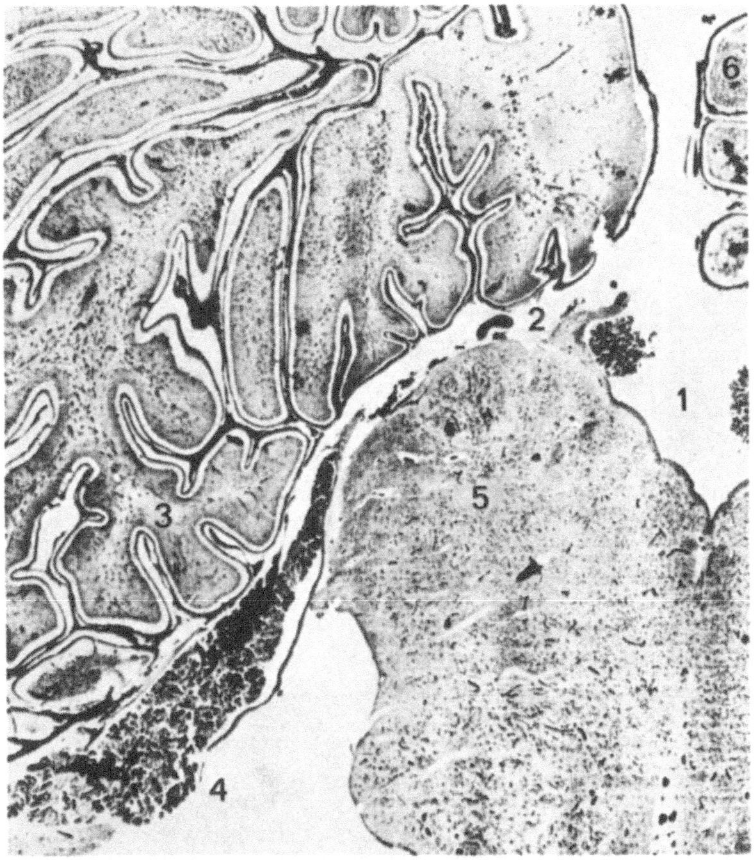

Figure 61

Coronal section of lateral recess of fourth ventricle (according to TOURNADE)

1 Fourth ventricle
2 Vein of lateral recess of fourth ventricle
3 Flocculus
4 Choroïdal plexus
5 Restiform body
6 Vermis

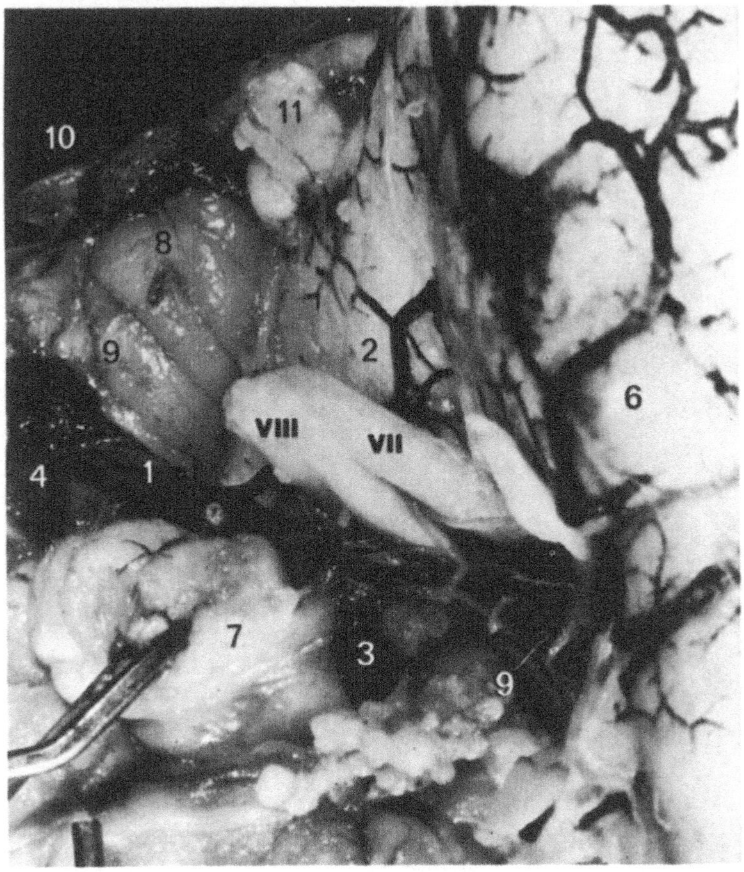

Figure 62

Photo of anatomical specimen showing superficial veins of pontocerebellar angle (according to TOURNADE)

1 Lateral medullopontine vein
2 Inferior transverse pontine vein
3 Vein of lateral recess of fourth ventricle
4 Vein of the great horizontal fissure of Vicq d'Azyr
VII Facial nerve
VIII Acoustic nerve
5 Medulla oblongata
6 Protuberance
7 Flocculus
8 Cerebellar hemisphere
9 Inferior petrosal vein
10 Superior petrosal vein
11 Vermis

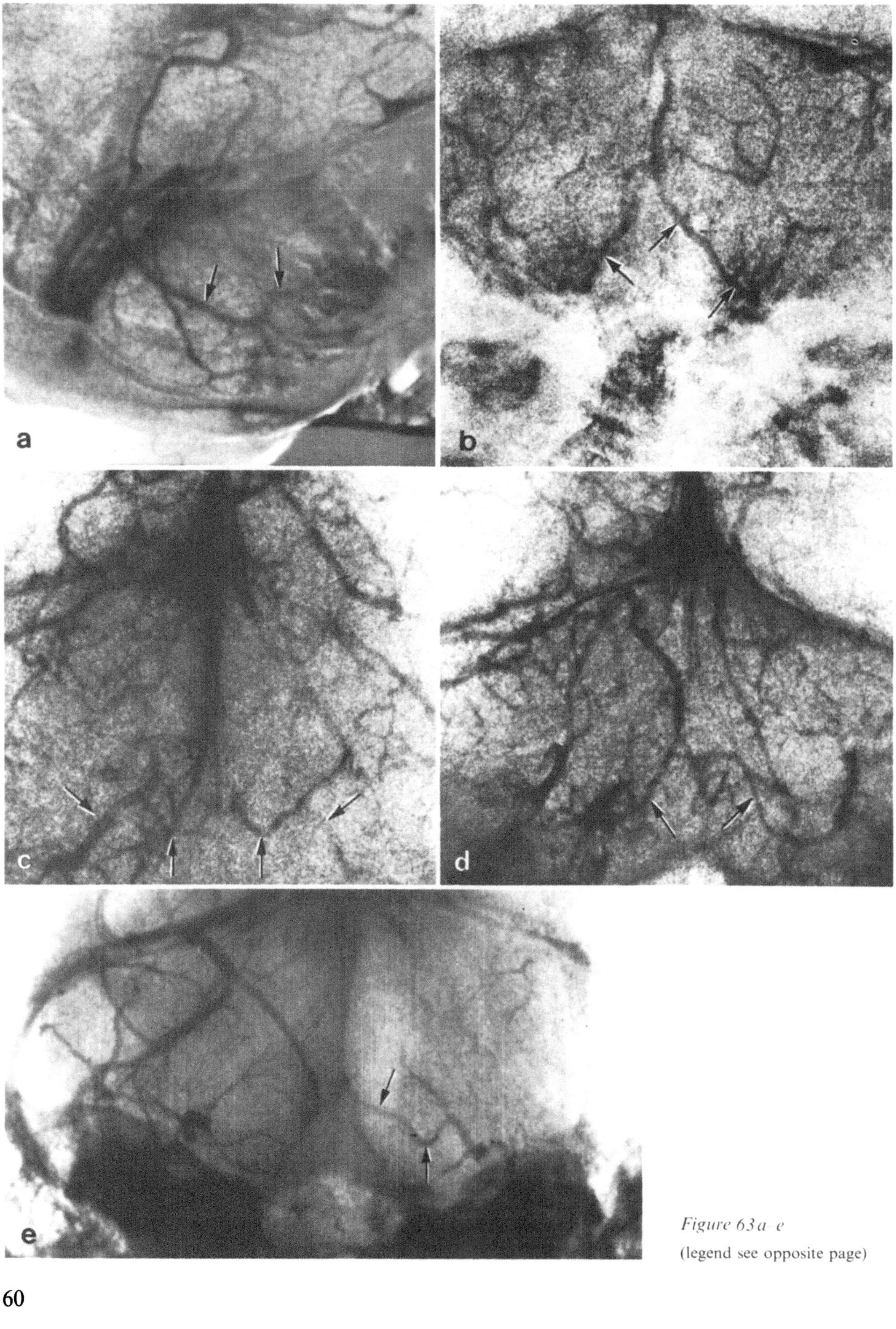

Figure 63a–e
(legend see opposite page)

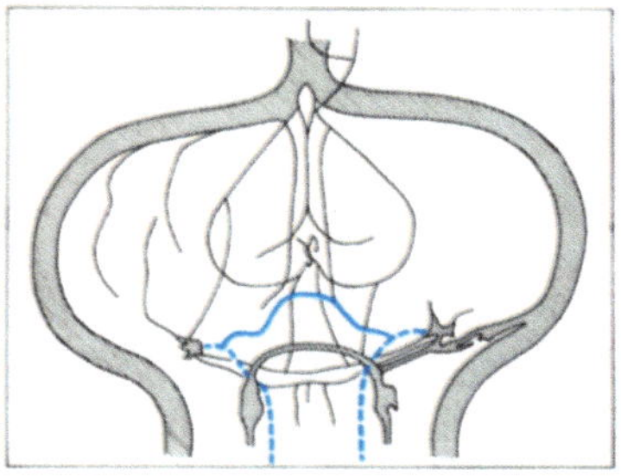

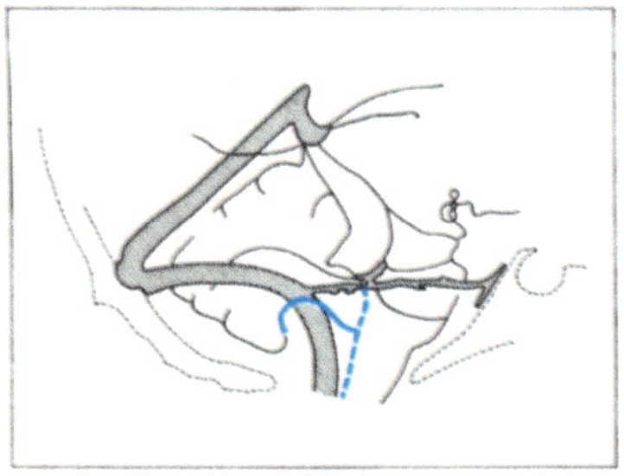

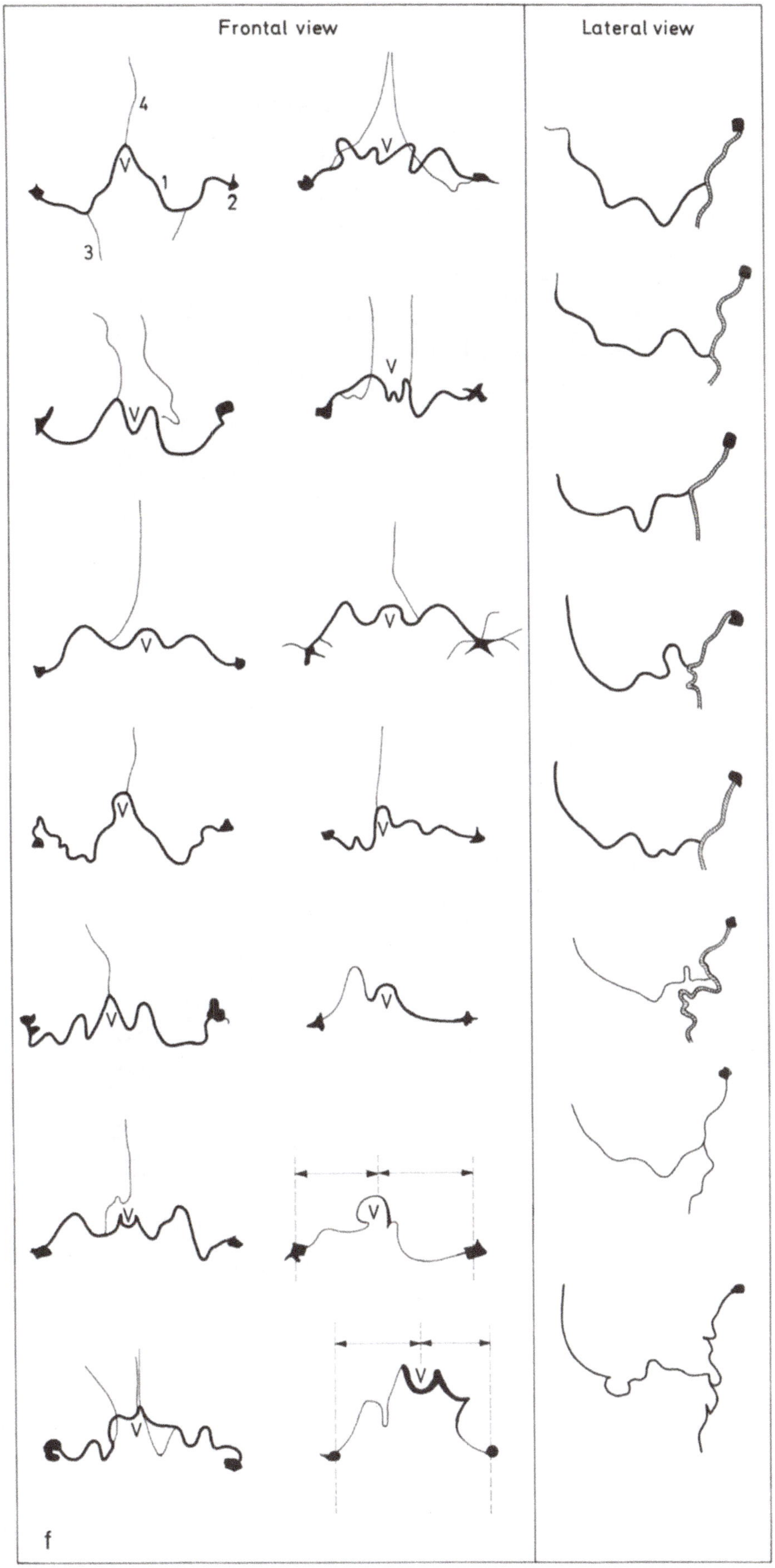

Figure 63 a–f

Radiographs and drawings depicting course of vein of lateral recess of fourth ventricle during normal vertebral angiography in frontal and lateral projection

V Vermis
1 Vein of the lateral recess of fourth ventricle
2 Superior petrous vein
3 Lateral bulbopontine vein
4 Inferior vermian vein

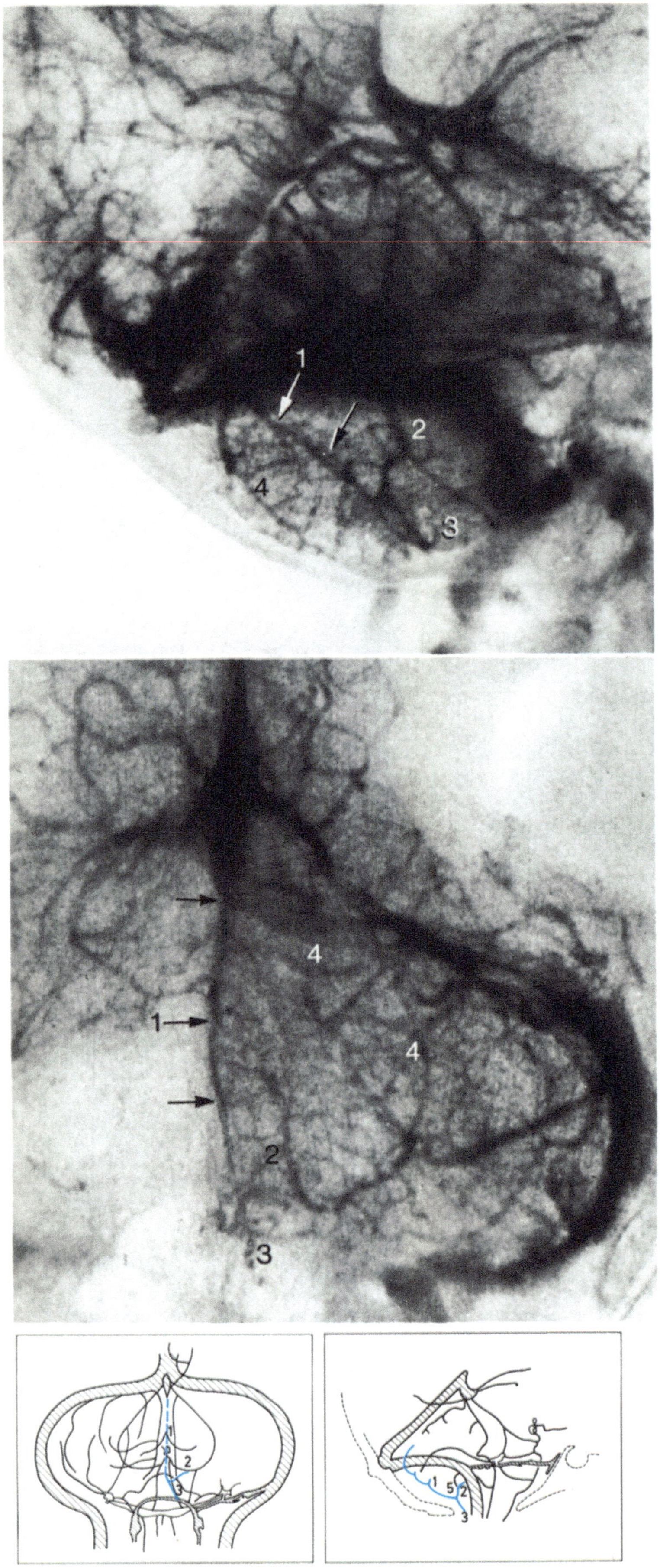

Figure 64

Inferior vermian vein in strictly medial course

1 Inferior vermian vein
2 Superior tonsillar vein
3 Inferior tonsillar vein
4 Hemispheric cerebellar vein
5 Tonsillovermian angle measuring roughly 90°

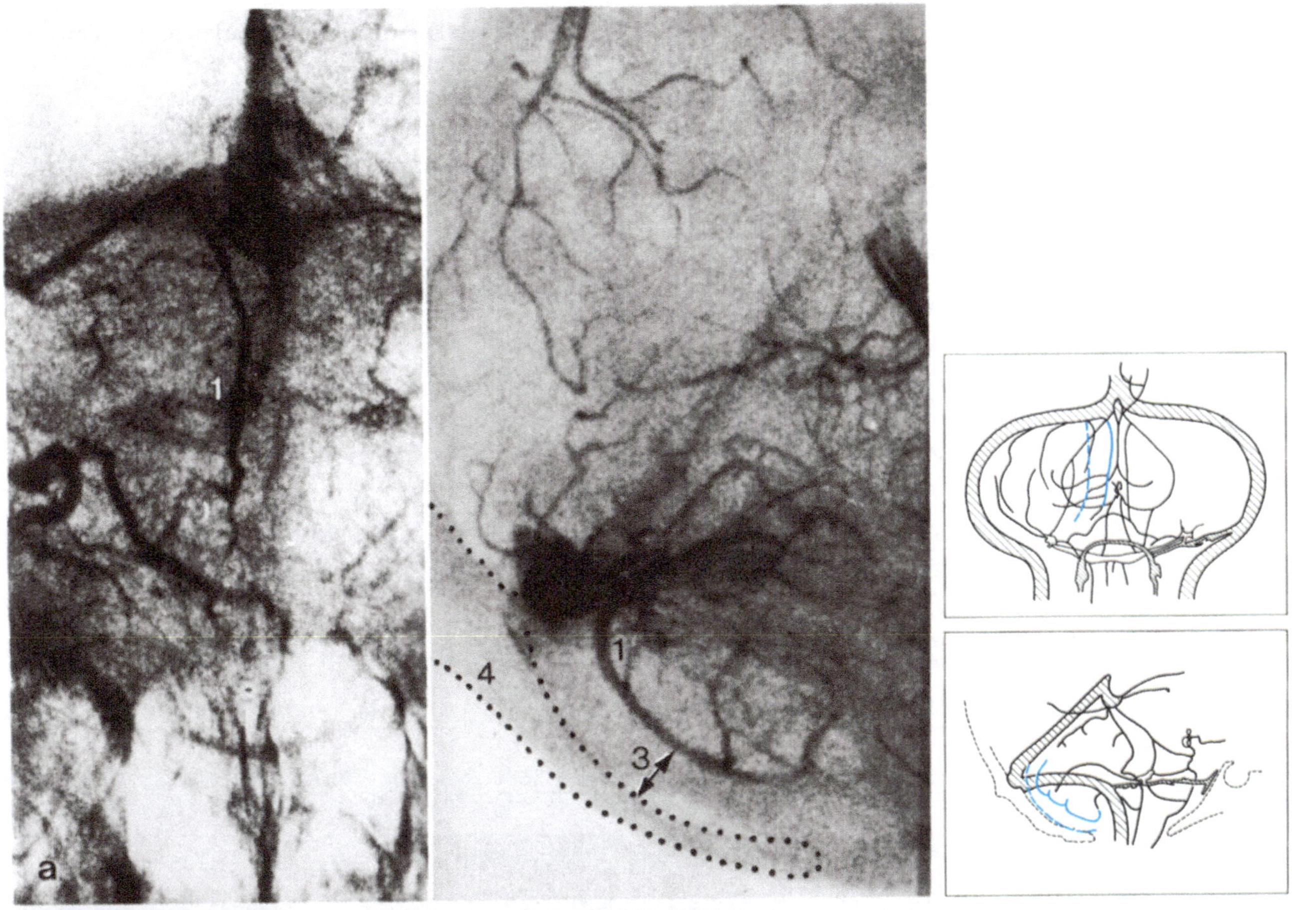

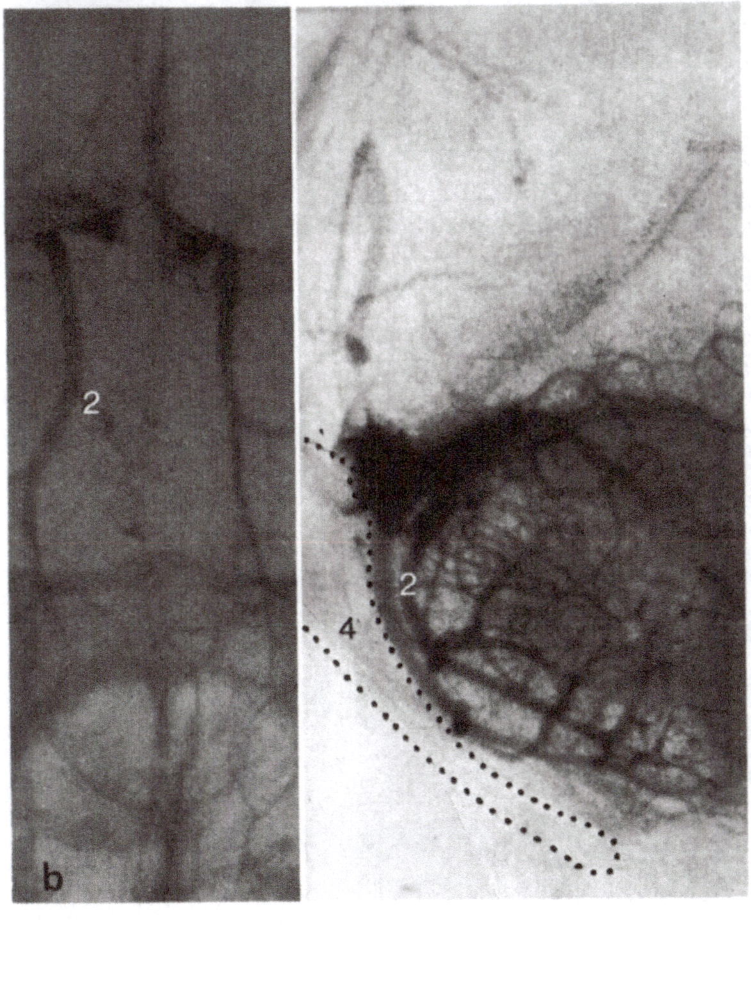

Figure 65 a and b

Median or paramedian (a) and lateral (b) course of inferior vermian vein. The more the vein courses laterally, the more it is posteriorly located in lateral view. Then it (b) has no value as a landmark for anterior aspect of cisterna magna. See also Figure 4

1 Inferior vermian vein (median or paramedian course)
2 Inferior vermian vein (lateral course)
3 Cisterna magna
4 Occipital bone

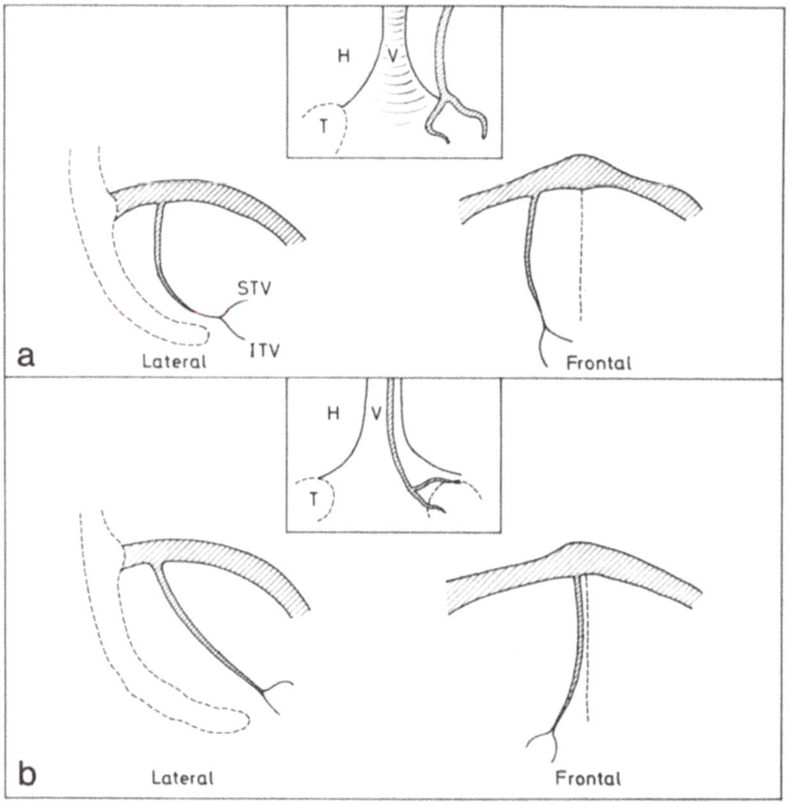

Figure 66 a and b

Drawing of main variation of inferior vermian vein

a Lateral location of vein
- *H* Cerebellar hemisphere
- *V* Cerebellar vermis
- *T* Tonsil
- *STV* Superior tonsillar vein
- *ITV* Inferior tonsillar vein

b Medial or paramedial location of vein

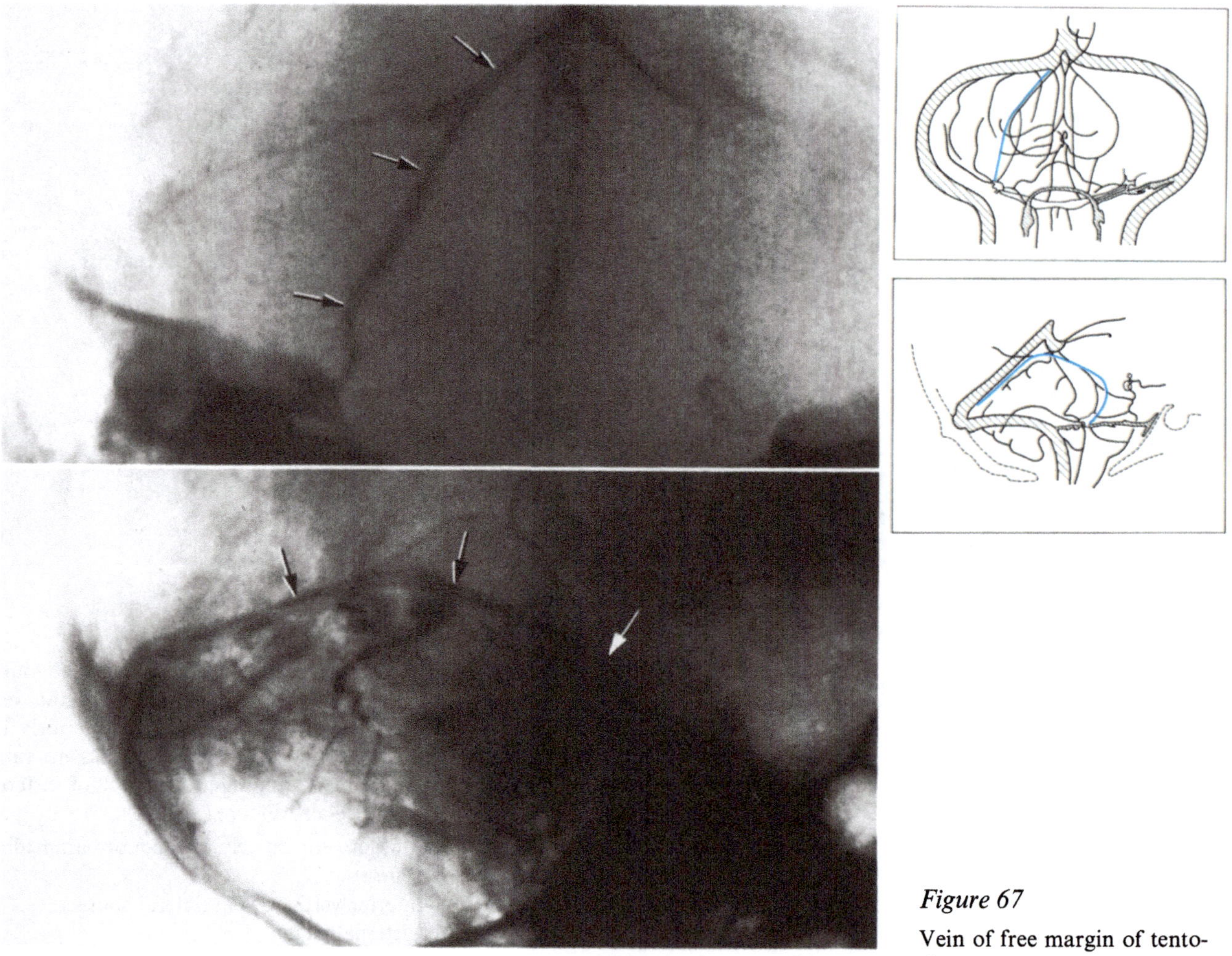

Figure 67

Vein of free margin of tentorium

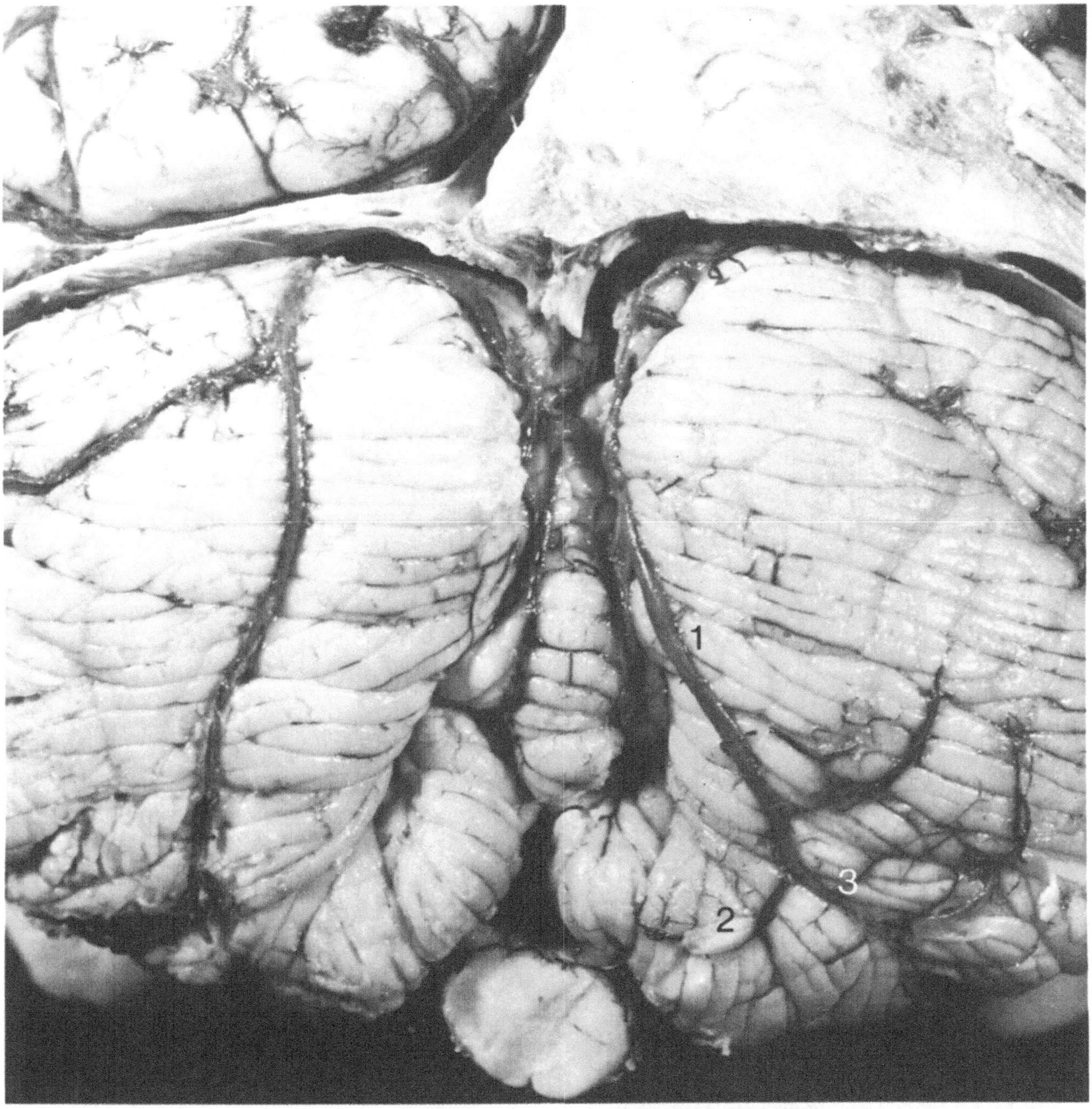

Figure 68

Photography of posterior aspect of cerebellum

1 Inferior vermian veins in a paramedian course
2 Inferior tonsillar vein
3 Superior tonsillar vein

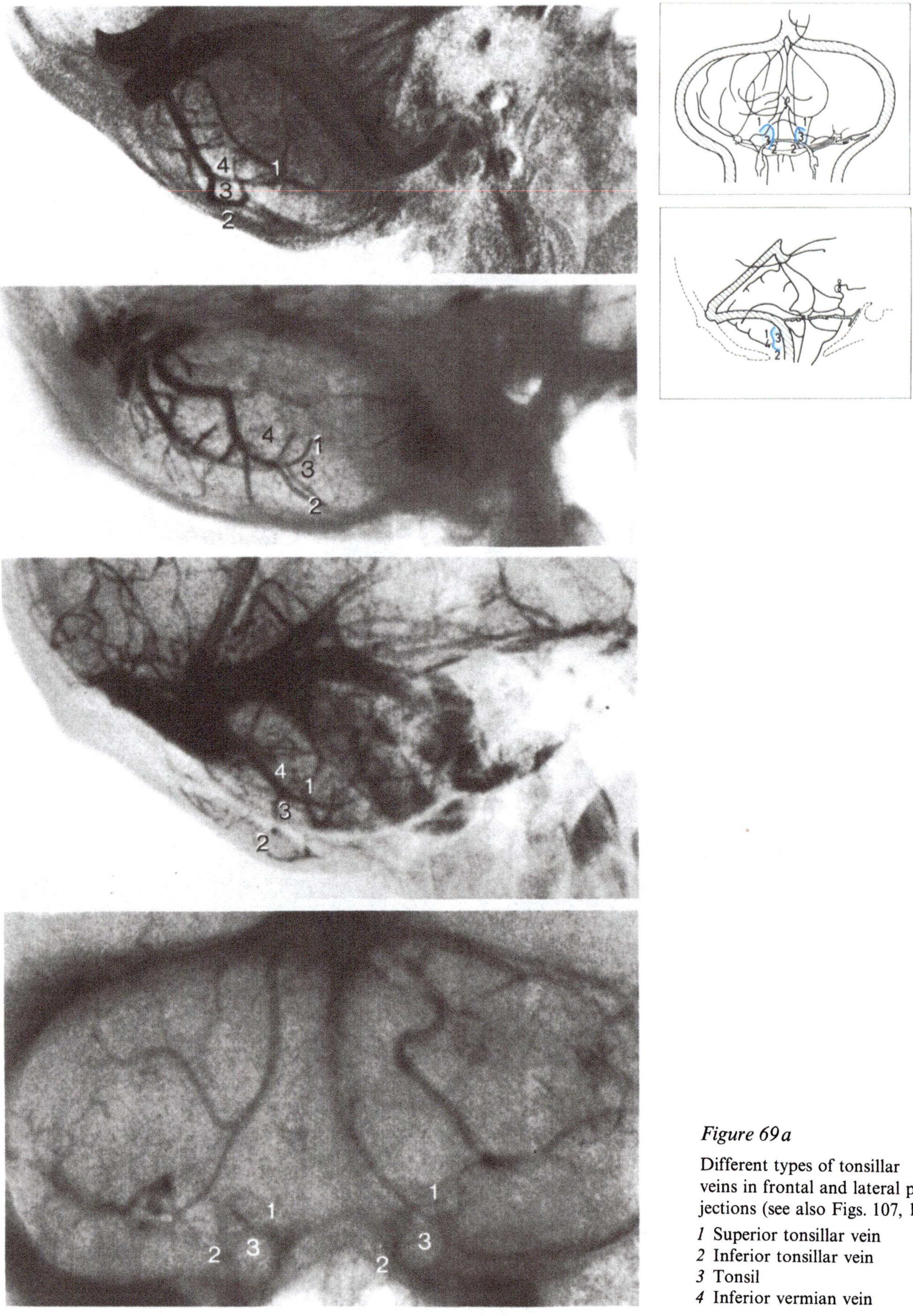

Figure 69a

Different types of tonsillar veins in frontal and lateral projections (see also Figs. 107, 125)

1 Superior tonsillar vein
2 Inferior tonsillar vein
3 Tonsil
4 Inferior vermian vein

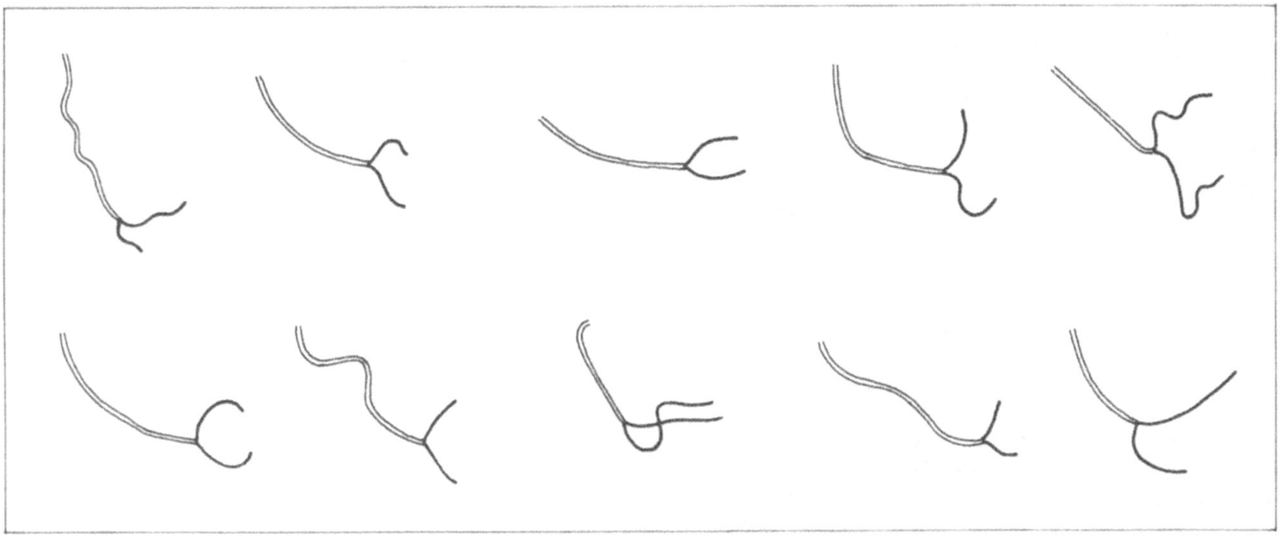

Figure 69 b

Drawing of different types of tonsillar veins and copular point in lateral projection

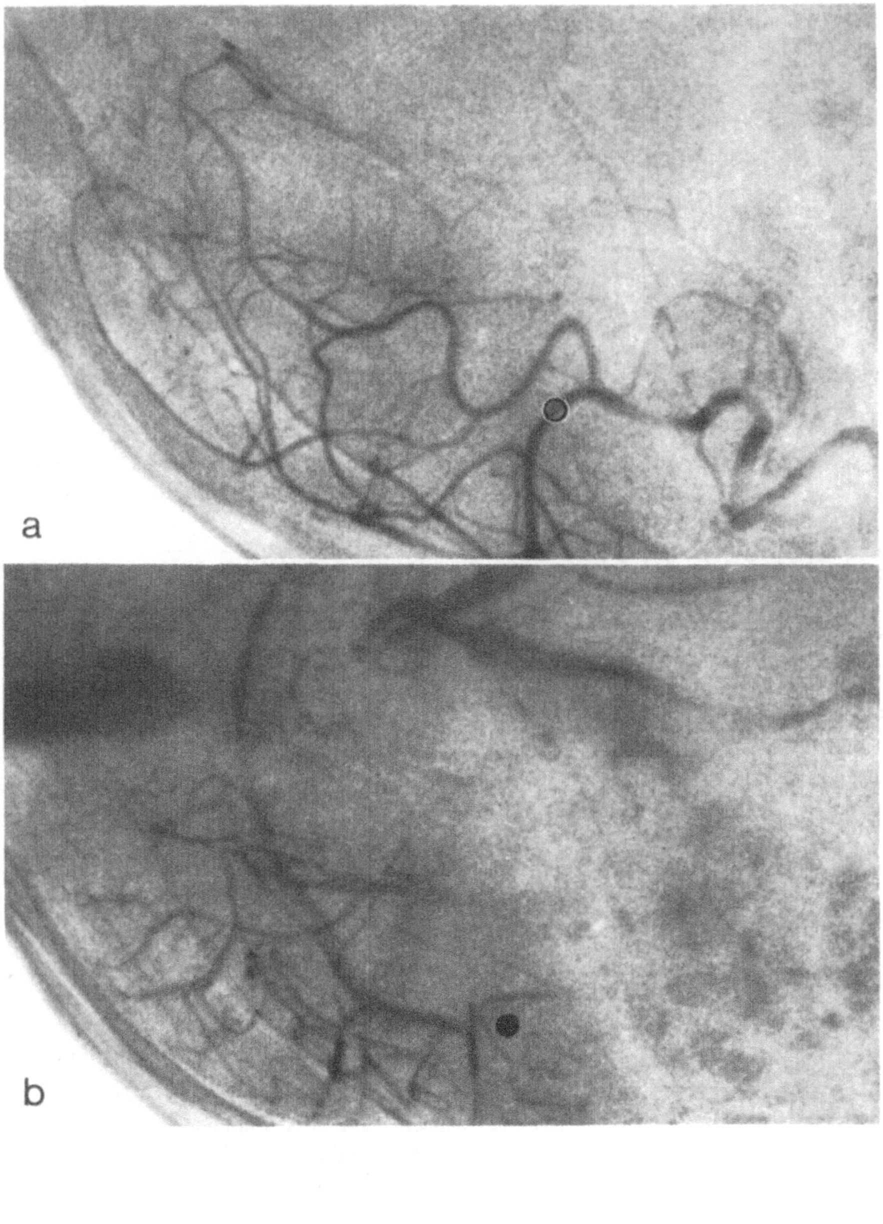

Figure 70 a and b

Tonsillar vessels and location of tonsils in arterial and venous phase

a Arterial phase
b Venous phase

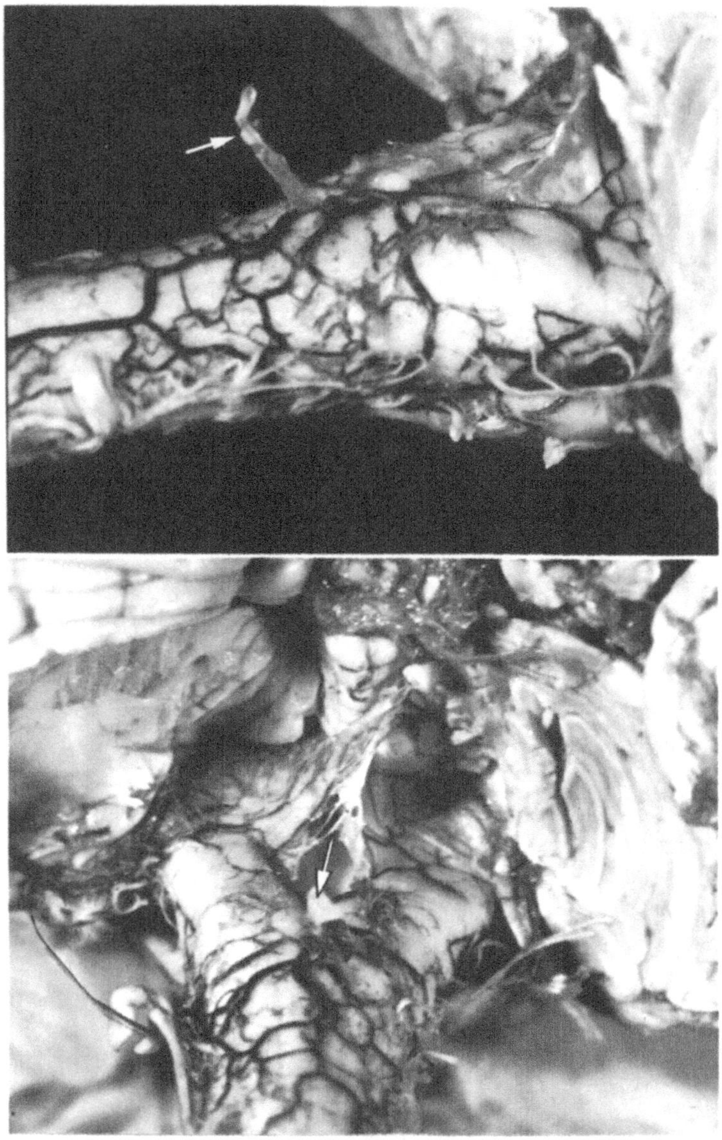

Figure 71

Specimen demonstrating posterior venous network of bulb with, in this case, an emissary vein (arrows) joining bulbar veins to occipital plexus

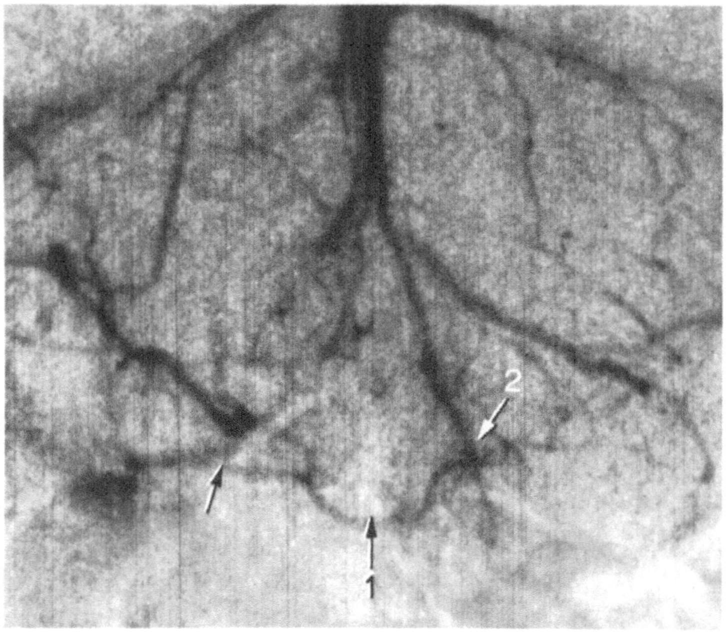

Figure 72

Anterior transverse veins of medulla oblongata

1 Transverse anterior vein delineating anterior aspect of medulla oblongata

2 Venous landmark of lateral limit of lateral groove of medulla oblongata

Figure 73a–d

Drawing of venous network around foramen magnum

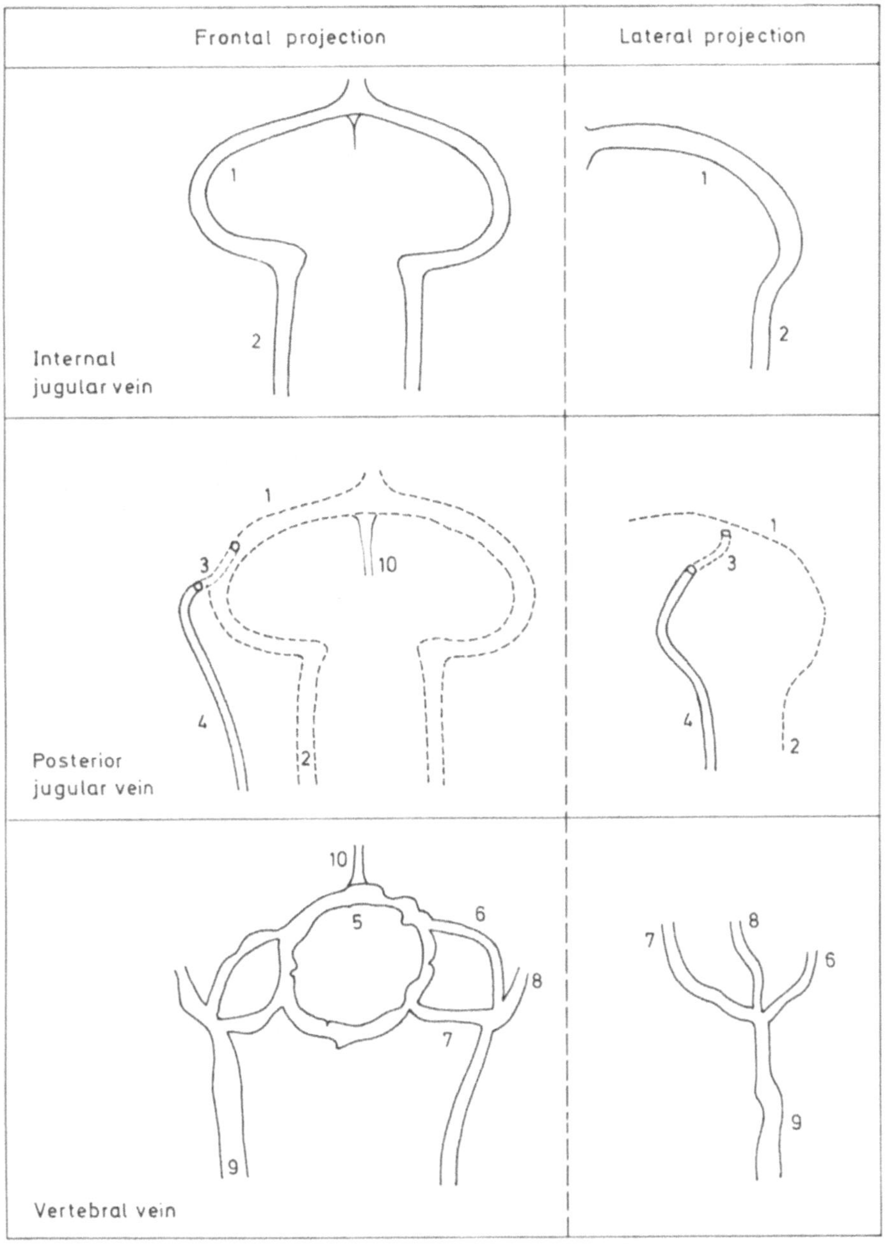

Figure 73a

Course of main venous trunk (semiaxial view)

1 Lateral sinus
2 Internal jugular vein
3 Mastoïd emissary vein
4 Posterior jugular vein
5 Perioccipital plexus
6 Posterior condylar vein
7 Anterior condylar vein
8 Posterior emissary condylar vein
9 Vertebral vein

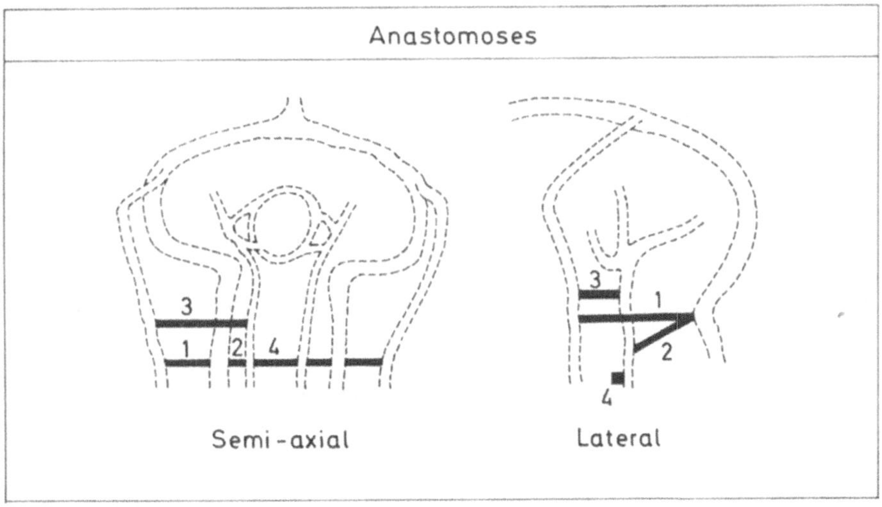

Figure 73b

Venous anastomic network between main venous trunk (semiaxial and lateral view)

1 Between posterior and internal jugular veins
2 Between internal jugular vein and vertebral vein
3 Between posterior jugular and vertebral veins
4 Between vertebral veins: vertebral plexuses

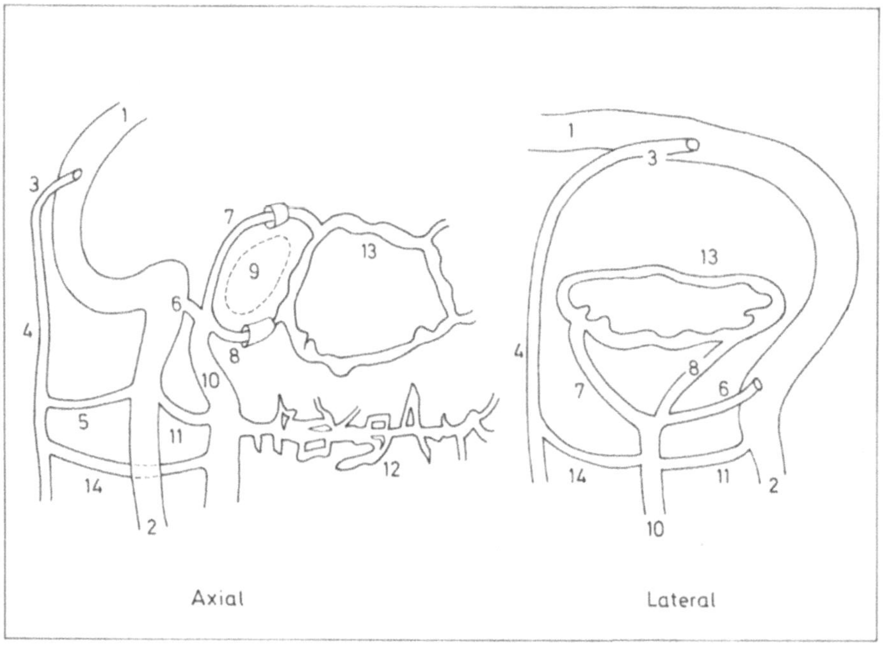

Figure 73c

Connections of condylar veins (axial and lateral view)

1 Lateral sinus
2 Internal jugular vein
3 Mastoïd emissary vein
4 Posterior jugular vein
5 Venous anastomosis between posterior jugular and internal jugular vein
6 Emissary condylar vein
7 Posterior condylar vein
8 Anterior condylar vein
9 Occipital condyle
10 Vertebral vein
11 Anastomosis between internal jugular and vertebral veins
12 Plexus between vertebral veins
13 Plexus around foramen magnum
14 Anastomosis between posterior jugular and vertebral veins

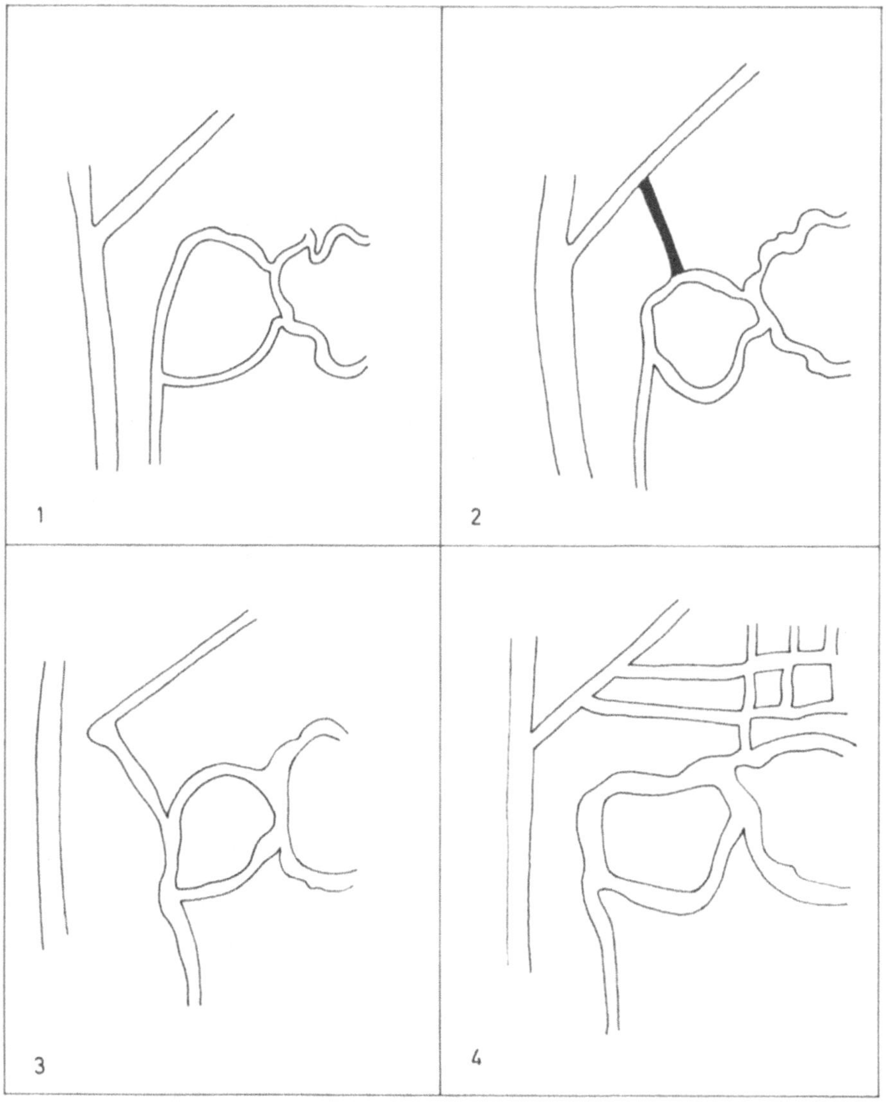

Figure 73d

Variations in condylar veins (axial view)

1 Absence of condylar emissary vein
2 Anastomosis between inferior petrous sinus and anterior condylar vein
3 Confluence of inferior petrous sinus and anterior condylar vein
4 Anastomosis between plexus of the clivus and petrous sinus

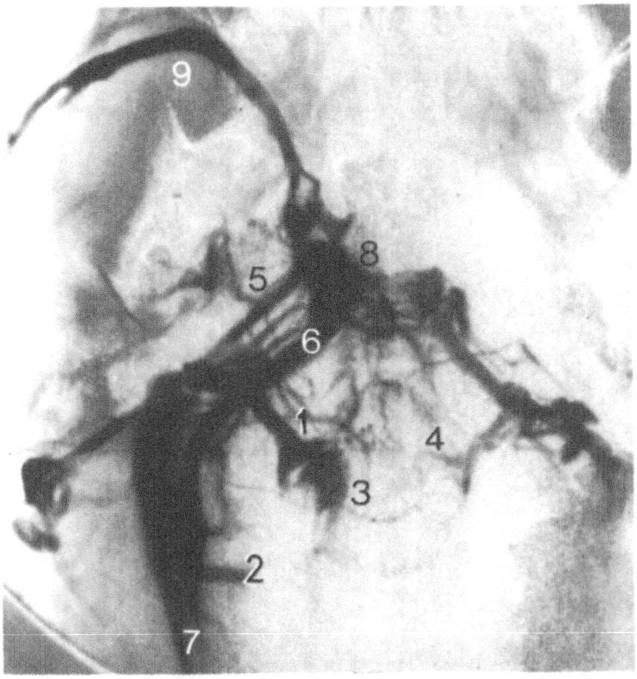

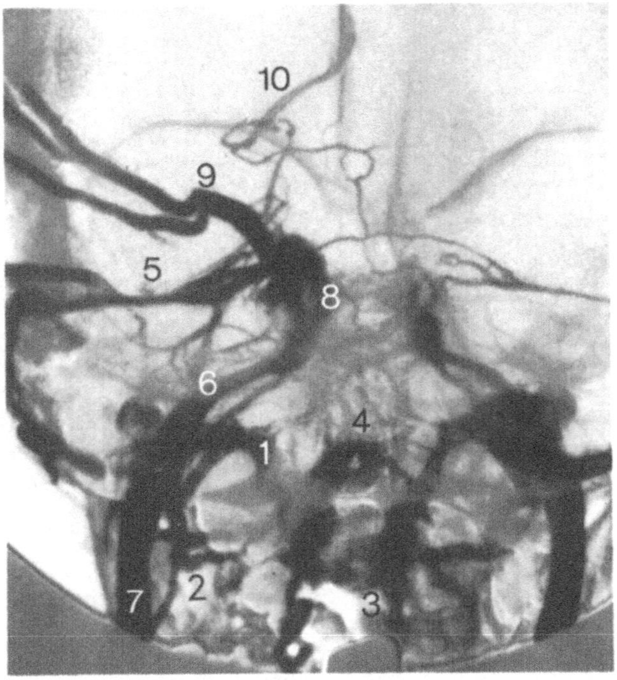

Figures 74 and 75

Veins of the craniovertebral region: jugulography in axial (Fig. 74) and frontal view (Fig. 75)

1 Anterior condylar vein
2 Vertebral vein
3 Anastomotic venous plexus
4 Perioccipital plexus
5 Superior petrous sinus
6 Inferior petrous sinus
7 Internal jugular vein
8 Cavernous sinus
9 Sphenoparietal sinus
10 Basal vein

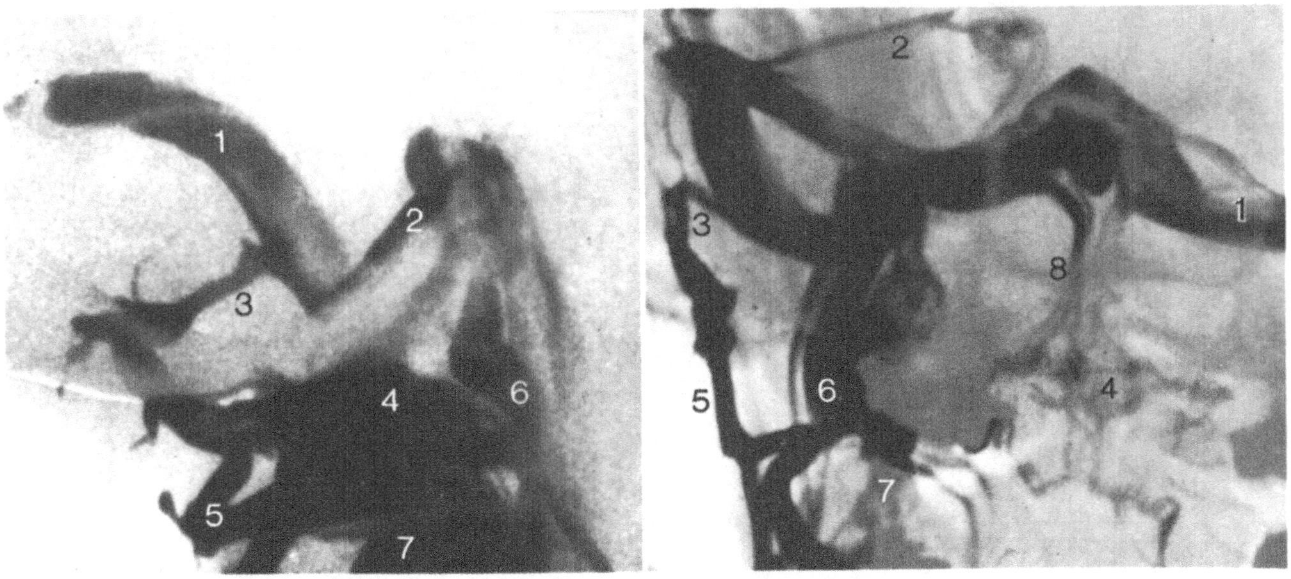

Figure 76

Perioccipital plexus: jugulography in lateral (above) and frontal (below) projection

1 Lateral sinus
2 Inferior petrous sinus
3 Mastoïd emissary vein
4 Perioccipital sinus
5 Posterior jugular vein
6 Internal jugular vein
7 Cervical venous plexus
8 Occipital sinus

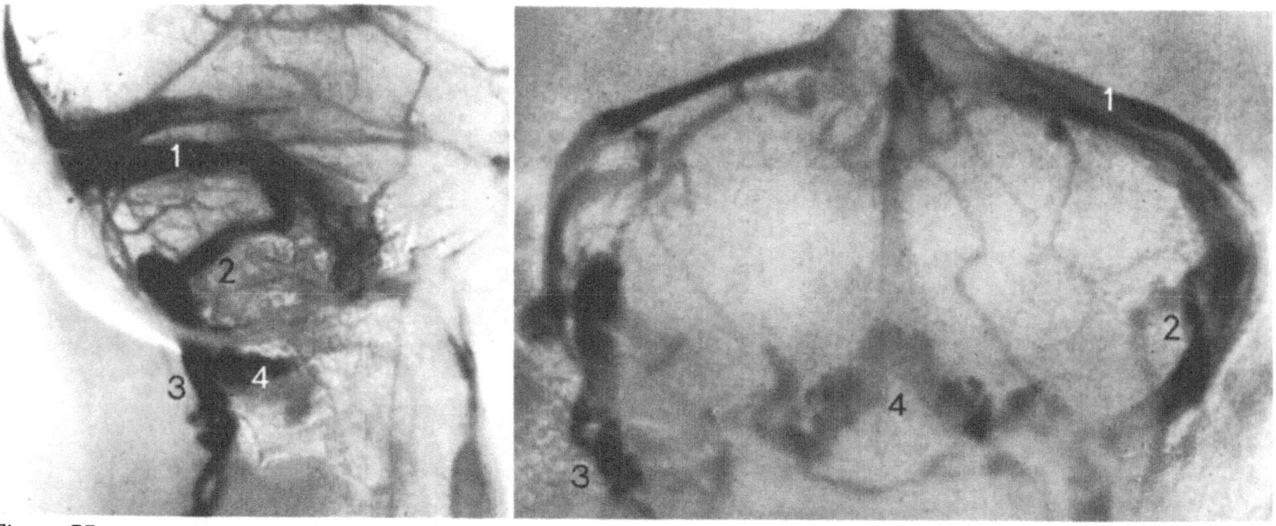

Figure 77 Perioccipital plexus (jugulography)

1 Lateral sinus
2 Mastoïd emissary vein
3 Posterior jugular vein
4 Perioccipital plexus

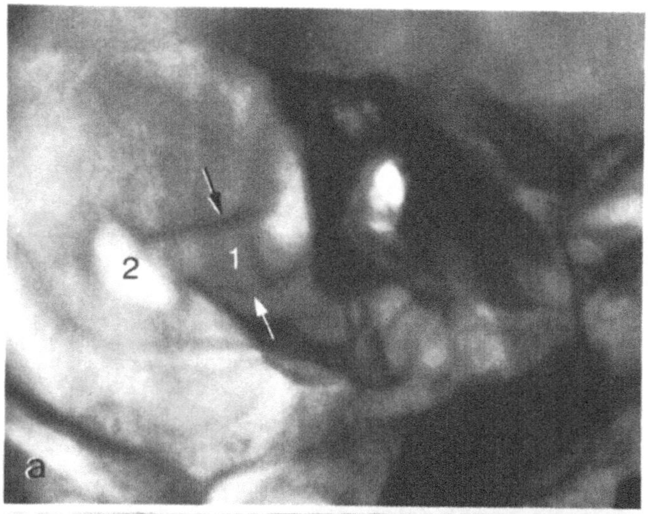

Figure 78

a The osseous canal of mastoïd emissary vein (*1*) with its foramen (*2*) on plain film
b Exocranial (*1*) and endocranial foramen (*2*) of retromastoïd canal
c Veins of craniovertebral junction during venous phase of vertebral angiography
_1_ Mastoïd emissary vein
_2_ Posterior jugular vein
_3_ Perioccipital plexus
_4_ Anterior condylar vein
_5_ Posterior condylar vein
_6_ Vertebral artery surrounded by venous network
_7_ Cervical anastomosis veins

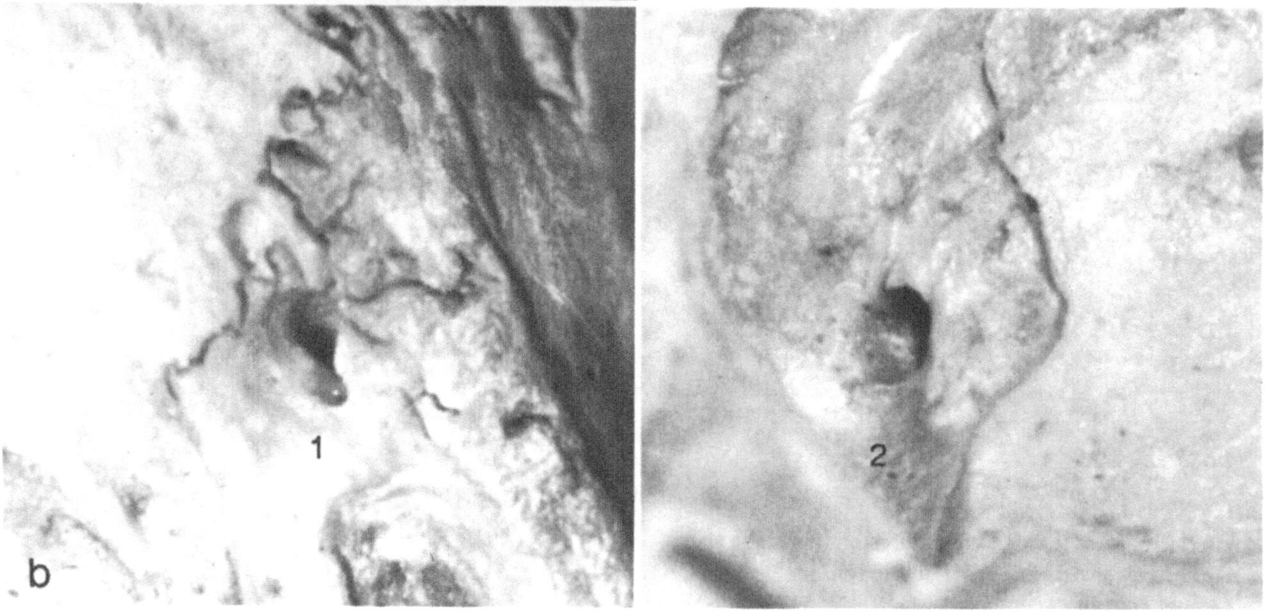

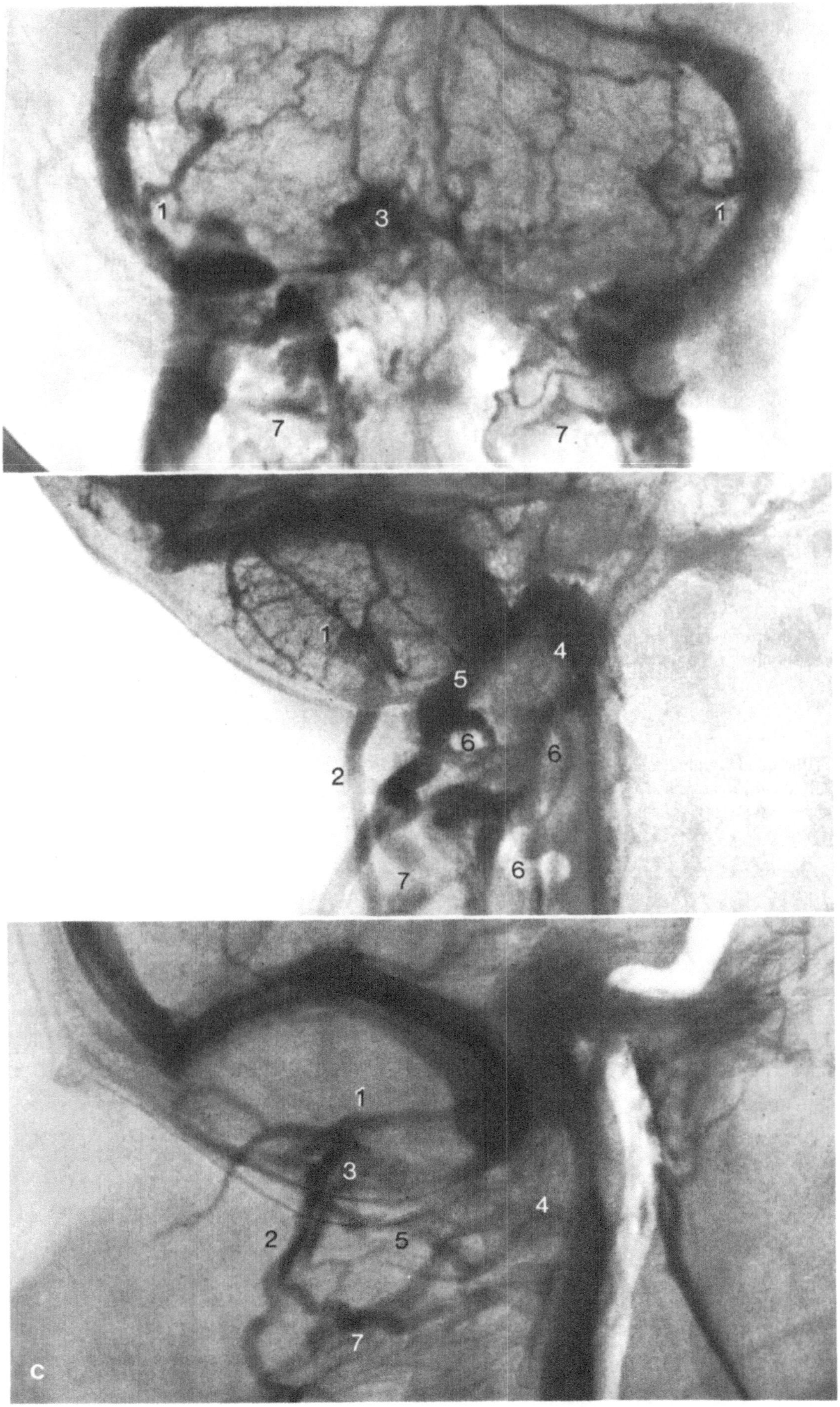

Figure 78c
(legend see p. 72)

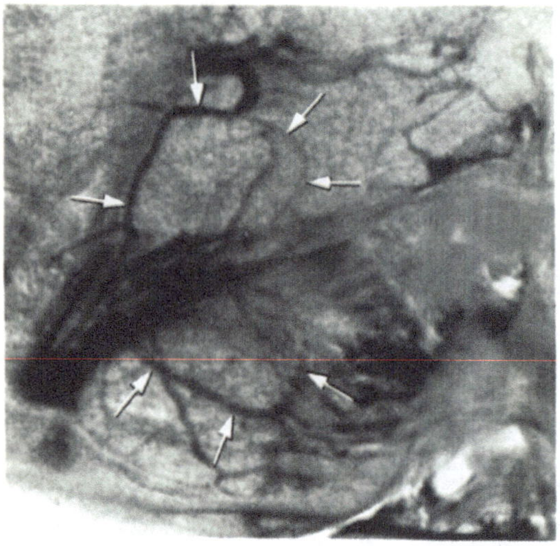

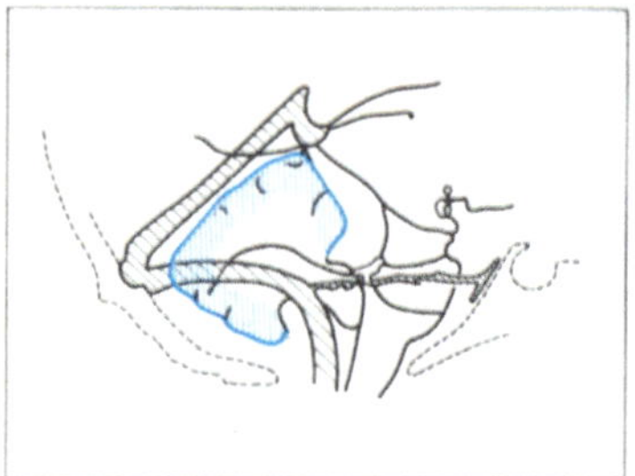

Figure 79

The joined superior and inferior vermian veins and precentral vein form a landmark for practically the whole cerebellar vermis

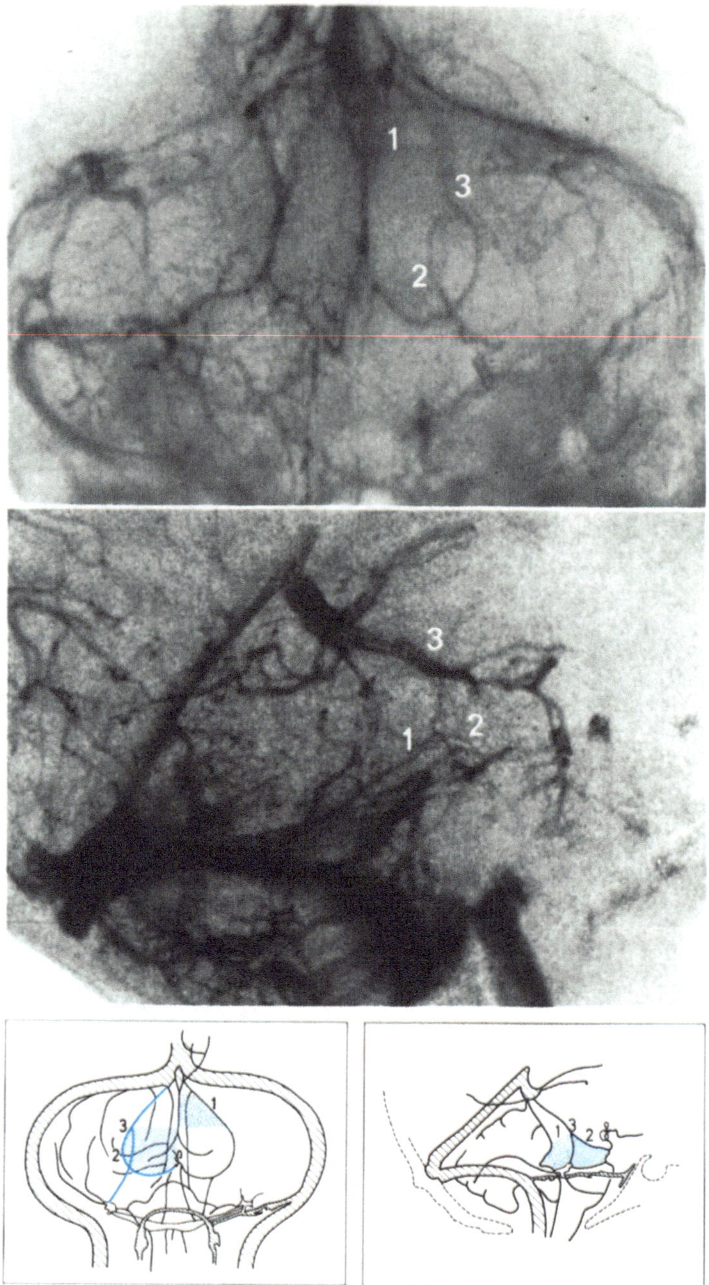

Figure 80 ▷

Venous landmarks of the quadrigeminal plate (*1*), pes pedunculi (*2*). These two areas are separated by lateral mesencephalic sulcus (*3*) which corresponds with course of vein

Figure 81 a and b

Venous landmarks of pontocerebellar angle

1 Superior petrous vein
2 Lateral medullopontine vein
3 Vein of lateral recess of fourth ventricle

▽

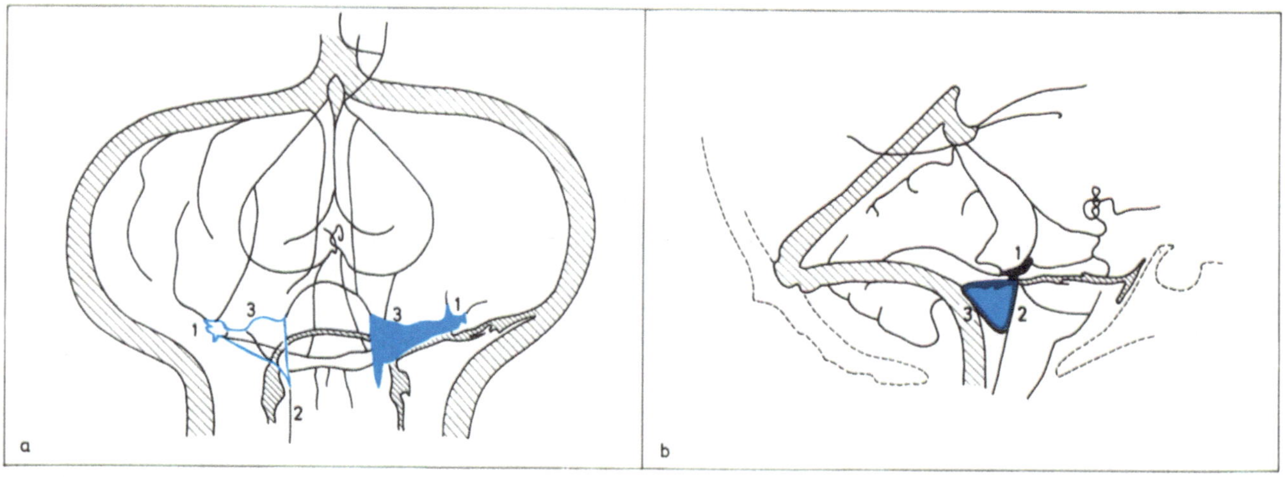

E. Pathologic Findings

I. Pathology of the Venous Sinuses

Conventional vertebral arteriography in adults generally results in good opacification of the lateral sinuses and at times of the jugular venous system as well. In the case of children, direct sinusography is necessary (Figs. 6–8). Quite often retrograde jugulography is also utilized to achieve a visualization of the venous network of the posterior fossa in its entirety. In any case, the radiologic study of the lateral venous sinuses, when needed, is easy and the difficulties brought up by KINAL fall away with retrograde jugulography [77]. From a practical point of view the following chapters can be taken into consideration:

1. Thromboses
2. Extracerebral tumors touching the sinuses:
   Cranial tumors
   Meningiomas of the tentorium
   Tumors of the jugular glomus

1. Thromboses of the Lateral Sinus

The conditions of flow into the lateral sinuses are quite variable and are characterized particularly by a predominance of flow into the right lateral sinus. The diagnosis of thrombosis of the lateral sinus should be rendered with extreme caution. We feel that a combination of three simultaneous signs should permit the discernment of a thrombosis:

a) The dye stop, which may have a variable morphology (Fig. 82).

b) The visualization in negative shape of the thrombus proper. We submit an image of typical "snake head"-shaped thrombosis (Fig. 83).

c) The visualization of a venous derivation network. This circulation can be visualized at levels not usually opacified during the venous phase in vertebral angiography. The same is applied to the mastoid emissary vein which joins the lateral sinus to the vertebral venous plexuses, or to the veins which connect the lateral sinus to the superior petrous sinus.

When these three signs are combined there can be no doubt as to thrombosis of the lateral sinus. Absence of one or the other should call for extreme caution in rendering such an interpretation, all the more so that the case of an isolated thrombosis of the lateral sinus is rare. This is not the case when there exists a general clinical and radiologic syndrome of an extended cerebral thrombophlebitis whose symptomatologic wealth facilitates evaluation.

2. Extracerebral Tumors Touching the Sinuses

a) *Cranial Tumors*

The most demonstrative example of an invasion of the sinuses by a craniodural tumor is illustrated by Figure 85, where on the one hand a bony hypertrophying lesion can be observed and, on the other, signs of invasion of the lateral sinus. These signs are quite typical. Several tumoral digitations may also be identified. The veins of the posterior fossa, on the other hand, permit the detection of an expansion in the posterior hemifossa with a characteristic deviation of the inferior vermian vein.

b) *Meningiomas*

The meningiomas of the posterior fossa encroach upon the lateral sinus during their development in its walls. An example is given in Figure 86. The venous signs are very important not only in the elaboration of a dura mater tumor diagnosis, but for presurgical data as well.

c) Jugular Glomus Tumors

Concerning tumors of the jugular glomus, may we suggest Professor Castan's excellent monograph on the subject [30]. We illustrate such a tumor in Figure 87. Conventional vertebral angiography is usually inadequate to provide the topography of the venous circulation of a jugular glomus. This is why it is advisable to complete the picture by jugular retrograde phlebography.

II. Venous Signs of Hydrocephalus in Children

Venous anomalies due to simple ventricular dilatation will not be described. These anomalies are known and require no special commentaries. Nonetheless we shall concentrate our efforts on distinguishing venous anomalies characteristic of one of the four following cases producing an inordinate accumulation of cephalospinal fluid in the intracranial space:

1. Noncommunicating hydrocephalus
2. Communicating hydrocephalus
3. Hydranencephalus
4. Hydroma

On several occasions in this chapter we recall the possibility of defining intracranial volumes by planimetry. This method, applied during the venous phase, furnishes the topography of the straight sinus which delimits the tentorium. It is of prime necessity to distinguish "harmonious" hydrocephalus involving an increase in volume of the sub- and supratentorial space and "disharmonious" hydrocephalus with predominant dilatation of the supratentorial space. The determination of projection surface relations of these spaces provides an objective element allowing an index to be evaluated.

1. Noncommunicating Hydrocephalus

We shall take as an example stenosis of the aqueduct of Sylvius, be it direct via obliteration of the aqueduct due to a localized process at its level, or secondary through compression or indirect kinking.

a) Venous Signs of Congenital Stenosis of the Aqueduct of Sylvius

Depending on the precociousness and size of the stenosis, it is possible to distinguish two cases (Fig. 88):

α) Considerable Noncommunicating Supratentorial Hydrocephalus. The straight sinus is rectilinear, upright, pushed backward and has a definitely retrobregmatic axial orientation. The great vein of Galen is lowered and forms a right or obtuse angle with the straight sinus.

The internal cerebral vein is flattened, stretched, and rectilinear. Its morphology depends, in any case, on the relation of the ventricular dilatations. The vein is all the more lowered and rectilinear in that the dilatation of the lateral ventricles is predominating, that is, that the third ventricle is less dilated.

The posterior mesencephalic basal and peripeduncular veins, seen in a frontal view, describe the contours of the cerebral trunk, which appears small due to bitemporal expansion. This anomaly can be discerned further, in the same view, through an external concavity of the basal and posterior mesencephalic veins. These signs of bitemporal expansion appear equally in a lateral view under the form of a lowering and of an anterosuperior concavity (hammock) of the posterior mesencephalic, basal, optopeduncular, and precentral veins. Summing up, there exists a venous syndrome of a small posterior fossa and of an increase in the volume of the supratentorial space. This anomaly in volume has already been detected by former neuroradiologists following simple and precise analyses of a lateral cranial radiography. The analysis of the veins, however, permits not only a more reliable appreciation of the disturbances, but also the establishment with greater accuracy of their variants and stages.

An endeavor to objectify the foregoing by means of planimetric measurements is shown in Figures 99 and 100. The surface relation supratentorial space is slightly higher than 3 in normal cases. In the case of a noncommunicating hydrocephalus, the relation is frequently plus 4 and sometimes plus 5.

β) Late or Slow Hydrocephalus. Topographic anomalies of the right sinus and of the great vein of Galen can be totally absent or only slightly outlined. Venous symptomatology is at such time limited to the signs of bitemporal expansion, that is, to the slightness in diameter of the cerebral trunk in a frontal view. In a lateral view, the optopeduncular, basal, posterior mesencephalic, and precentral veins are lowered and pushed backward and diminished in certain temporal herniations. The planimetric relation of the supra- and subtentorial spaces is not modified, or only slightly, in the case of a young child or of hydrocephalus dating back to infancy.

b) Venous Signs of a Tumor Responsible for Compression of the Aqueduct of Sylvius

We shall consider three examples, one of a posterior thalamic tumor, one of a pinealoma, and one of a tumor of the superior cerebral trunk.

α) In the case of a posterior thalamic tumour or of peduncular expansion, the venous signs at the level of the posterior fossa are quite characteristic and always include a considerable posterior pushing back of the precentral vein (Fig. 90).

β) In the case of a pinealoma (Fig. 91), the venous signs are analogous to those of a posterior thalamic tumor, but predominate at the level of the angle formed by Galen's great vein and the basal vein. In numerous cases, pinealomas invading the thalamus or posterior thalamic tumors cannot be distinguished one from the other through venous or arterial symptomatology. When the tumor shifts laterally it can be considered to be of thalamic origin.

γ) In the case of a peduncular glioma viewed frontally, the venous symptomatology is characteristic, thus permitting the recognition beyond doubt of the increase in the transversal and vertical diameter of one or both of the cerebral peduncles. In a lateral projection, the veins in the peduncular region are more or less developed and stretched. In all cases of enveloping mass expansions of the aqueduct of Sylvius, the increase in volume of the supratentorial space depends on the age of the child, the speed of growth, and the degree of compression of the aqueduct. It is only during the first years of life that one can follow the changes in volume analogous to those of a tight congenital stenosis of the aqueduct. In the majority of cases the direct, specifically mesencephalic signs are the most distinct and the most useful in rendering a diagnosis. Thus it can be said as well that at such time as disturbances of the volumetric relation between the posterior fossa and the supratentorial space predominate, congenital stenosis of the aqueduct is usually present.

c) Direct Venous Signs of Vascular Malformation Localized in the Aqueduct of Sylvius

We have observed this malformation in a single case shown in Figure 91. A venous budding of an arterial-venous aneurysm within the aqueduct causing a progressive and tardy stenosis through an increasing dilatation of this drainage vein.

2. Communicating Hydrocephalus

While it is possible to assert that precocious stenosis of the aqueduct of Sylvius entails a large skull having a small posterior fossa, communicating hydrocephalus determines a large relatively harmonious skull having an equally enlarged posterior fossa. Let us say at the beginning that the increase in volume predominates slightly in the supratentorial space in such a way that not all communicating hydrocephalus are perfectly harmonious. Our planimetric method permits easy recognition (Fig. 95). The supra- and subtentorial planimetric index is in effect relatively high, exceeding 3 but not going up to 4. In such cases of communicating hydrocephalus the basal vein is only slightly displaced. In fact, there is no truly temporal herniation and one can observe only a slight decrease in the angle of point L with a normal localization of the origin of the vein. A posterior displacement of the precentral vein becoming concave anteriorly is frequently observed. The colliculocentral point is shifted backward and the tonsillovermian angle is more or less closed due to the dilatation of the fourth ventricle. Venous anomalies in communicating hydrocephalus indicate a certain disharmony between the supra- and subtentorial levels. Thus in a frontal view a certain degree of bilateral temporal herniation can be found having characteristic venous signs at the level of the cerebral trunk. However, these coexist with a large posterior fossa which distinguishes these cases definitely from those of stenosis of the aqueduct.

3. Hydranencephaly

Numerous forms are shown in this chapter, depending on the importance of supratentorial hemispheric agenesis, its topography, and especially of the volume of dilated hydrous cavities. This explains why we shall describe here an average case of agenesis localized in the mean and anterior part of the cerebral hemispheres. The phlebogram of hydranencephaly is characterized by three orders of signs (Fig. 97):

a) Supra- and subtentorial volumetric relations such as those in precocious stenosis of the aqueduct, that is to say, a large skull with a small posterior fossa, with or without bilateral temporal herniation. The index of the supra- and subtentorial surfaces is always heavily increased.

b) The number and the diameter of the supratentorial cortical veins is greatly modified depending on the topography of the cerebral ageneses. Thus the frontal and parietal cortical veins are absent or slight while relatively well visualized in the posterior supratentorial area.

c) Both the capillary and venous phase in angiography show a heavy density of contrast material in the region of the basal ganglia, an area which appears hypertrophied. This peculiarity is not observed in cases of precocious stenosis of the aqueduct of Sylvius.

4. Hydroma

The most common form of pericerebral hydroma is temporosylvian in nature. The cystic or pseudocystic liquid collection in this region determines a group of characteristic venous signs (Fig. 98) made up of, on the one hand, an avascular zone edged by normal veins and, on the other hand, a displacement of the basal vein. These temporosylvian hydromas are relatively slow in their evolution so much so that they result on the one hand in a cranial deformation (a blowing-up of the temporal fossa), and on the other hand, in an increase in the volume of the supratentorial space, as in the case of a precocious stenosis of the aqueduct of Sylvius. Thus we are in the presence of a cranial deformation which:

Takes on the same characterics as an early stenosis of the aqueduct of Sylvius, when the hydroma is precocious and evolutive (large skull with a small posterior fossa having an increased planimetric index).

Takes on the same characteristics as a late stenosis of the aqueduct of Sylvius or a pinealoma, when the hydroma is itself tardy or slowly evolutive.

As to the deformation of the basal vein, this is illustrated in Figure 98: a straightening and backward-shifting of the entire anterior ensemble of the vein, which is deformed and enlarged.

In conclusion, let us recall that the planimetric method elaborated in our department by Ludwiczak enables to put into concrete form the relations between the posterior fossa and the supratentorial space. In this method three zones are distinguished: the supratentorial, basal, and subtentorial (Figs. 99 and 100). We report the normal figures obtained in different age brackets. For each of these zones we have noted that the index of the supratentorial space (*A* and *B*) over the subtentorial space (*C*) is approximately 3.

III. Intracranial Hypertension and Herniations

Multiple venous signs premonitory of herniation are observed in diffuse hypertension of the posterior fossa (Fig. 102).

In a lateral view:

The basal, posterior mesencephalic, precentral, lateral mesencephalic, and optopeduncular veins are raised.

The course of the vein of the lateral recess of the fourth ventricle and the hemispheric cerebellar veins is stretched.

A decrease in the venous tonsillovermian angle.

The inferior vermian vein shows a stretched course and is shifted more or less backward against the occipital [155].

The superior vermian vein occupies the venous supravermian triangle [152].

The pontine veins are pushed toward the clivus.

In a frontal view:

The course of the hemispheric cerebellar veins is stretched, as is the vein of the lateral recess of the fourth ventricle.

The tonsillar veins, when visible, are lowered.

These venous alterations, due to diffuse hypertension in the posterior fossa, are forerunners of herniations.

1. Herniations

Expanding processes cause cerebral herniation through the displacement of one intracranial part into another. Sometimes the displacement is accompanied by a strangulation occurring at its usually narrow opening. However, even through the foramen magnum is wider than the cervical canal, tonsillar herniation occurs through it. Consequently there is no strangulation proper.

The term "herniation", which is large in scope, leads us to propose it be used to designate the so-called superior vermian herniation as well [152].

In the posterior fossa, four varieties (Fig. 103) of herniation are distinguishable. The veins in a vertebral angiography are also modified in cases of temporal herniation so that we have to take this type of herniation into account.

a) The cerebellar herniation
b) The superior vermis herniation
c) The tonsillar herniation
d) The temporal herniation
e) Associated herniations.

a) Cerebellar or Transtentorial Herniation (Fig. 104)

The diagnosis of this hernia of the cerebellum is easy to establish and is characterized by venous anomalies. Indeed, the antero-superior area of the posterior fossa in its entirety is displaced toward the supratentorial space. The venous signs:

An anterior and superior displacement of the pontine and optopeduncular veins.

Anterior and superior displacement of the lateral and posterior mesencephalic veins and of the precentral vein.

b) Hernia of the Superior Vermis or Hernia of the Superior Vermis in the Subdural Supravermian Space (Figs. 105, 106)

Hernia of the superior vermis is a special type of hernia. We reported on this form in 1973 at the European meeting in Frankfurt [152]. It is based on the existence of a subdural supravermian space. This subdural space is located above the culminating point of the superior vermis and is not shown by pneumoencephalography. It is bounded above and behind by the straight sinus, laterally by the tentorium, in front by the great vein of Galen and below by the arachnoid which covers the superior cerebellar vermis and the cerebellar hemispheres (Figs. 36–39, 103–106). This area is filled with air only in cases of pneumosubdurography.

We are inclined to believe that the air injected by the lumbar route enters this area by flowing along the trunk of the superior vermian vein (Figs. 38, 39). In practice, the veins outline the subdural supravermian area, the straight sinus behind, the great vein of Galen in front, and the superior vermian vein below. A lateral projection of these venous elements fixes the boundaries of a triangular surface which corresponds to the subdural supravermian area. Its form and size vary depending on the general shape of the skull while experience helps us to recognize with ease the normal course of the superior vermian vein and the void in the triangular area. Obviously one must consider the fact that the height of this space depends in part on the more or less medial course of the superior vermian vein (Fig. 38). When its course is strictly medial, the space is less high than when it runs more laterally.

When the superior cerebellar vermis enters into the subdural supravermian space, the lateral projection of the superior vermian vein is modified. In an early stage of hernia the vein is distinctly convex (Fig. 106), stretched and displaced upward. However, a triangular area is still distinguishable. When hernia of the superior vermis is larger, the supravermian subdural area disappears completely (Figs. 104, 105, 108, 110, 114). In a very late stage, hypertension is great enough to compress completely the supravermian vein so that it is not opacified during the venous phase in vertebral angiography.

Hernias of the superior cerebellar vermis are not always present in cases of cerebral expanding masses. There are cases (Figs. 111, 141) where a hernia of the inferior pole of the cerebellum (tonsillar herniation) seems to exercise traction on the cerebellum in such a manner that, paradoxically, the subdural supravermian space is enlarged. When a hernia of the superior vermis is present there can be no doubt as to the pathologic localization of the superior vermian vein (Figs. 93, 102–104 and 106). The association of different herniations is very variable. Superior vermian herniation can be found with other hernias and this association is seen in cases of both supra- and subtentorial tumors (Figs. 104, 110, 112, 113). Superior vermian herniation can be absent, especially in cases of supratentorial tumors which have caused a temporal and tonsillar hernia with a general craniocaudal displacement of the brain stem (Figs. 110, 111, 141).

c) Tonsillar Herniation

With reference to the chapter concerning the copular point and its practical value in angiography, its study reveals a double disadvantage: the localization of the copular point is not constant on the one hand, and on the other, is subject, as all veins, to compression and in consequence to angiographic exclusion. In practice, however, the clearly lowered copular point, tonsillar and inferior vermian veins may be indispensable (Figs. 108, 111). More precocious and reliable venous information is given by the value of the angle of junction of the superior retrotonsillar vein with the inferior vermian vein: under normal conditions this angle measures approximately 90° or more. When the inferior pole of the posterior fossa is subject to pressure in an anteroposterior or posteroanterior direction, we observe the progressive narrowing of this angle and propose calling it "the venous tonsillovermian angle". The narrowing of this angle is the premonitory stage in tonsillar herniation (Figs. 107, 114). The lowering of the inferior pole of the posterior fossa in the cervical canal under the influence of intracranial hypertension causes two displacements in the inferior vermian vein and superior tonsillar vein: a backward displacement on the one hand, illustrated by Figure 107b and a downward displacement on the other hand, illustrated in Figure 107c and 111. These venous displacements are the equivalent of early signs of herniation of the tonsils, i.e., of a lowering of the posteroinferior cerebellar artery. The venous signs of herniation are difficult to recognize in a frontal view since geometric projection varies the topography of the veins with reference to the foramen magnum. Only major displacements are identifiable in a frontal view.

In a lateral projection one may use the foramen magnum line and then drop a perpendicular on this line through the posterior margin of the foramen magnum. The copular point is normally located in the angle formed by these two lines. When displaced, it touches one of the two lines (Figs. 107, 164).

d) Temporal Herniation

Arterial signs of temporal herniation are well known in neuroradiology. The veins provide the complement to this angiographic pattern. In a frontal vertebral angiography, the basal and posterior mesencephalic veins are compressed and pushed toward the midline. The course of these veins becomes concave outward. In a lateral view, the venous displacements are characterized by the lowering of the posterior segment of the lateral mesencephalic vein, of the posterior mesencephalic vein, and of the optopeduncular veins. Furthermore, the precentral vein is more or less pushed backward (Figs. 109–113).

e) Associated Herniations

The study of the phlebogram often shows a multiplicity of hernias (Fig. 110). Thus it is possible to find in association:

A temporal hernia and a tonsillar hernia (Fig. 111).

A cerebellar hernia and a hernia of the superior vermis.

A temporal hernia and a hernia of the superior vermis (Fig. 112).

A triple hernia: temporal, vermian, and tonsillar (Fig. 113).

IV. Tumors of the Fourth Ventricle

The concept, tumor of the fourth ventricle, can certainly be criticized as such. However, it can be embraced since it corresponds to clinical language. The phlebogram of the posterior fossa is certainly not solely competent in the evaluation of such tumors and nowadays numerous methods are utilized in establishing the diagnosis.

At the moment it is impossible to outline in detail the different varieties of tumors of the fourth ventricle and we shall limit ourselves to the principal anomalies. Huang's copular point has been set forth and criticized in detail on page 20. Picard [106] has also published a critical study. In order to establish the modifications of the copular point, we shall report the case of an important dilatation of the fourth ventricle which can be considered as a mass expansion and not as an infiltration of the area (Fig. 116). Thus, it appears logically that such an expansion of the fourth ventricle does not lower the copular point but reclines the superior retrotonsillar vein in such a manner as to narrow the "tonsillovermian venous angle" illustrated in Figure 107. This pushing backward is its principal sign. Such an anomaly, easily demonstrated by the dilatation of the fourth ventricle, is found in other cases. An example is found in Figure 117, which represents a case of sarcoma of the floor of the fourth ventricle. In this case of infiltrating tumor, other venous signs are visible, such as the dislocation and the stretching of the vein of the lateral recess of the fourth ventricle and the presence of abnormal slender veins in its area. The vein of the lateral recess of the fourth ventricle, which covers a large area, can be rather easily identified in both a frontal and side view and must be looked for in all cases of suspected tumor. In the cases illustrated in Figure 125 one can observe not only a lowering, but also a stretching of this vein. Finally, it must be pointed out that this vein acts as a drainage vein, as in the case illustrated by Figure 123, where its hypertrophy is caused by the arteriovenous shunt of an ependymoma. As to the lateral medullopontine vein, it is usually modified, particularly stretched, and displaced forward. Other indirect venous signs are to be taken into consideration, such as the stretching of the hemispheric veins, their abnormal curve under the influence of tumoral growths, the raising of the precentral vein, the herniation of the superior vermis in the subdural supravermian space or the herniation of the tonsils.

For example, the latter, both in a frontal and a lateral view, may show modifications of the copular point, as seen in Figure 121.

In conclusion, it is possible to divide the phlebographic symptomatology of tumors of the area of the fourth ventricle into direct and indirect signs.

Direct signs:

Narrowing of the tonsillovermian angle.
Abnormal venous network in the area of the fourth ventricle.
Various alterations of the vein of the recess of the fourth ventricle.
Various alterations of the lateral medullopontine vein.

Indirect signs:

Herniations: tonsillar and in the subdural supravermian space.
Stretching and displacement of various cerebellar veins more or less removed from the tumor and more especially of the precentral and lateral mesencephalic vein.

V. Tumors of the Brain Stem

In the topographic study of mass expansions of the brain stem, we shall distinguish tumors of the cerebral peduncles, of the pons and of the medulla oblongata. The main venous changes are schematized in Figure 126.

General Remarks

In tumors of the brain stem, tonsillar herniation occurs tardily, thus few changes are generally found in the tonsillar veins and the inferior vermian veins. However in Figure 137 we illustrate the case of a large pontine tumor with tonsillar herniation; Figure 139 is a case of compression of the posterior tonsillar veins by a bulbar tumor as an example of direct involvement of the tonsillar veins by a tumoral process. A superior vermian herniation is as frequent as a tonsillar one. Modifications in the course and location of the superior vermian vein with respect to the great vein of Galen and the straight sinus (Figs. 131, 136, 137) provide information on the herniation of the superior vermis. The absence of a herniated superior vermis is the result of a tonsillar herniation corresponding to a general craniocaudal displacement of the brain stem (Fig. 141). Whatever the peduncular localizations, pontine or bulbar, a mass expansion is characterized by typical venous displacements in a frontal view: the lateral veins of the brain stem such as the peripeduncular veins, the lateropontine veins, the lateral and posterior mesencephalic veins as well as the laterobulbar veins are pushed externally (Figs. 129, 130, 132, 133). Brain stem tumors are rarely localized in one of the three main topographic regions. Pontine tumors are often found evolving toward the peduncles in particular or toward the bulb. We shall omit from this chapter the signs of expansion toward the cerebellar hemispheres of fourth ventricle tumors.

Brain stem tumors can be adherent and compact, but they can also be multilobular, forming different nuclei thus explaining the paradoxical topographic angiographic alterations.

Whatever their topographic localization, the veins are displaced in the following manner on a lateral projection:
The optopeduncular veins are displaced upward and forward in a pontine localization, or downward and backward in a peduncular localization.

The lateral mesencephalic veins and the precentral vein are displaced upward and backward.

The posterior segment of the basal vein and of the posterior mesencephalic vein are displaced upward in a pontine tumor and downward in a peduncular tumor.

The pontine veins show a rectilinear course sometimes angulated but generally forward-displaced against the clivus, especially in pontine tumors.

The tonsillar veins are lowered and pushed backward while the tonsillovermian angle is narrowed.

The arteriovenous dissocation [156] which we shall present is valid for pontine tumors, as the superficial veins are more easily incorporated in the tumor than the arteries (Figs. 134, 135).

Three principal signs should be looked for in a lateral view:

a) A thickening of the brain stem characterized by an increase in the distance which separates the collicular point and the pontine veins (Fig. 126b).

b) A compression of the aqueduct of Sylvius, which corresponds to the line which joins the collicular point to the point of confluence of the lateral mesencephalic vein with the posterior segment of the basal vein (Fig. 126d).

c) A kind of modification of the precentral vein (Fig. 127), which gives a certain idea about the location of the tumor in one of the three main topographic areas of the brain stem, the peduncles (Fig. 127a), the pons (Fig. 127b), and the bulb (Fig. 127c).

1. Tumors of the Cerebral Peduncles

The signs of peduncular involvement of a thalamic tumor invading a right cerebral peduncle were described by one of us [149] in 1971 (Fig. 128). We refer also to Figure 129 and 130 which are very explicit on this point. The posterior segment of the basal veins and the peduncular veins outline the cerebral peduncles so that an increase in the volume of a peduncle becomes apparent in a frontal view. In a lateral projection we shall describe two signs of peduncular localization:

A displacement of optopeduncular veins either forward and upward (Figs. 133, 137) or downward and backward (Figs. 131, 132).

A displacement of the lateral mesencephalic vein backward.

2. Tumors of the Pons

This is the most frequent localization and thus the best known from a venous semeilogic viewpoint. In a lateral view, we describe:

A lifting or lowering, but especially a marked tautness of the optopeduncular veins, indicates the degree of alteration and the extension of invasion or of edema of the cerebral peduncles.

Various displacements, sometimes paradoxical, of the pontine veins.

The uplifting of the lateral mesencephalic vein, forward or backward as well as externally.

The uplifting of the precentral vein, which produces a rigid course followed by a concave one, either forward or backward, depending on whether the tumor develops toward the superior brain stem (anterior concavity of the precentral vein (Fig. 133) or toward the inferior brain stem (inferior concavity of the precentral vein) (Fig. 134).

The shifting of the aqueduct of Sylvius, which is the main sign in pneumoencephalography, may be suspected thanks to the venous anomalies. The aqueduct landmark (Fig. 126d) indicates this displacement and is a useful pattern.

The venous sign of thickening of the brain stem (Fig. 126b).

A displacement upward and backward of the landmark of the aqueduct of Sylvius (Fig. 126d).

Frontally, an analysis of the venous topography is more difficult, but it is easy to distinguish a compression, an exclusion, or an external displacement of the posterior segment of the basal vein and of the posterior mesencephalic vein (Figs. 126a, c).

Arteriovenous dissociation: in Figure 134 we present two examples giving the normal disposition of the prepontine veins and arteries. The localization of the prepontine veins is more posterior than the topography of the basilar artery; the veins indicate, in consequence, the depth of the prepontine cistern. Figure 135, already published [156], illustrates the arteriovenous dissociation: thus an intrapontine tumor can incorporate veins which appear very much removed from the basilar artery, which is compressed forward. On the contrary, an extracerebral prepontine tumor pushes back the arteries as well as the veins (see legend to Fig. 135).

3. Tumors of the Medulla Oblongata

Tumors of the medulla oblongata are infrequent and characterized by a typical venous syndrome difficult to find evidence of. This kind of tumor needs a peculiarly precise angiographic technique using magnification and subtraction.

Three signs can be noted (Fig. 139):

a) Stasis in the superior medullar venous system, which is well opacified for this reason
b) A pushing backward and downward of the tonsillar veins
c) A spreading out of the lateral medullopontine veins. When the tumor has evolved, total compression of these veins, which then do not opacify, results.

VI. Tumors of the Mesencephalon and the Thalamus

We treated the subject of the angiography of the mesencephalon in a monograph [154]. Here we are mainly interested in the characteristics of the veins, which are particularly informative in the domain of tumors.

We have seen the masses of cerebral peduncles (Figs. 129, 131–133, 136). Anomalies in veins are quite characteristic when the phlebogram is sufficiently contrasted. Direct enlargement in this case is of great help. Pineal tumors involve an increase in the angle formed by the internal cerebral vein and the basal vein in a lateral projection and, quite precociously, a posterior pushing backward of the precentral vein (Fig. 144). As to thalamic tumors, these present a varied arterial and venous symptomatology which can be summed up as follows:

1. The phlebogram of a vertebral angiography may reveal the direct signs of opacification of a tumor [26–156], a condition illustrated in Figures 128 and 141. It is useful to note that the veins seem to follow the real contours of the tumor. In the observation illustrated by Figure 141, the "venous" dimensions of the mass expansion are definitely larger than the "arterial" dimensions. Naturally one may wonder whether this difference is due to a peritumoral edema.

2. In any case, the phlebogram of a vertebral angiography shows important alterations in the veins of the thalamic region, namely:

Of the superior choroidal vein (Figs. 141, 142, 144, 146).
Of the superior thalamic vein (Figs. 141–144, 146).
Of the basal or posterior mesencephalic vein (Figs. 141–146).
Of the pineal veins (Figs. 142, 144–146).

It should be noted that in one of our illustrations (Fig. 143), the two superior thalamic veins are sufficiently well opacified so as to distinguish a unilateral uplift.

3. The phlebogram of a vertebral angiography also furnishes information on the repercussion of a thalamic tumor on the posterior fossa proper and indicates remarkably well the signs of a cone of temporal pressure, or even of temporal herniation.

Figs. 141, 142, 144, 145, and 146 illustrate venous signs which include:

A posterior pushing back of the precentral vein.
A compression with hammock-type lowering and possibly the dislocation of the posterior segment of the basal or posterior mesencephalic vein.
At times an important lowering and dislocation of the venous optopeduncular system.
Modifications of the lateral mesencephalic vein either in the direction of those of the precentral vein (pushed backward), or in the direction of those of the optopeduncular veins (lowered).

The venous symptomatology of thalamic tumors appears especially in lateral projections. Naturally, frontal projections can provide information on the lateralization of a blush, but indicate only a few discrete modifications of the basal and posterior mesencephalic veins except when the cerebral peduncles are encroached upon (Figs. 128–133). They complement the data furnished by the lateral projection, but alone do not provide sufficient information to render a diagnosis of thalamic tumor. On the contrary, venous anomalies seen in profile are usually characteristic of this variety of tumor.

Lastly, it should be noted that thalamic tumors seem to involve a precocious temporal herniation, with or without tonsillar herniation. This craniocaudal displacement does not produce a superior vermian herniation, and even appears to increase the domain of the triangular projection of the supravermian, subdural space (Figs. 111, 144–146, and the diagram Fig. 110b).

VII. Tumors of the Cerebellar Hemispheres

These tumors present relatively early abundant venous signs which render their diagnosis more or less easy on the indications of the veins alone. There exists a category of vermian tumors which expand in the two cerebellar hemispheres in a rather symmetrical way showing venous signs of diffuse hypertension of the posterior fossa illustrated in Figure 102. We submit such a case of voluminous vermian tumor with bilateral hemispheric extension in Figure 147. This is a case of an anterior vermian tumor.

As opposed to true hemispheric tumors or to vermian tumors predominantly lateralized in one hemisphere, anterior vermian tumors with bilateral hemispheric encroachment involve a posterior pushing back of the precentral vein. In effect, in all other forms of hemispheric masses an uplifting of the precentral vein is observed (Figs. 148–152). Indirect venous signs of stretching of the vein of the lateral recess of the fourth ventricle and of forward displacement, rarely backward, of the point of junction of the vein of the lateral recess with the lateral medullopontine vein (Fig. 147) appear at the level of the inferior pole of the posterior fossa.

These displacements of the vein of the lateral recess of the fourth ventricle, of the lateral medullopontine vein, and of their point of junction with relation to Twining's midline are schematized in Figure 125b.

Well-lateralized hemispheric tumors, be they vermian tumors having invaded a single hemisphere or purely hemispheric tumors, have a homogeneous and fairly characteristic venous symptomatology. Such tumors are illustrated by Figures 148–152 and involve all of the following signs:

1. Forward and upward displacement of the entire venous system of the superior part of the posterior fossa, that is of:

The precentral vein
The lateral mesencephalic vein
The posterior mesencephalic vein
The optopeduncular veins
The prepontine veins
The superior vermian vein.

2. A stretching and dislocation of the cerebellar hemispheric veins, a possible blush, and an avascular zone. A special place must be set aside for the inferior paravermian vein among the hemispheric veins. This vein is generally well visible and can have a course so far away from the midline so as to take on the value of a hemispheric vein. Its alterations in curvature are extremely important in the analysis of a hemispheric cerebellar tumor (Figs. 149, 150).

3. Variable signs at the level of the inferior part of the posterior fossa. Among these signs one can distinguish:

a) Signs of tonsillar herniation: tonsillovermian angle, copular point, inferior vermian vein (Figs. 149–151).
b) Various alterations of the vein of the lateral recess of the fourth ventricle: stretching in a lateral projection (Figs. 147, 149), asymmetry of the inferior branch in a frontal projection (Fig. 149), with or without displacement of the junction point of the vein of the lateral recess with the lateral medullopontine vein (Fig. 125b).
c) Controlateral deviation of the medial veins: Of the inferior vermian vein (Figs. 148–150). Of the whole of the superior cerebral trunk (Fig. 152).

VIII. Tumors of the Cerebellopontine Angle

Acoustical neurinomas, meningiomas, occasional epidermoid or other rare tumors are the principal mass expansions with which we are concerned in the following pages. In this connection, phlebography plays an important role in angiographic diagnosis. A schematized analysis follows.

Intermediate or Capillary Phase

In approximately one-third of the cases a blush or an avascular zone is observed. A blush on neurinomas is usually limited and "heterogeneous" in appearance in one part of the tumor (Figs. 155, 156). A blush on meningiomas on the other hand, is more outspread and more homogeneous (Figs. 159, 161). It is very hard to depend on such differences, all the more so in that certain tumors are histologically difficult to distinguish. Avascular zones can be very coherent radiographically, furnishing the boundaries of the tumor, especially when bordering veins exist (Figs. 156, 160, 164). Finally, it is the association of a small blush and of an avascular zone which gives the best diagnostic sign (Fig. 156) of an acoustic neurinoma in the cerebellopontine angle. The more extensive the blush, the more open to doubt the neurinoma and the more possible the meningioma. However, a clear-cut distinction of either is controversial, no absolute rule existing.

Drainage Veins

Apart from the superior petrosal vein, various drainage channels can be observed, characterized by sinuous and dilated venous trunks. These apparently normal veins correspond most frequently to the normal veins of the posterior fossa, such as the lateral mesencephalic vein, the vein of the lateral recess of the fourth ventricle, and the superior petrosal vein. These drainage veins are obviously very useful in that their considerable vascularization of the tumor draws attention to it and delimits its boundaries (Figs. 155, 157, 159).

Changes in the Superior Petrosal Vein

This principal venous element is subject to different types of modifications. It is to be remembered that its normal opacification above and beyond the porus of the internal auditory meatus, near the external wall of the cerebellopontine cistern, is a good criterion of normality. Thus various types of modifications of the superior petrosal veins in cases of tumors of the angle can be schematized:

1. The vein's great axis normally obliques upward and inward. An early sign consists in a straightening and even an inversion of the obliquity (Fig. 165b). Figure 153 and 158 illustrate a case of obliquity upward and outward on the side of the tumor.

2. The superior petrosal vein is generally displaced laterally. The difference in the distance with reference to the midline measures an average of 10 mm. The central point of this vein, which is a venous segment, may be difficult to localize exactly, but generally one can choose a similar left and right point of the vein so as to measure its distance to the midline (Fig. 165a–e).

3. An elevation of the superior petrosal vein is certainly infrequent but is a reliable sign of a tumor originating in the auditory meatus, spreading into the cerebellopontine cistern (Figs. 155, 157, 162, 165d, e). This elevation, when accompanied by a dilatation of the affluents of the vein, takes on an umbrella shape which covers the superior pole of the tumor (parasol sign and stellar sign). This elevation is not constant but when it exists (in 30% of cases), its mean value is 5 mm.

4. A large avascular zone due to the exclusion of the veins results from the compression of the venous system in the cerebellopontine angle. In these cases, it is mandatory to examine the phlebogram prudently and to look for other signs (Fig. 156).

5. Finally, it must be pointed out that we have observed a deviation of the superior petrosal vein toward the midline in a case of epidermoid tumor of the cerebellopontine angle (Figs. 164, 165f). At present such a displacement has not been observed in our cases of neurinomas and meningiomas.

Changes in the Precentral Vein

Two types of displacements can be observed: the precentral vein is usually shifted backward in such a manner that it is deformed and its course presents a more or less important anterior concavity. This displacement is due to a tumor involving the paramedian area (Figs. 154, 156, 163). On the other hand, when the precentral vein is elevated or shifted forward, the tumors responsible for such a displacement have a more lateral expansion (Figs. 155, 162, 164). This latter form was observed especially in the case of the epidermoid tumor illustrated in Figure 164.

Changes in the Lateral and Posterior Mesencephalic Veins and in the Optopeduncular Venous System

The shifting of the posterior mesencephalic and optopeduncular vein indicates a supratentorial extension of a tumor in the pontocerebellar angle (Figs. 154, 163, 164). As to the lateral mesencephalic vein, it presents the same changes as the precentral vein. However, it should be remembered that the lateral mesencephalic vein is less often opacified and that its course lies higher and more laterally than that of the precentral vein. A frontal projection reveals the precise change in course of the lateral mesencephalic vein, especially when it concurs in the tumoral drainage.

Changes in the Pontine Veins

These veins lie flat against the basisphenoid when the tumor does not expand into the prepontine cistern. On the contrary, the longitudinal pontine veins are shifted and dislocated during tumoral expansion into the prepontine cistern (Figs. 155, 162, 163).

Changes in the Vein of the Lateral Recess of the Fourth Ventricle

It is usually lowered and shifted backward (Figs. 156, 160, 163). It can be raised (Figs. 157, 162) or displaced toward the midline as in the case of an epidermoid tumor (Fig. 164). Specifically its role is important in venous drainage (Fig. 157).

Changes in the Lateral Medullopontine Vein

It is lowered, displaced inwardly or outwardly and sometimes takes part in the drainage of the tumor through multiple small affluents (Figs. 158, 160, 164).

Changes Surrounding and Involving the Cerebellopontine Angle (Fig. 167)

Different types of tumors can extend toward the cerebellopontine angle, such as hemispheric cerebellar masses, tumors of the fourth ventricle, pontine tumors, as well as sellar tumors. An example is given in Figure 167 whose venous signs are very characteristic: the modifications occur in the more or less medial veins such as the optopeduncular and precentral veins, as well as in the veins of the cerebellar pontine angle proper.

IX. Hemangioblastomas and Vascular Malformations

Hemangioblastomas of the posterior fossa are relatively frequent and are present under the form of several small rounded fleshy nodes or by a small fleshy area which opacify precociously during the arterial phase and show up extremely well during the capillary phase. These direct signs, easily revealed through subtraction, exist along with indirect signs of a cystic mass expansion, the liquid collection formed by the solid matter. Arterial symptomatology can be less characteristic, whereas venous anomalies are of major importance. These anomalies include two orders of signs (Fig. 169):

1. The extent of the drainage veins, much more numerous and of a greater caliber than in other cases of cerebral tumors, but less developed than in arteriovenous aneurysms.
2. Indirect venous signs of displacement due to the cyst.

Vascular Malformations do not present special characteristics in the posterior fossa. There exist however three cases:

1. Arteriovenous aneurysms of the lateral sinus resulting in the connection of the dural arteries with the sinuses (Fig. 166).
2. The paths of venous drainage followed by the veins of the posterior fossa, while the malformation proper resides in the spinal canal, or in the subtentorial space (Fig. 168). During vertebral angiography these drainage veins can be observed, thus giving ground for suspicion of an arteriovenous malformation.
3. Arteriovenous malformations: these types of malformation generally are made up of branches of the external carotid artery which is connected directly to the lateral sinus. This is depicted in Figs. 166a–b.

In the same case the fistula may be fed by numerous arteries: internal and external carotid artery, vertebral arteries. A recent report was given by Debrun [35b].

X. Congenital Malformations

The Dandy-Walker and the Arnold-Chiari malformations are well known and are described in the classical literature. Veins are of great use in the analysis of these malformations. In the Dandy-Walker malformation, they are characteristic at the level of the sinuses, whereas the small veins of the occipital cervical junction are modified in the Arnold-Chiari's syndrome. We have observed the various venous variants, notably the topography of the tonsillar veins, the lateral medullopontine vein, and the vein of the lateral recess of the fourth ventricle (Figs. 170, 171).

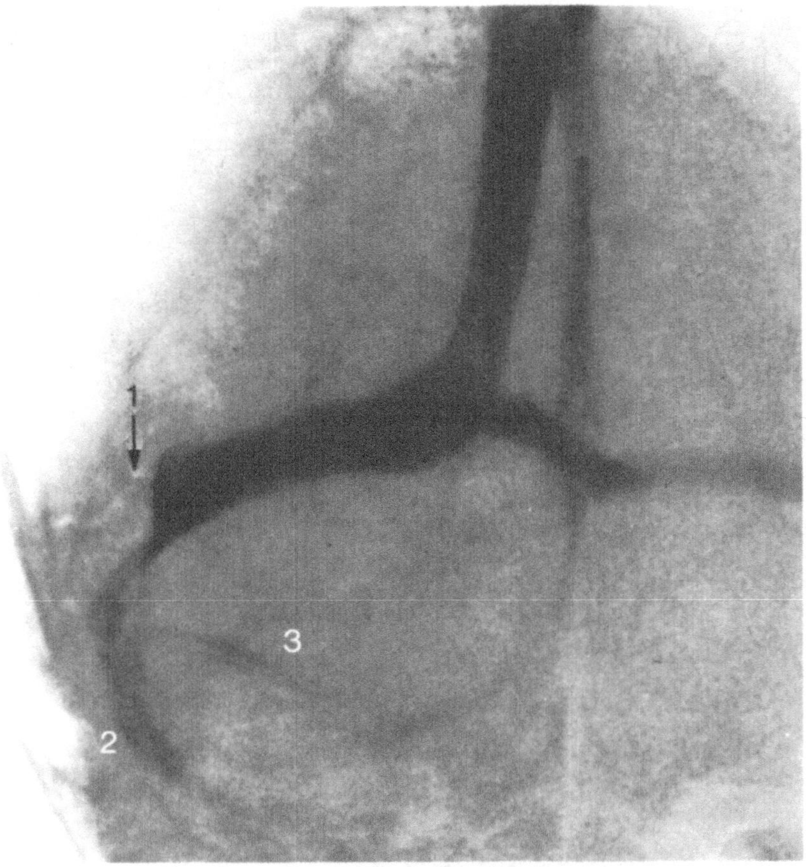

Figure 82

Thrombosis of right lateral sinus

1 Image of uncharacteristic hold-up
2 Derivation through mastoid emissary vein
3 Derivation through superior petrosal sinus

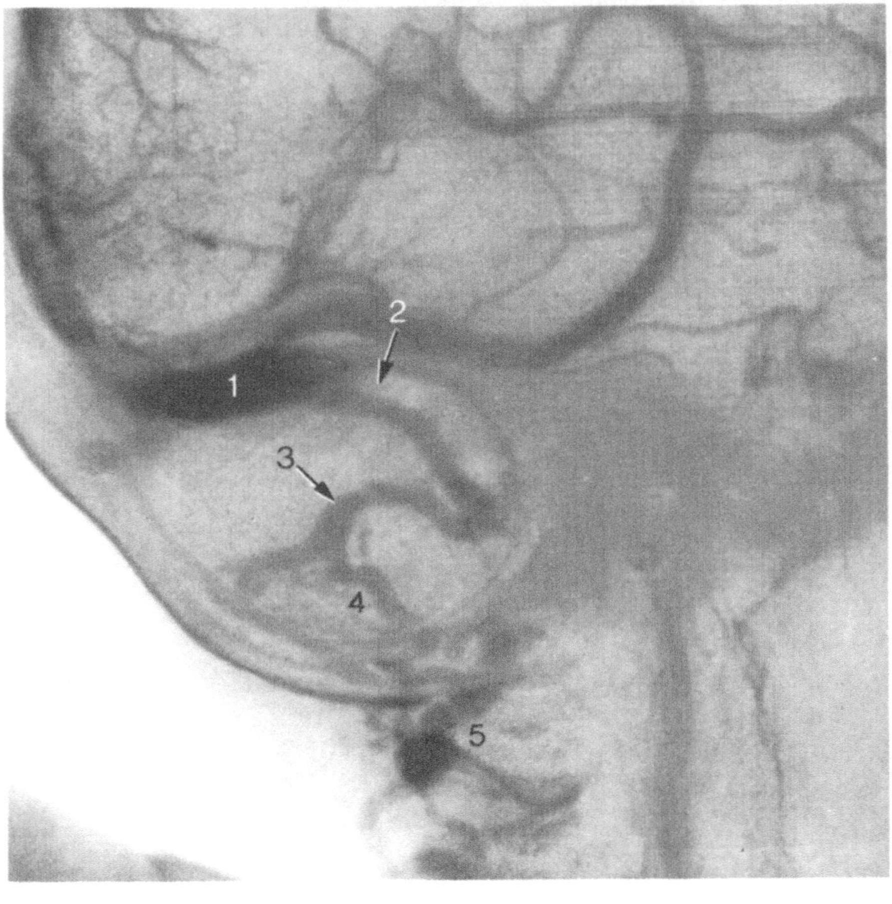

Figure 83

Thrombosis of right lateral sinus

1 Open segment of lateral sinus
2 Snake head-shaped sinus thrombosis
3 Mastoid emissary vein
4 Communicating vein between mastoid emissary vein and vertebral veins (*5*)

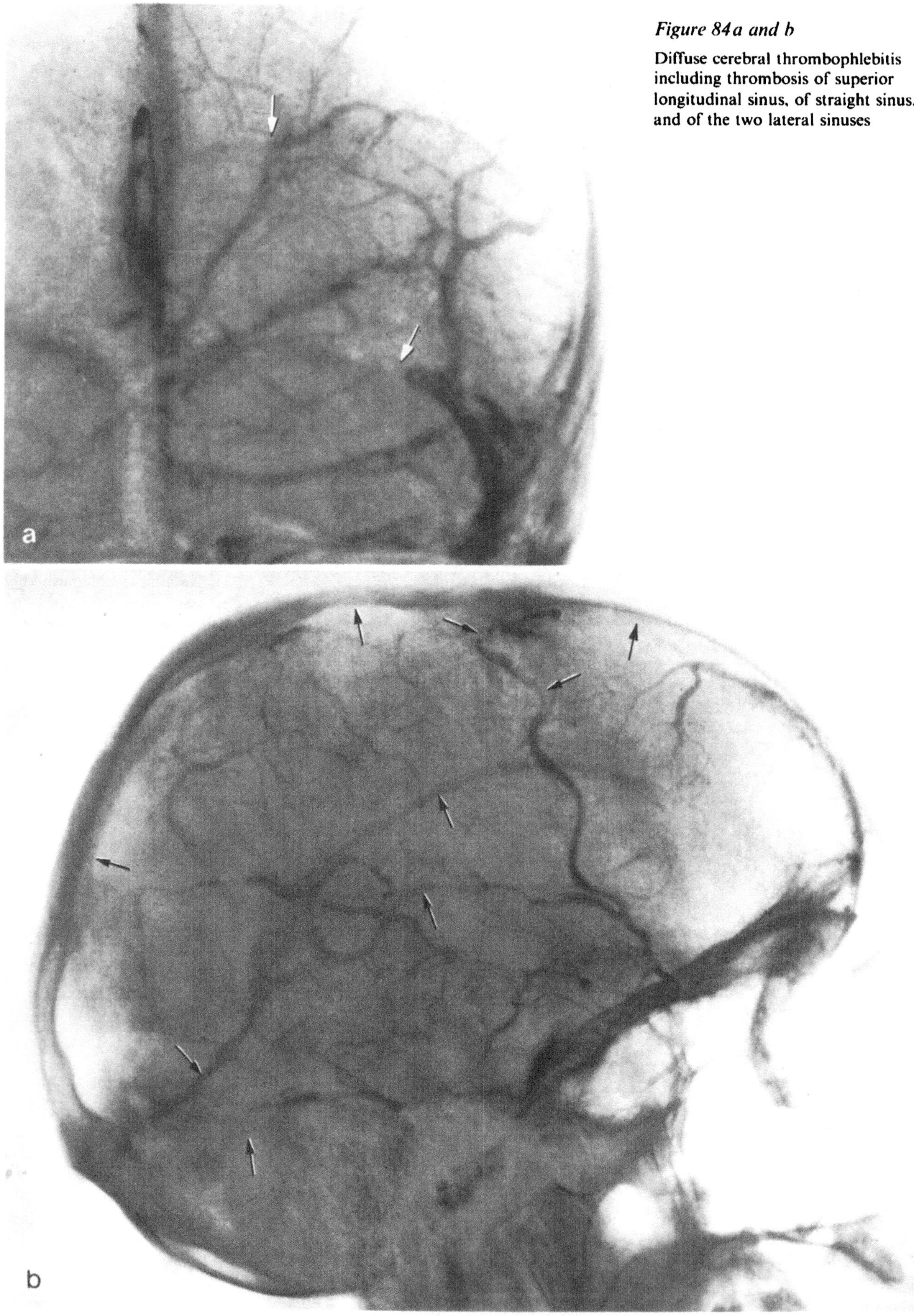

Figure 84 a and b

Diffuse cerebral thrombophlebitis including thrombosis of superior longitudinal sinus, of straight sinus, and of the two lateral sinuses

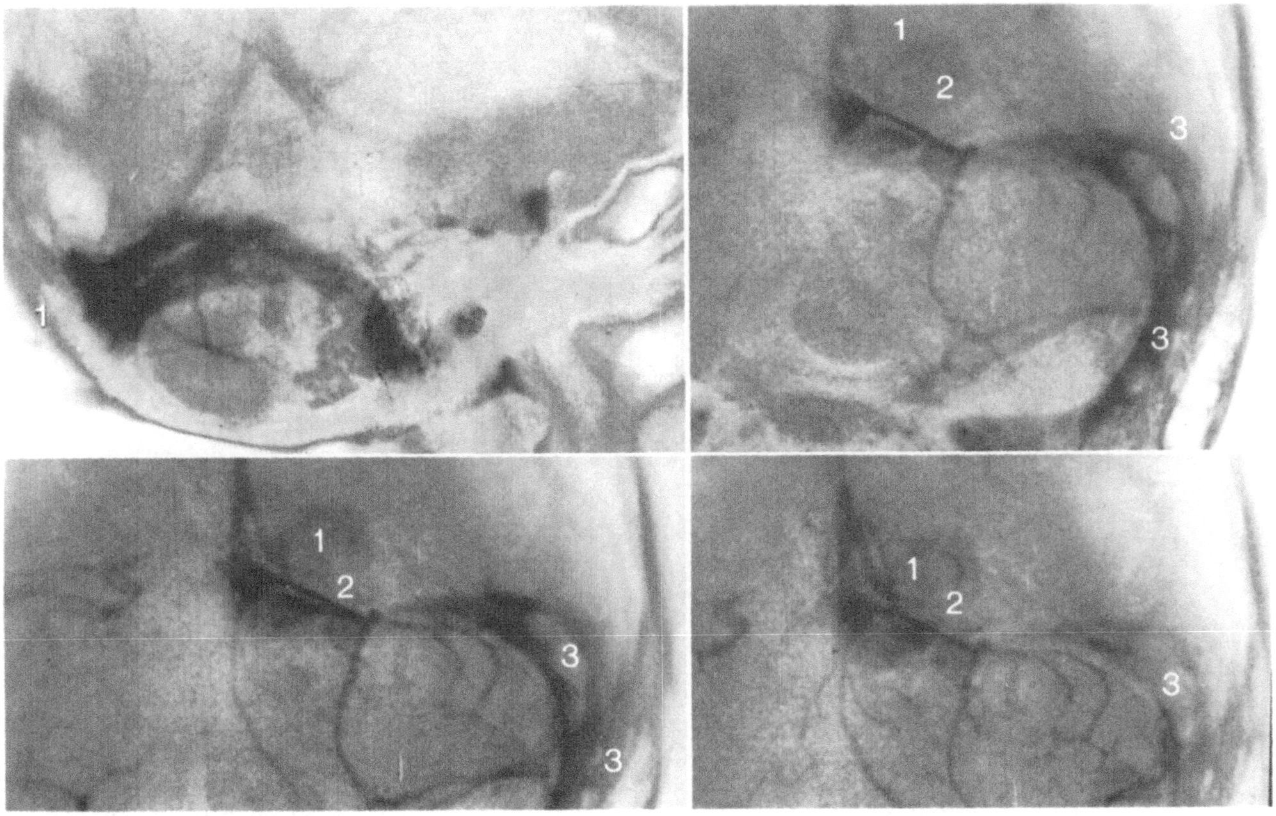

Figure 85

Posttraumatic diploic hydrocele of left occipital squama and dura mater involving lateral sinus (multiple arachnoïdal cysts). *1* area corresponding to a considerable translucency in left occipital paramedian area above and below lateral sinus; *2* affected area in lateral sinus; *3* multiple arachnoïdal cysts in lateral sinus

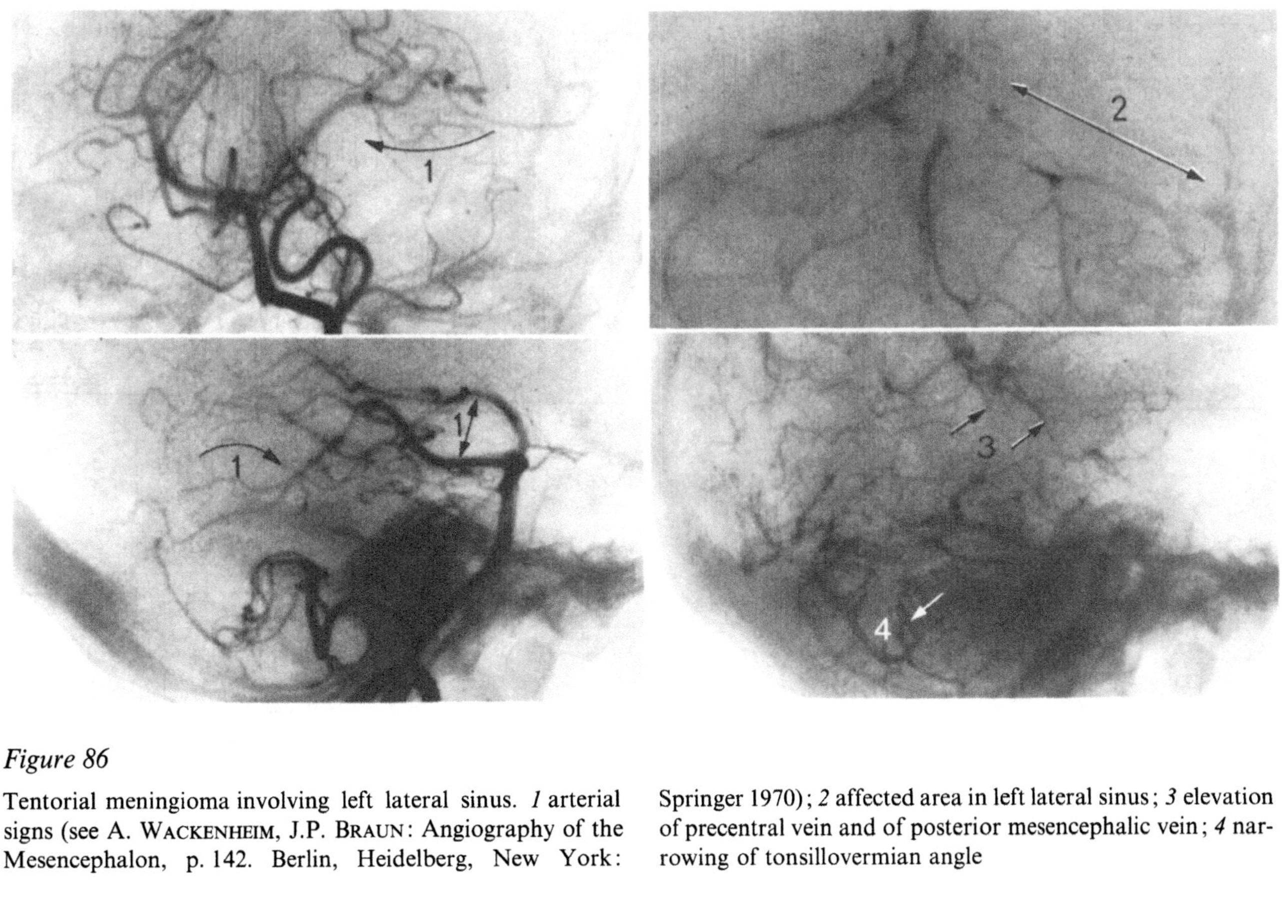

Figure 86

Tentorial meningioma involving left lateral sinus. *1* arterial signs (see A. Wackenheim, J.P. Braun: Angiography of the Mesencephalon, p. 142. Berlin, Heidelberg, New York: Springer 1970); *2* affected area in left lateral sinus; *3* elevation of precentral vein and of posterior mesencephalic vein; *4* narrowing of tonsillovermian angle

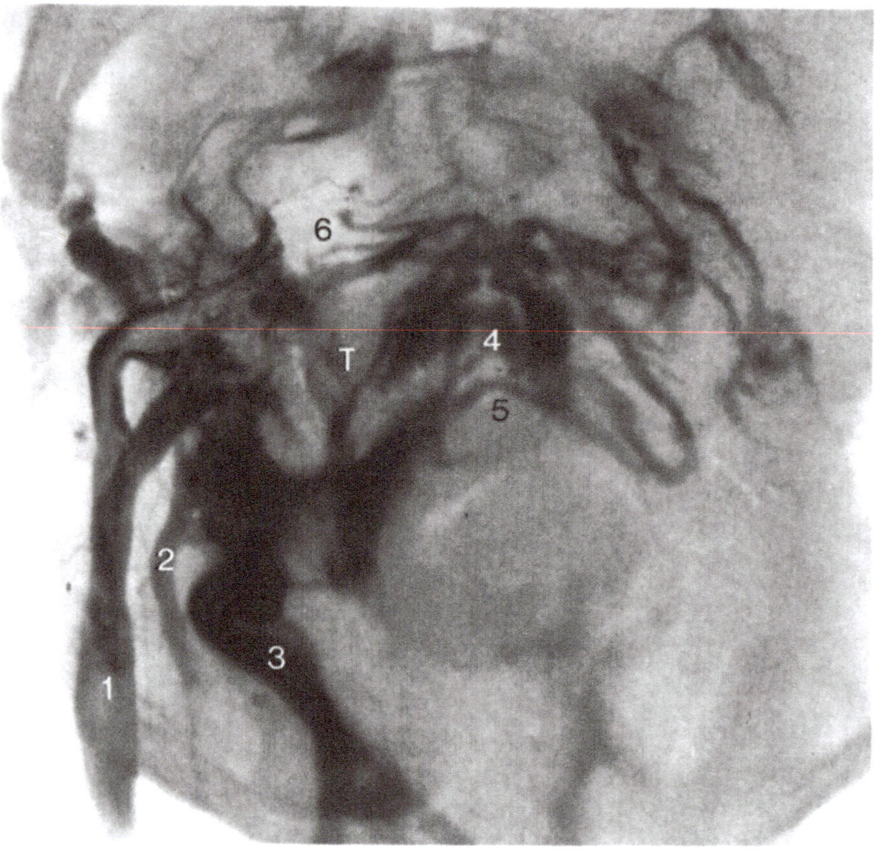

Figure 87

Jugular glomus tumor demonstrated by retrograde jugulography

1 Jugular vein
2 Condylar vein
3 Vertebral vein
4 Basal plexus
5 Perioccipital plexus (anterior part)
6 Inferior petrosal sinus
T Tumor of jugular glomus

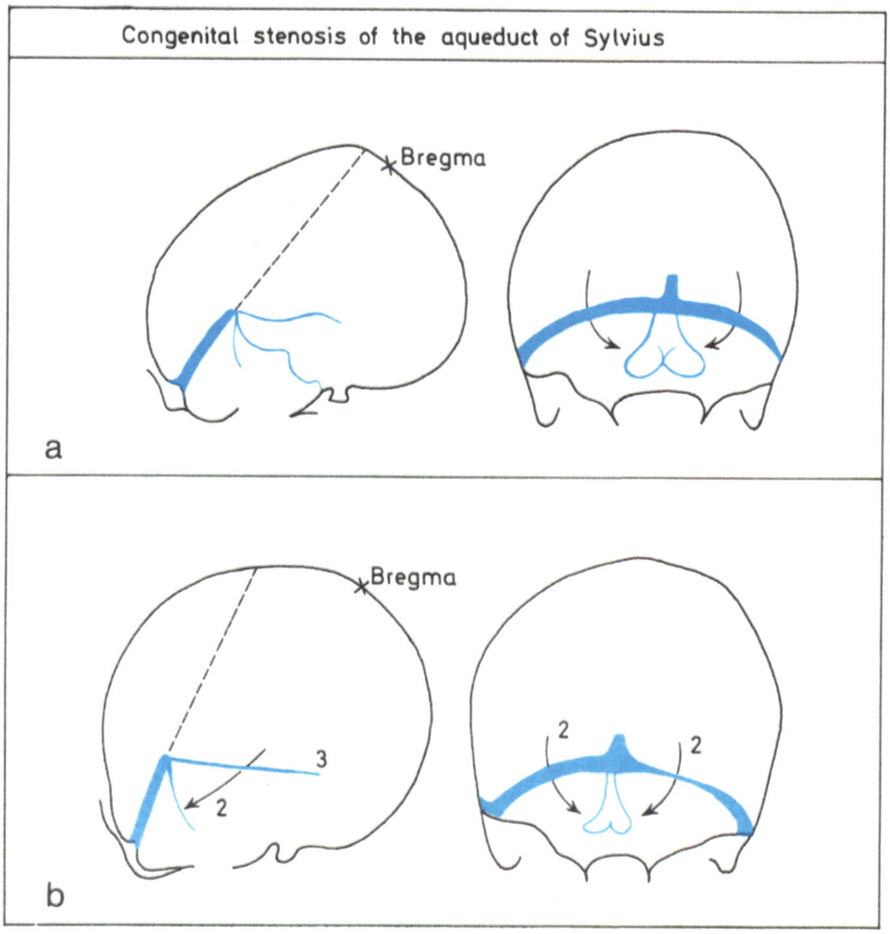

Figure 88 a and b

Congenital stenosis of aqueduct of Sylvius

a Minor or late form with slight straightening of straight sinus
b Severe or precocious form with marked straightening of straight sinus (*1*), signs of bitemporal herniation (*2*), and straightness of internal cerebral vein (*3*)

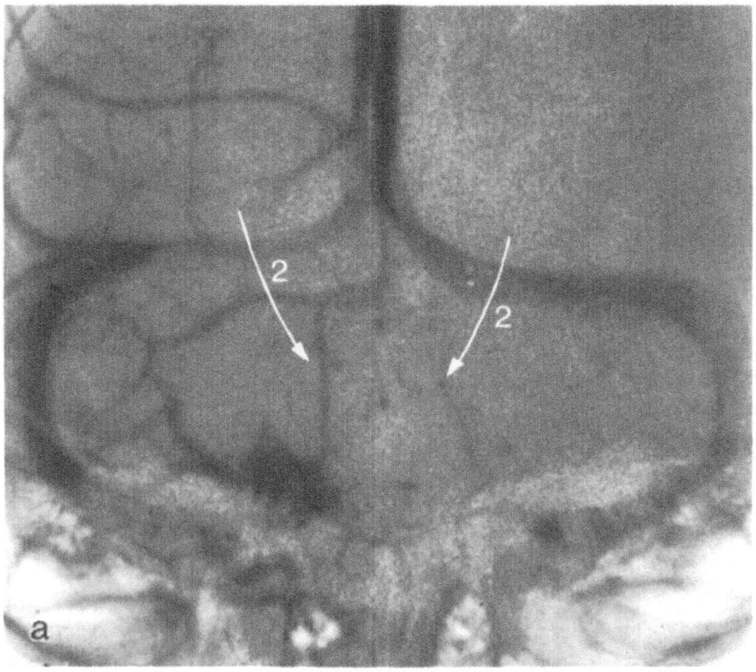

Figure 89a and b

Precocious congenital stenosis of aqueduct of Sylvius having characteristic venous anomaly in a 3-year-old child

1 Straightening of straight sinus
2 Sign of bitemporal herniation
3 Straightness of internal cerebral vein

Planimetry (see Figs. 99 and 100)

A	202.26	(increased)
B	7.02	(increased)
C	45.93	(normal)
$A+B+C$	255.21	(increased)
$A+B$	209.28	(increased)
$(A+B)/C$	4.5	that is a markedly abnormal relation indicative of an obstacle at level of aqueduct of Sylvius

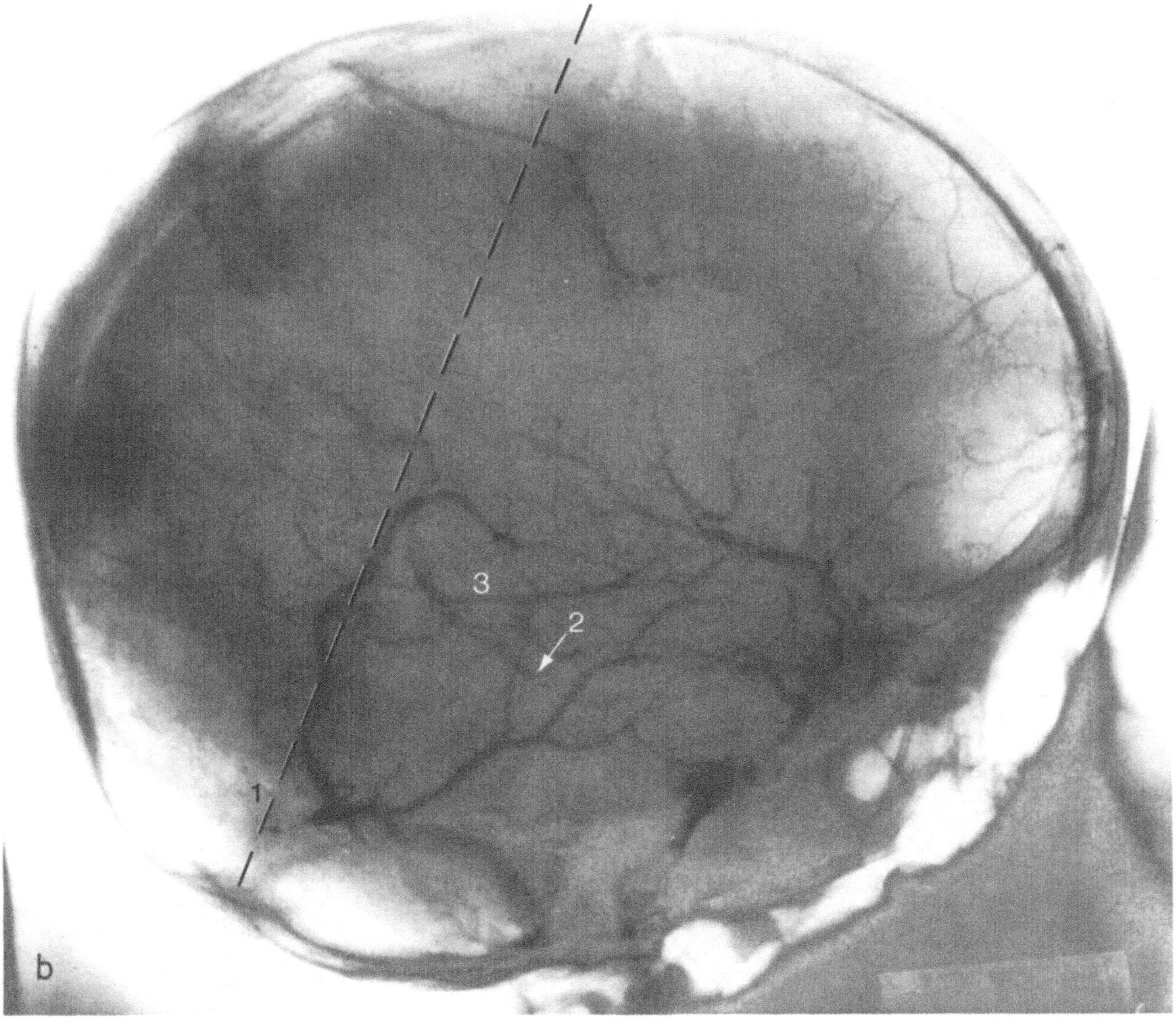

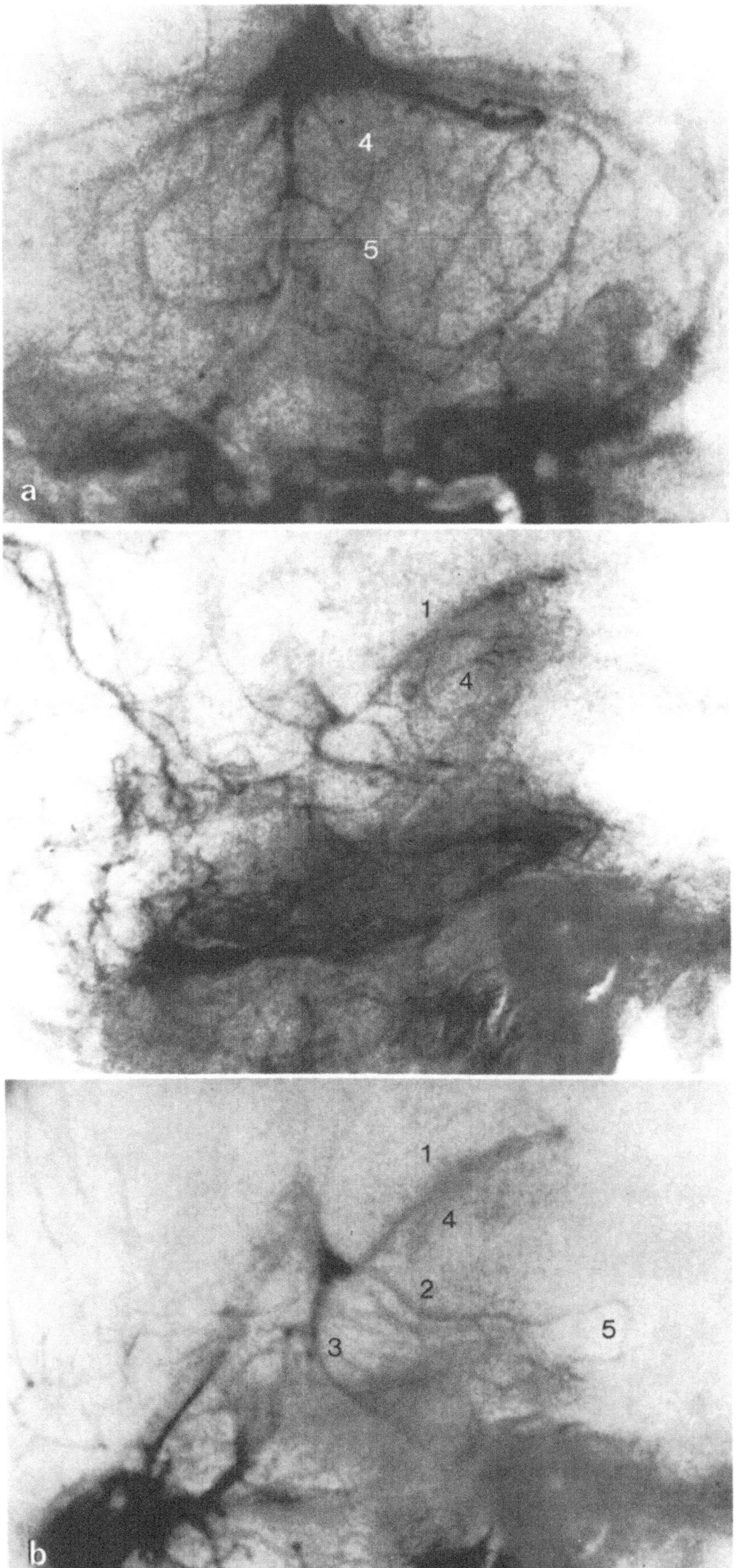

Figure 90 a and b

Posterior thalamic glioma causing an obstruction in aqueduct of Sylvius

1 Rigidity of superior thalamic vein
2 Hammock of posterior segment of posterior mesencephalic vein
3 Considerable pushing back of precentral vein
4 Tumoral blush
5 Invasion of cerebral peduncle

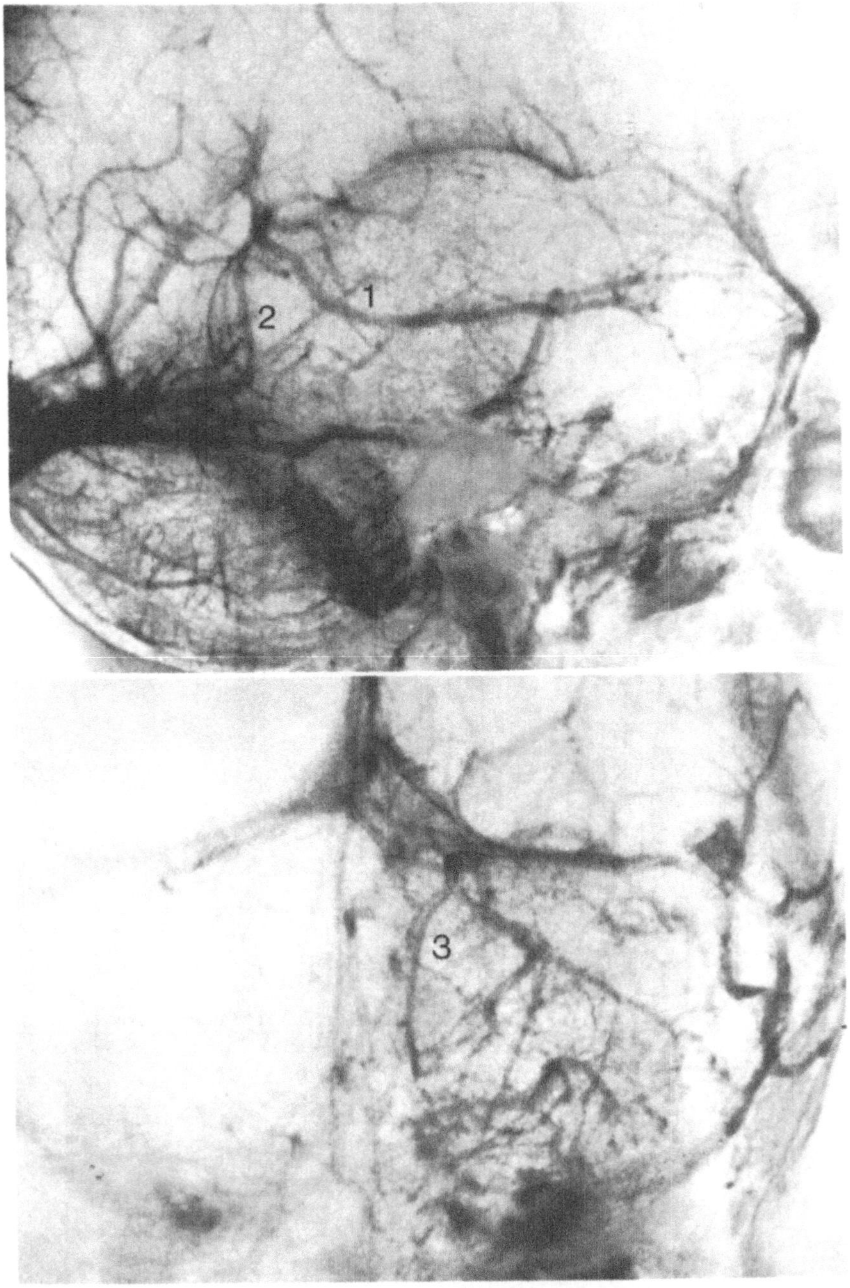

Figure 91

Pinealoma causing an extrinsic stenosis of aqueduct of Sylvius in adult

1 Lowering of posterior segment of basal vein
2 Considerable pushing back of precentral vein
3 Dislocation of veins of brain stem

Planimetry

A	146.83	(increased)
B	11.93	(normal)
C	51.33	(increased)
A+*B*+*C*	210.09	(increased)
A+*B*	158.75	(increased)
A+*B*/*C*	3.0	that is a relation within limits of normalcy in adult

Figure 92

Stenosis of aqueduct caused by tumor of superior brain stem

1 Bulging of posterior mesencephalic vein to left
2 Increase in volume of left cerebral peduncle
3 Contralateral deviation of interpeduncular veins characterizing increase in left cerebral peduncle
4 Exclusion of inferior segment of posterior mesencephalic vein to right
5 Hammock of posterior mesencephalic vein
6 Dislocation of optopeduncular veins
7 Elevation of collicular point

▽

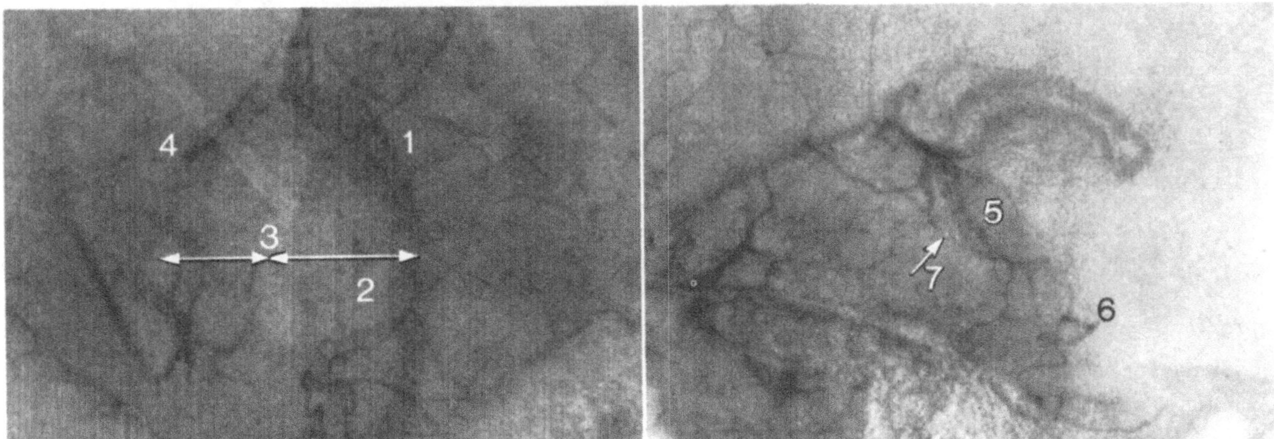

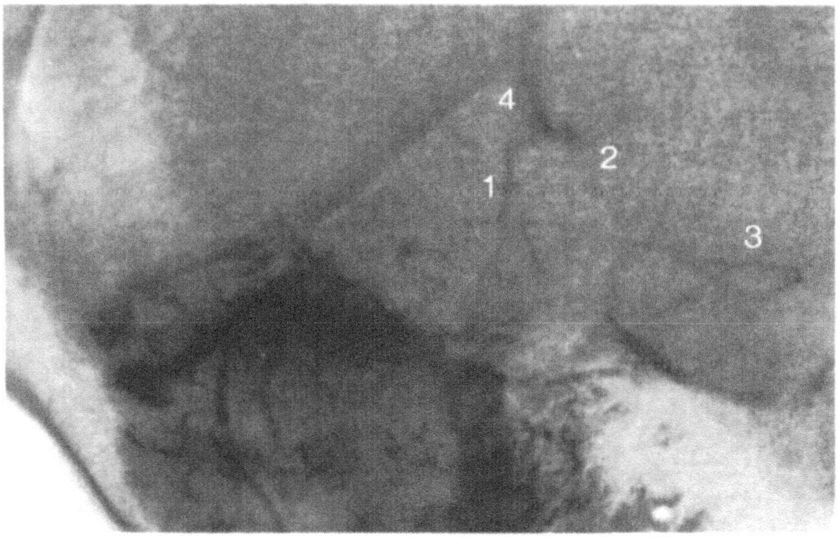

Figure 93

Tumor of mesencephalic area with stenosis of aqueduct

1 Pushing back of precentral vein
2 Marked irregularity of posterior mesencephalic vein
3 Flattening of optopeduncular veins
4 Herniation of superior vermis

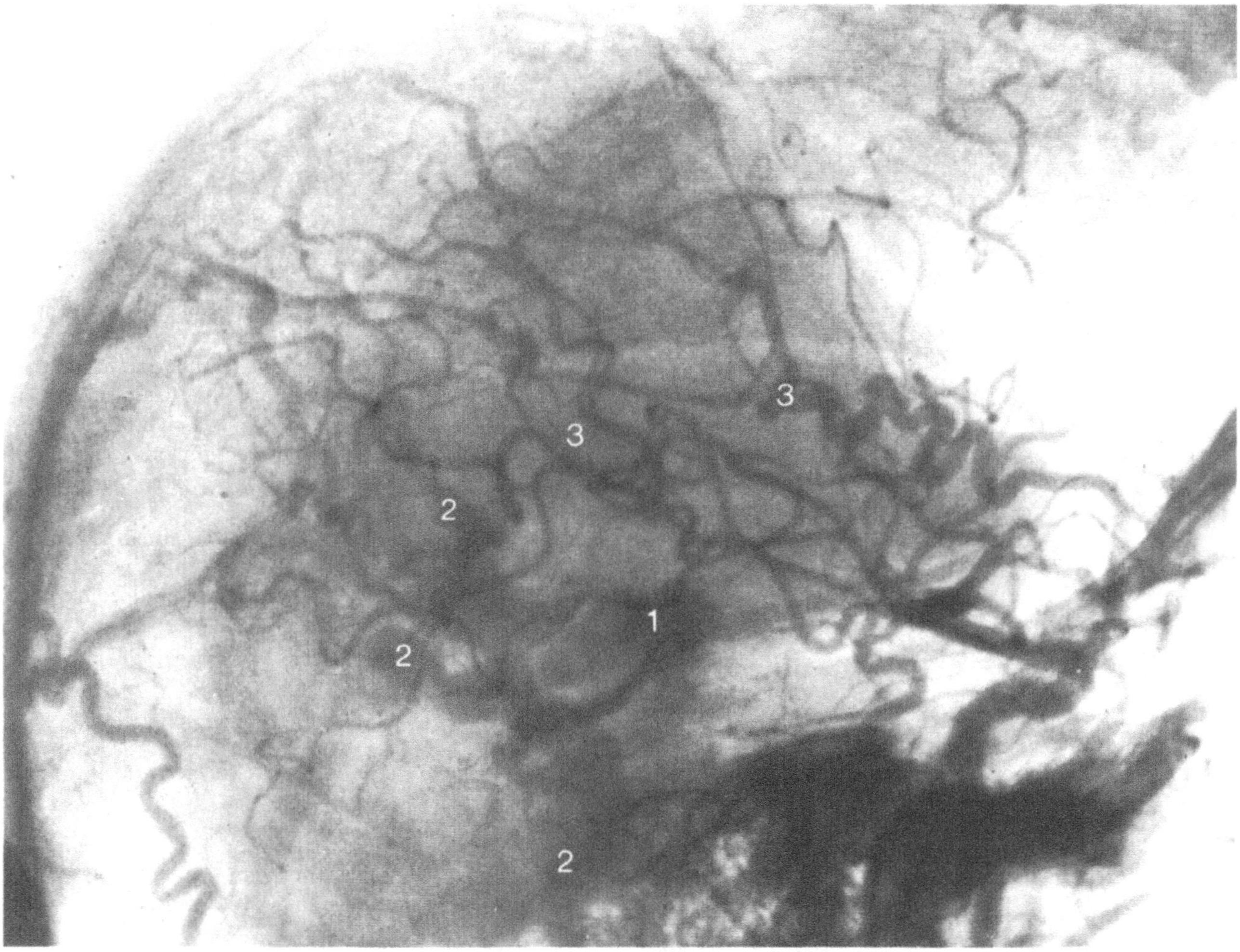

Figure 94

Arteriovenous aneurysm involving aqueduct of Sylvius

1 Veins in aqueduct
2 Various efferent veins
3 Various afferent arteries

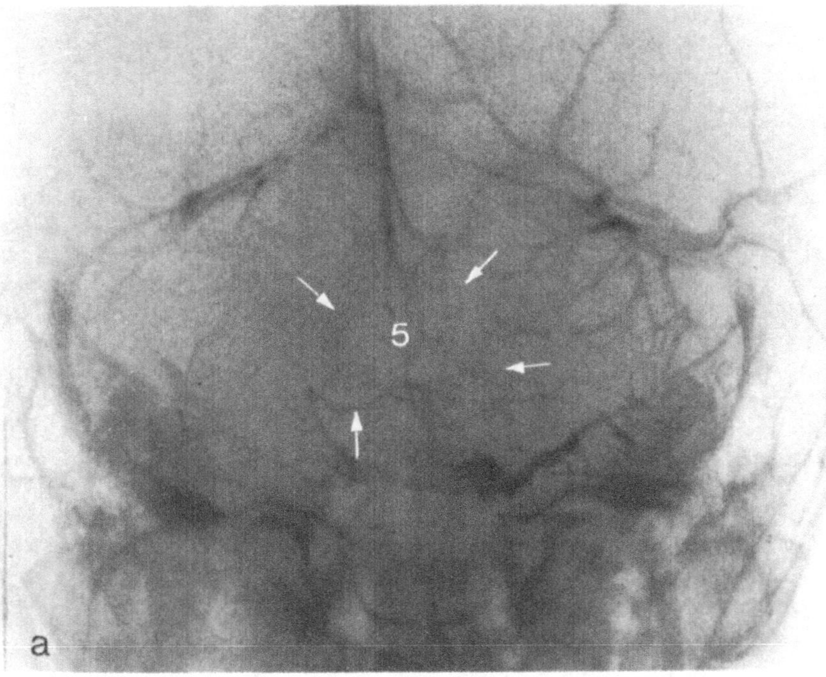

Figure 95a and b

Communicating hydrocephalus with slight bitemporal herniation and a normal-sized posterior fossa in 4-year-old child

1 Straight sinus
2 Lowered basal vein
3 Increase in straight sinus-great vein of Galen angle
4 Flattening of optopeduncular veins
5 Relatively small cerebral trunk

Planimetry

A	151.66	(increased)
B	11.86	(normal)
$A+B+C$	215.52	(increased)
$A+B$	163.52	(increased)
$(A+B)/C$	3.1	that is a normal relation

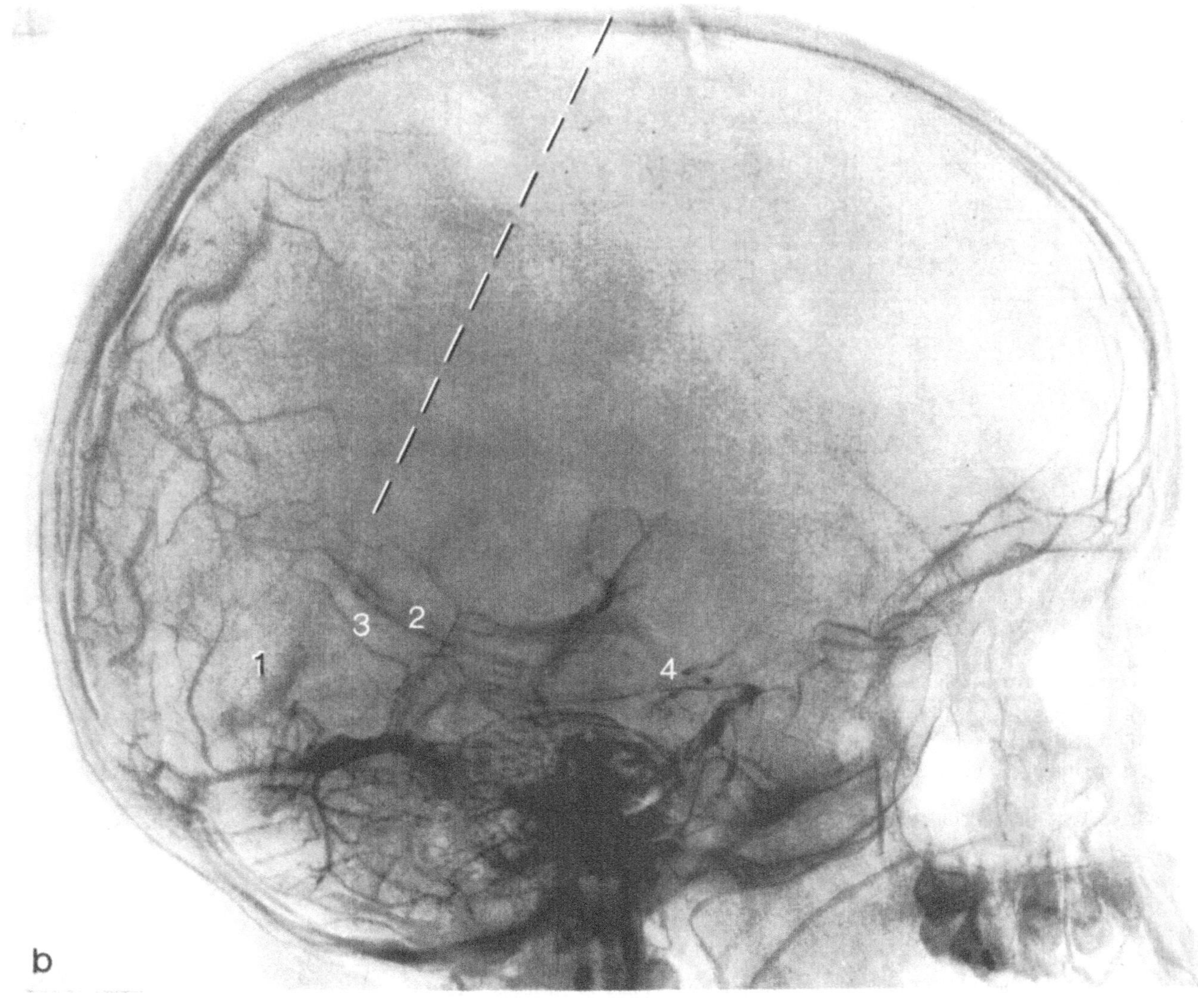

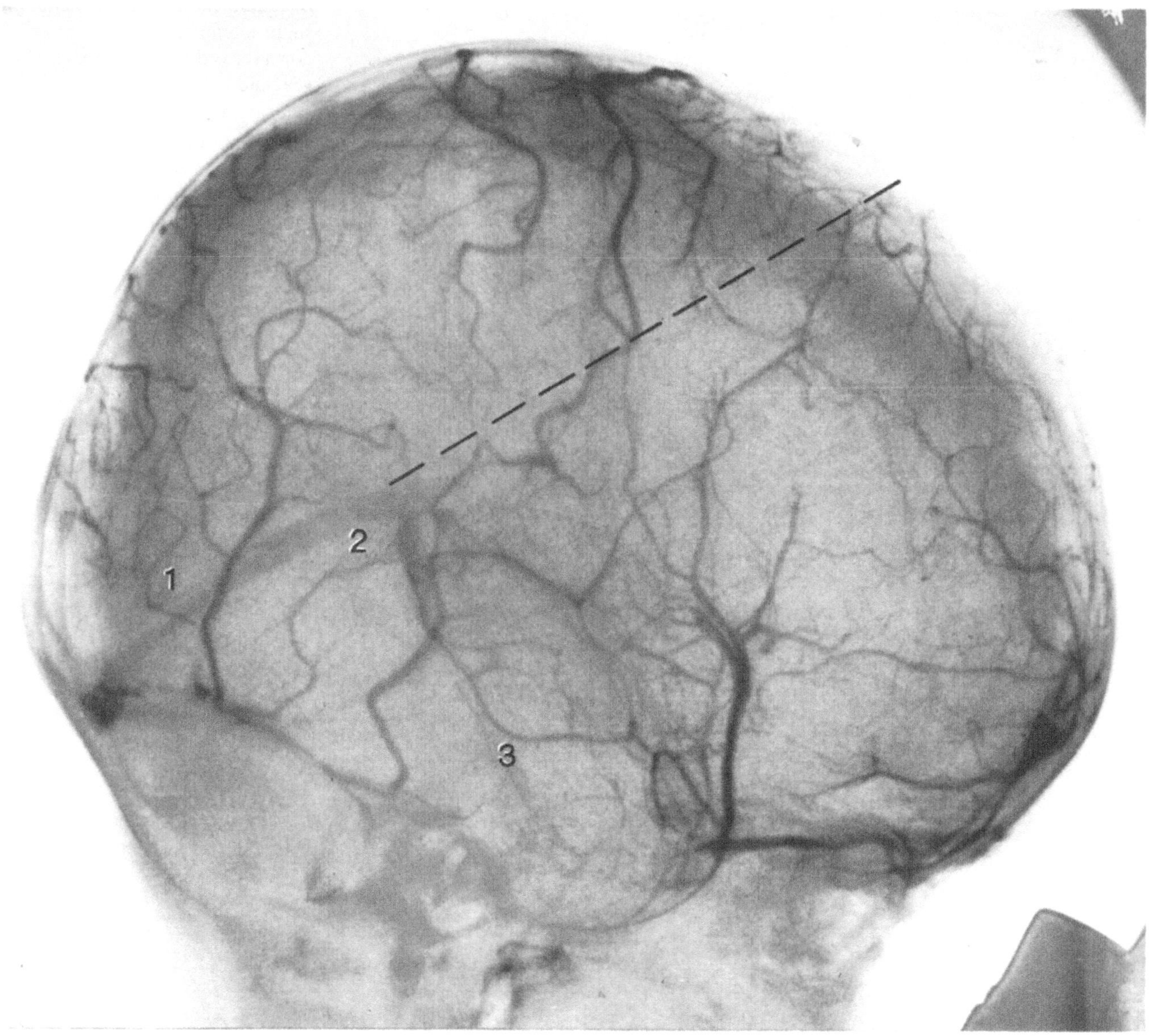

Figure 96

Communicating hydrocephalus with slight bitemporal herniation in 3-year-old child

1 Straight sinus in normal position
2 Straight sinus-great vein of Galen angle slightly open
3 Excessive angulation of point *L*

Planimetry (see Figs. 99 and 100)

A	167.00	(increased)
B	12.16	(normal)
C	51.20	(increased)
$A+B+C$	230.36	(increased)
$A+B$	179.16	(increased)
$(A+B)/C$	3.4	that is a normal relation in spite of increase in surfaces A and C

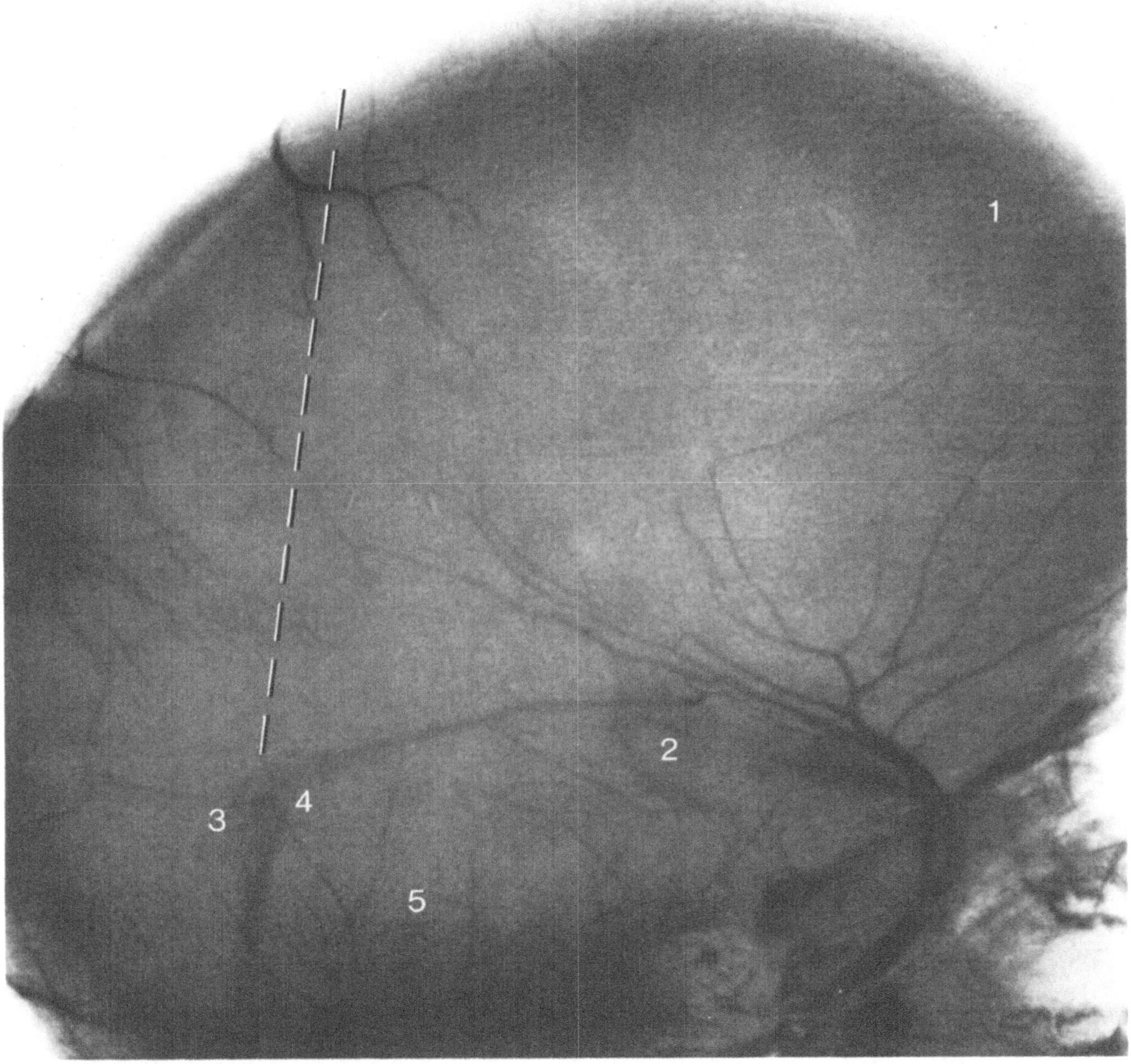

Figure 97

Hydranencephaly in 3-month-old child

1 Absence of frontal cortical veins
2 Increased opacity of basal ganglia
3 Straightening of the straight sinus
4 Straight sinus-great vein of Galen angle greatly increased
5 Small posterior fossa

Planimetry (see Figs. 99 and 100)

A	275.35	(greatly increased)
B	12.46	(normal)
C	51.20	(increased)
$A+B+C$	338.99	(increased)
$A+B$	287.77	(increased)
$(A+B)/C$	5.6	that is a considerably increased relation in spite of certain increase in surface of projection of posterior fossa

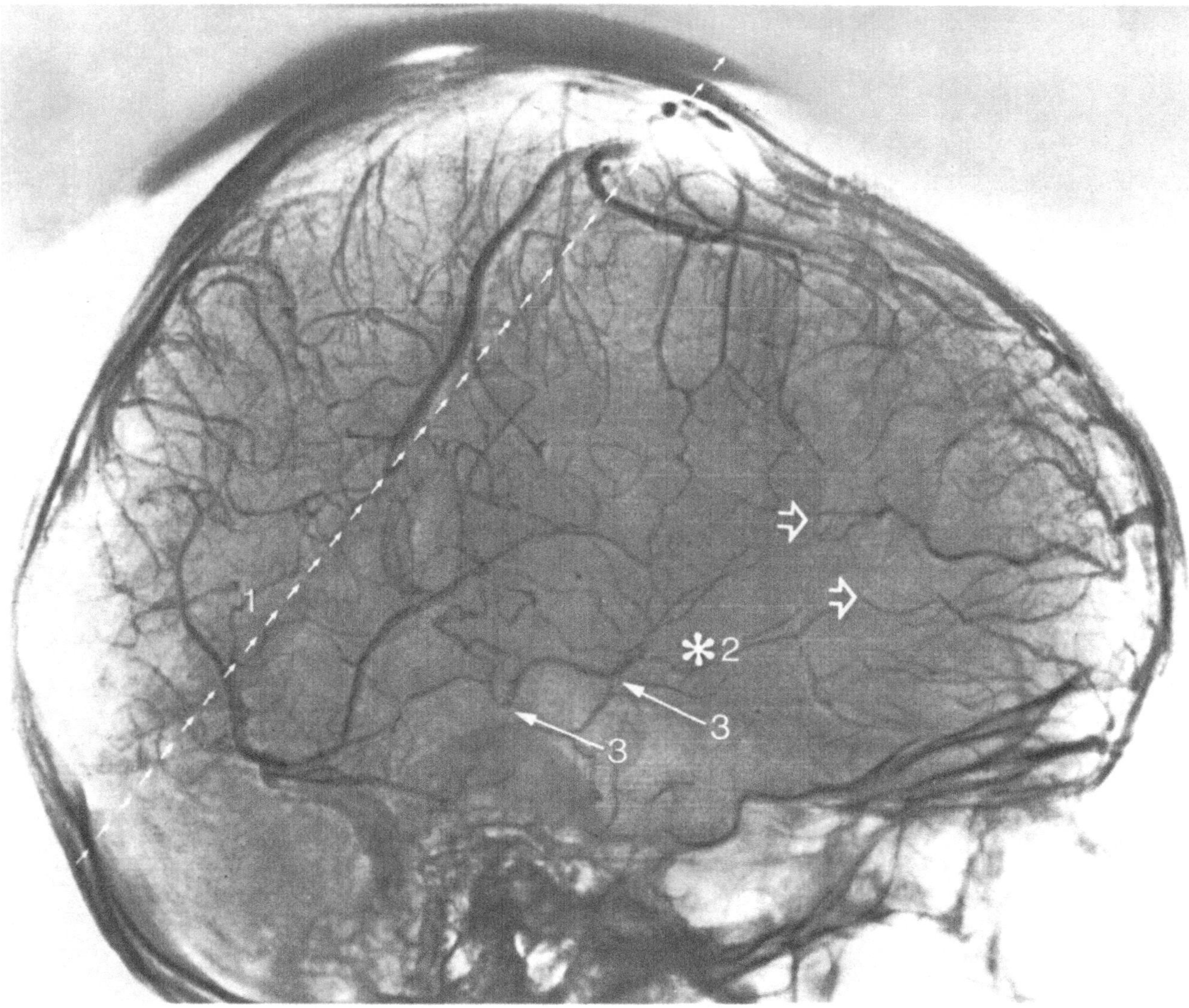

Figure 98

Temporosylvian hydroma in a 12-year-old child

1 Straight sinus slightly straightened
2 Avascular zone rimmed by normal veins
3 Considerably elevated and pushed backward (accordion-wise) basal vein
Double arrows: anterior limit of the hydroma

Planimetry (see Figs. 99 and 100)

A	157.46	(slightly increased)
B	16.26	(slightly increased)
C	59.30	(slightly increased)
A+*B*+*C*	233.02	(increased)
A+*B*	173.72	(slightly increased)
(*A*+*B*)/*C*	2.9	that is a normal relation in spite of increase in surfaces *A*, *B* and *C*

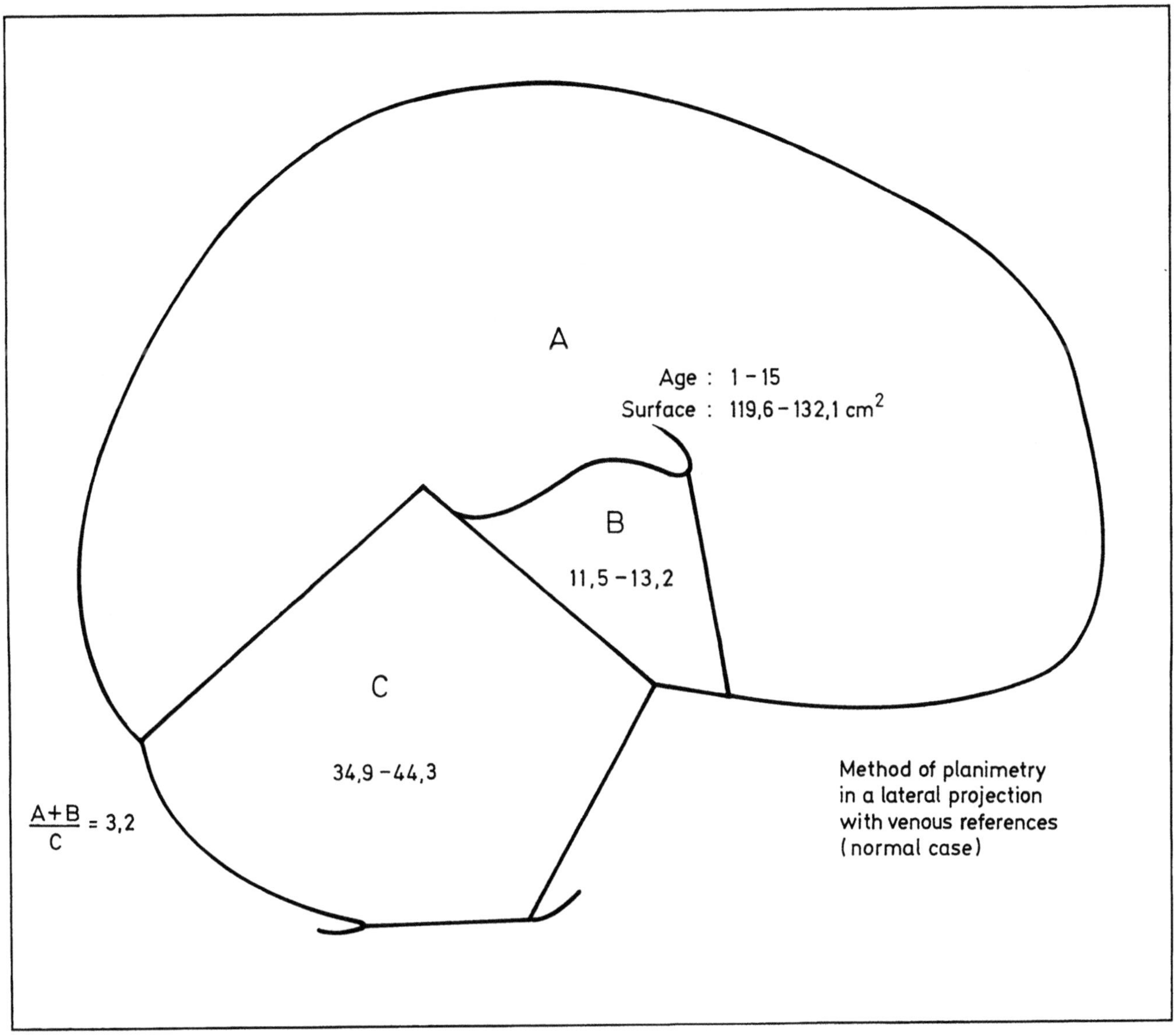

△

Figure 99

Diagram of planimetric method
A Supratentorial space
B Basal space
C Subtentorial space

Age	*A*	*B*	*C*	*A+B+C*	*A+B*	$\frac{A+B}{C}$	No of cases
0–1	119.60	11.52	34.90	166.02	131.12	3.7	4
2	106.78	13.11	31.19	151.08	119.89	3.8	2
6	123.83	12.86	44.33	181.02	139.69	3.0	3
7	129.15	12.35	43.65	185.05	141.50	3.2	3
8	148.43	13.19	52.08	213.70	161.62	3.1	2
9	119.20	13.93	44.43	177.56	133.13	2.5	1
10	131.68	11.88	46.09	189.65	143.56	3.1	5
11	135.95	13.90	46.42	196.27	149.85	3.2	3
12	130.44	14.38	47.51	192.33	144.82	3.0	4
13	127.87	13.09	46.04	187.00	140.96	3.0	7
14	133.54	14.66	45.50	193.70	148.20	3.2	12
15	132.15	13.23	44.30	189.68	145.38	3.2	4

Figure 100

Summary of planimetric results in 50 normal cases

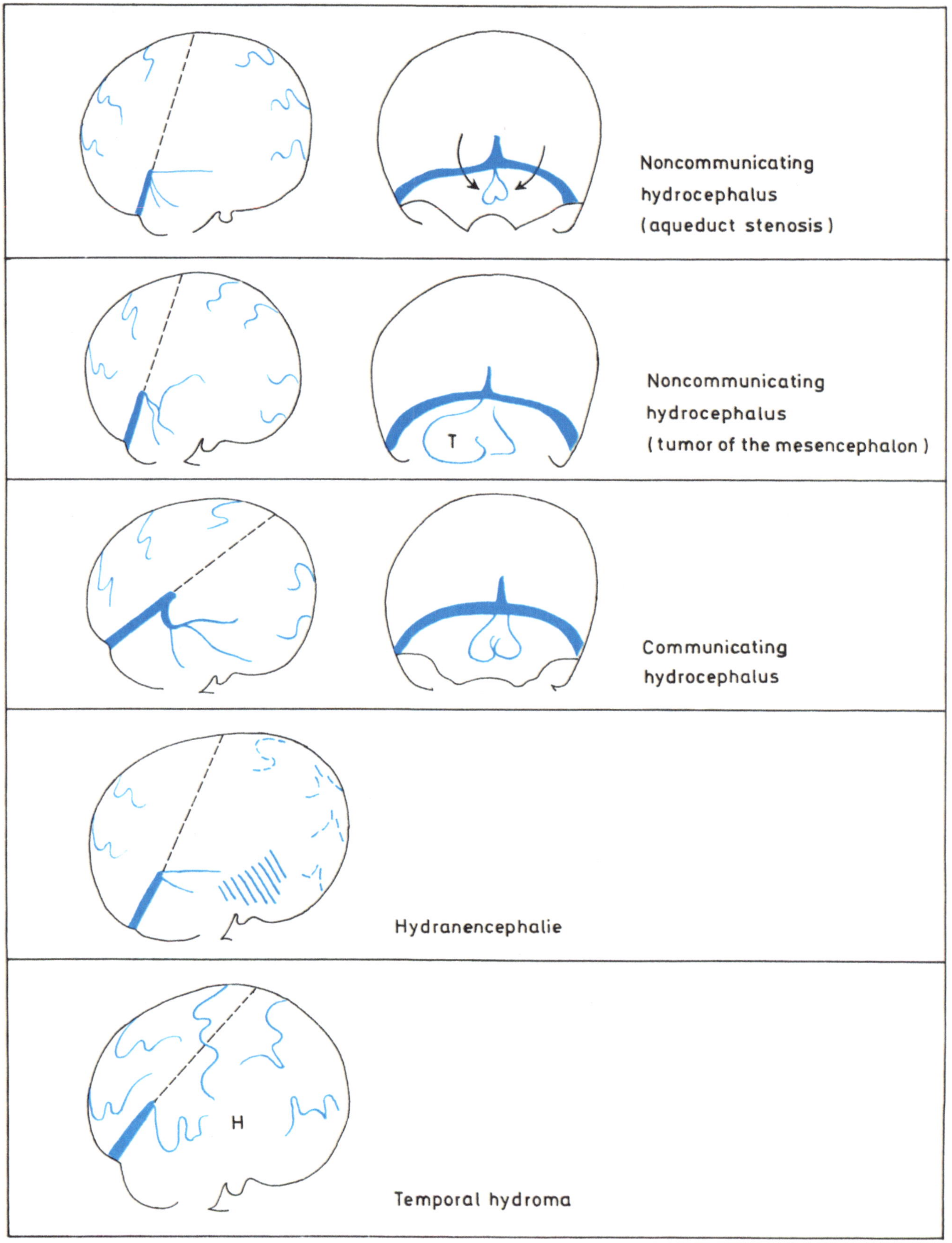

Figure 101

Schematic table of various varieties of venous anomalies found in cases of excessive accumulation of spinal fluid in intracranial space

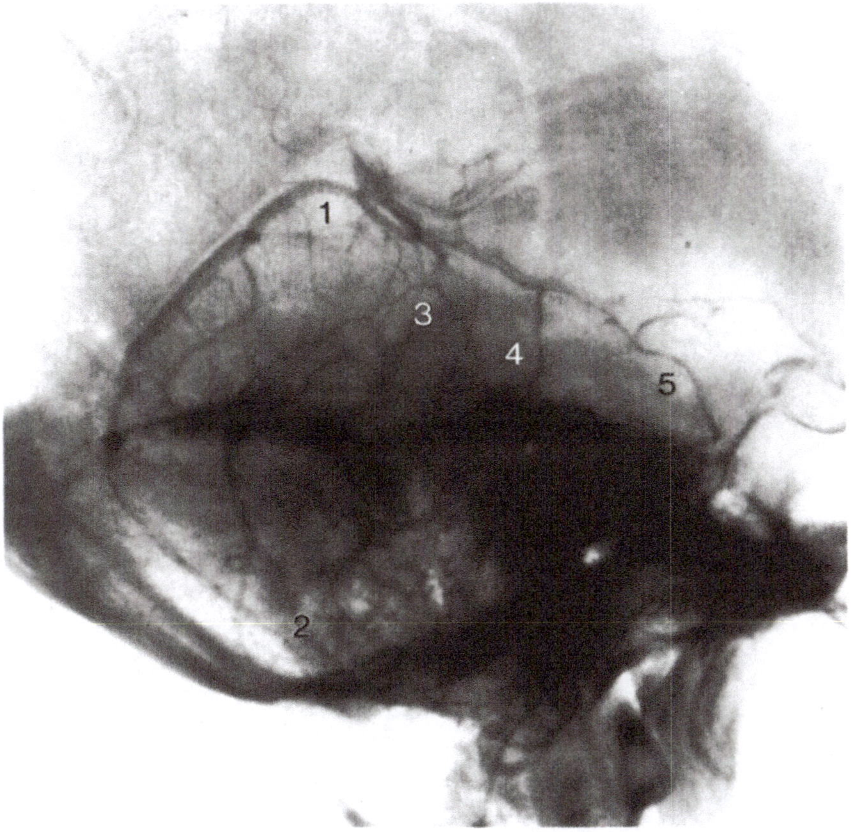

Figure 102

Diffuse hypertension in posterior fossa without specific herniations in case of multiple metastases

1 Raised superior vermian vein
2 Lowered inferior vermian vein
3 Raised precentral vein
4 Rigid and forward-displaced lateral mesencephalic vein
5 Raised and forward-displaced veins of prepontine and optopeduncular system

Figure 103a–d

Drawing of venous displacements in four main varieties of herniation in posterior fossa

a Cerebellar herniation
b Herniation of the superior vermis
c Tonsillar herniation
d Temporal herniation

▽

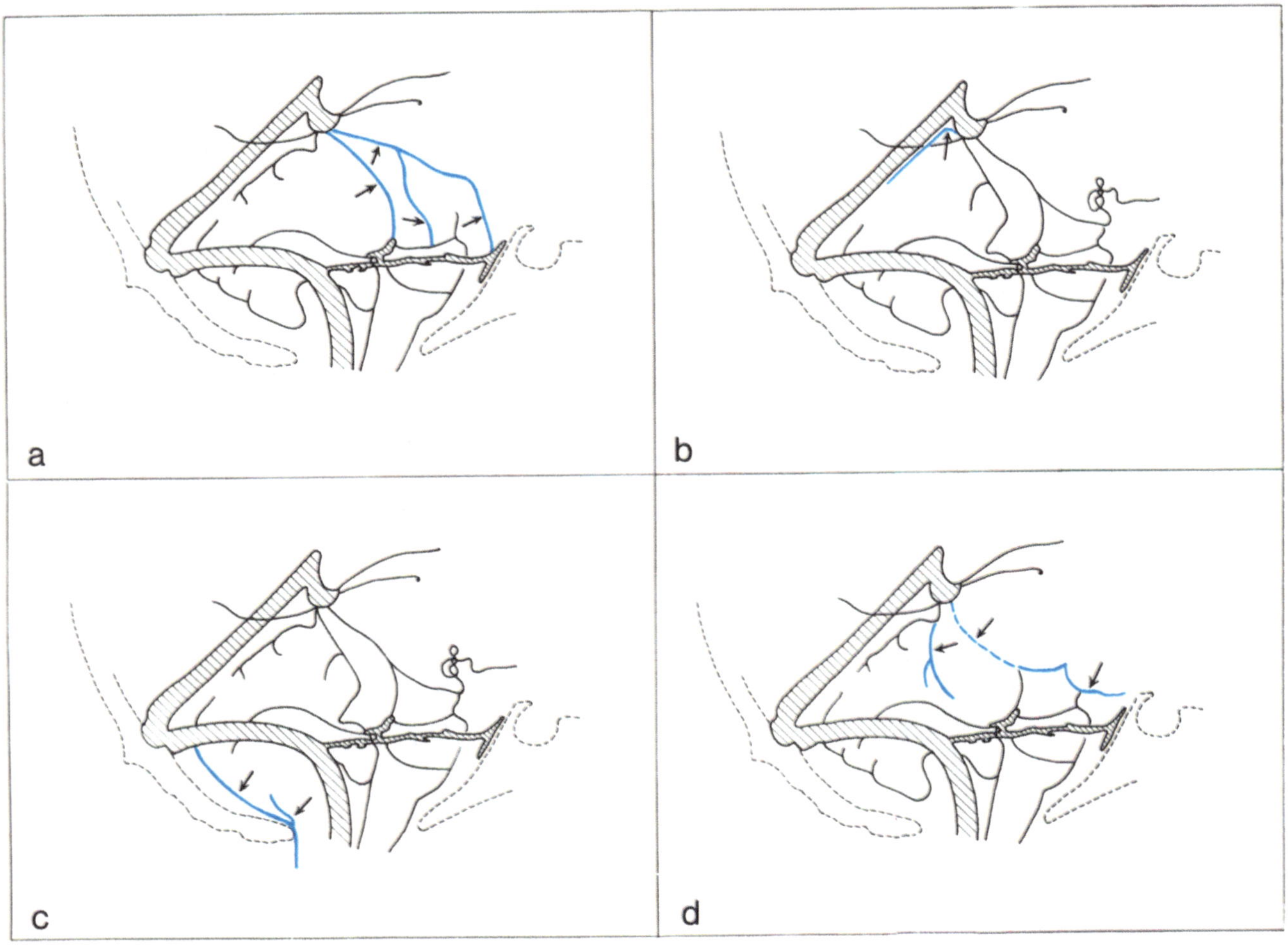

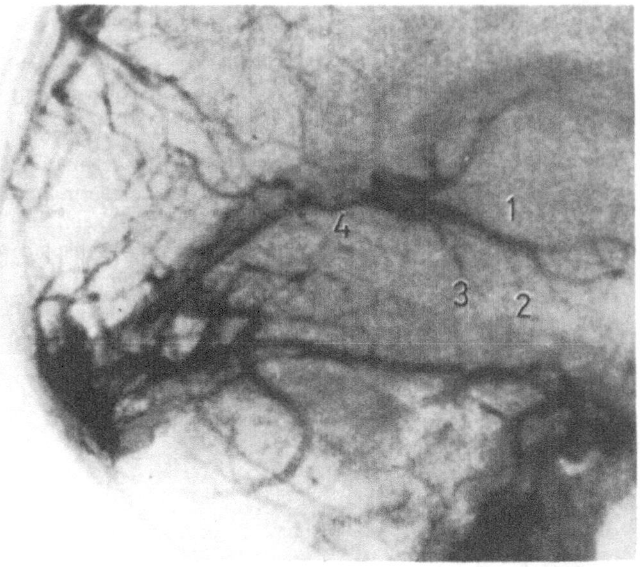

Figure 104

Cerebellar herniation by expanding mass in anterosuperior part of posterior fossa. Associated herniation of the superior vermis

1 Elevated basal vein
2 Elevated lateral mesencephalic vein
3 Elevated precentral vein
4 Herniation of the superior vermis

Figure 105

Superimposition of arterial and venous phases in case of supratentorial tumor (temporal glioma)

1 Herniation of superior vermis into subdural space
2 Small venous supravermian triangle
3 Superior vermian vein
4 Inferior vermian artery (p.i.c.a.)
5 Inferior vermian vein

▽

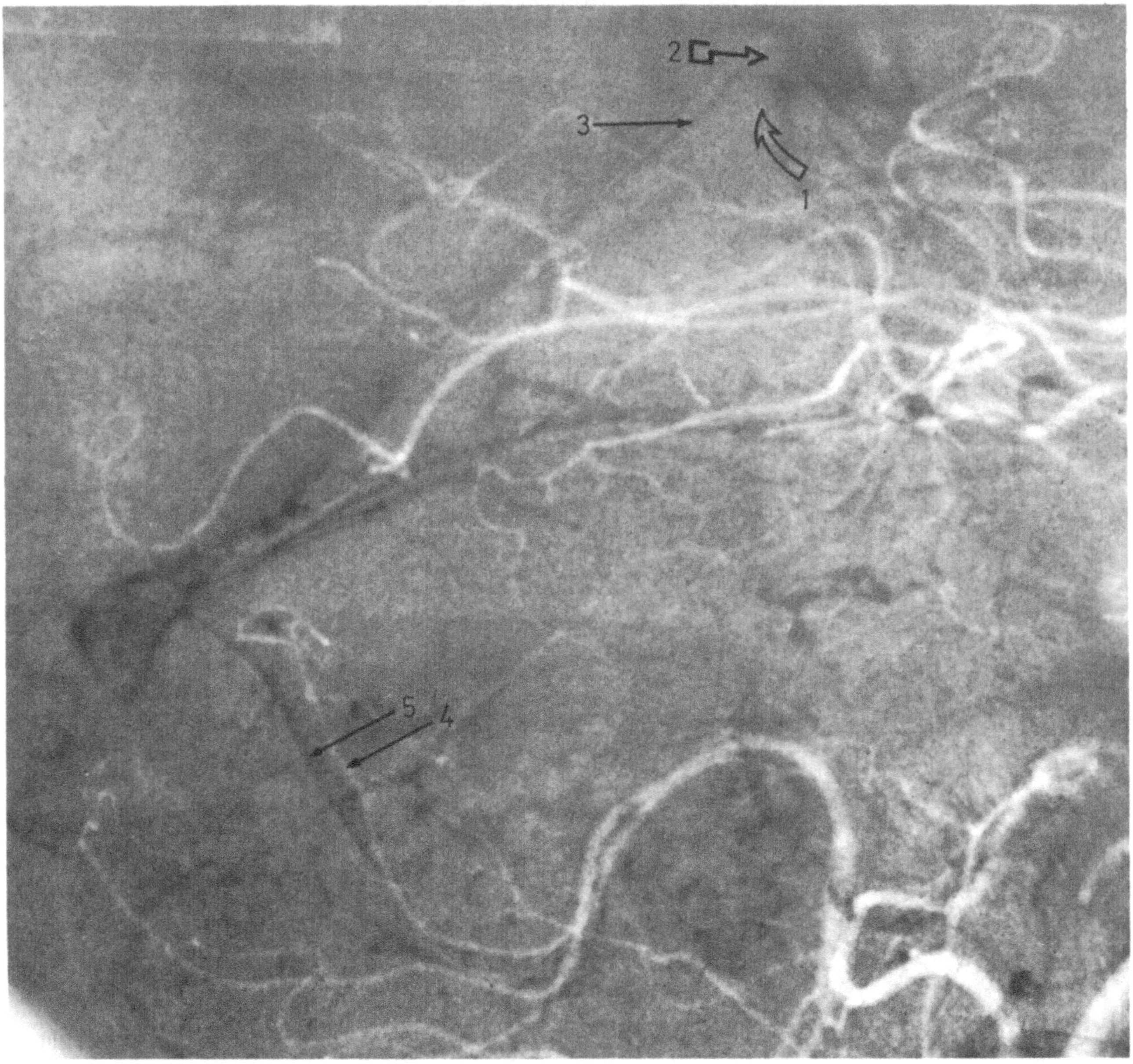

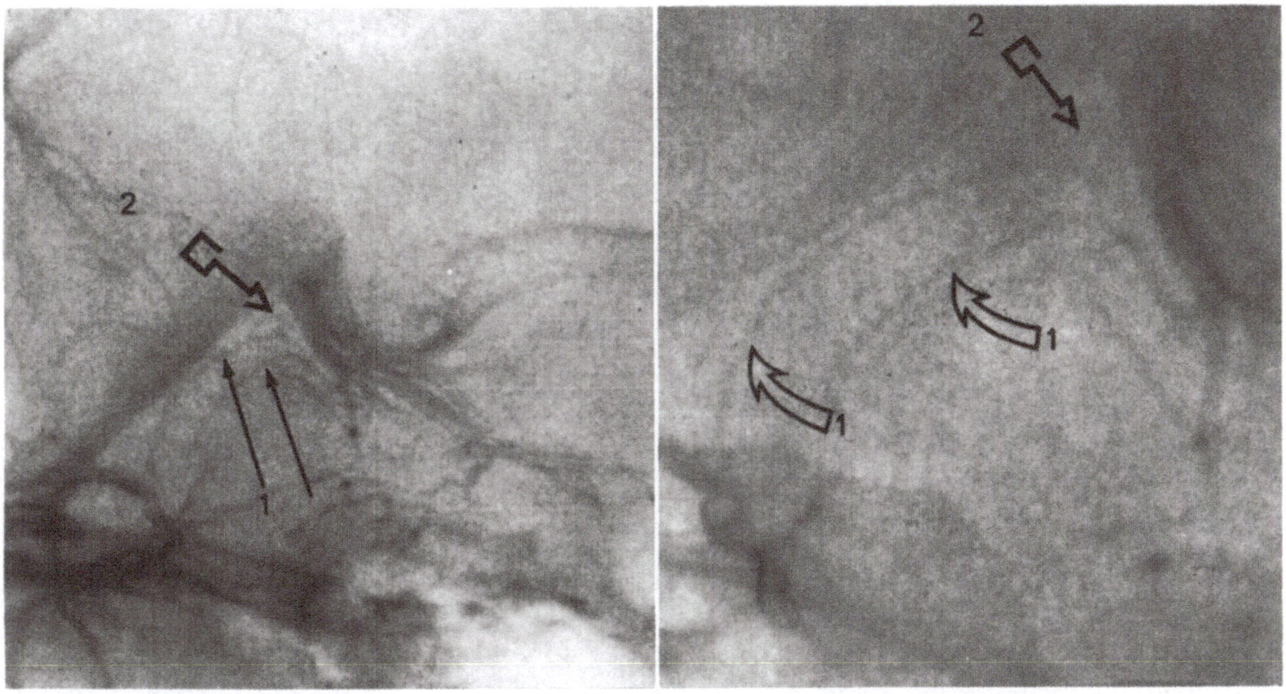

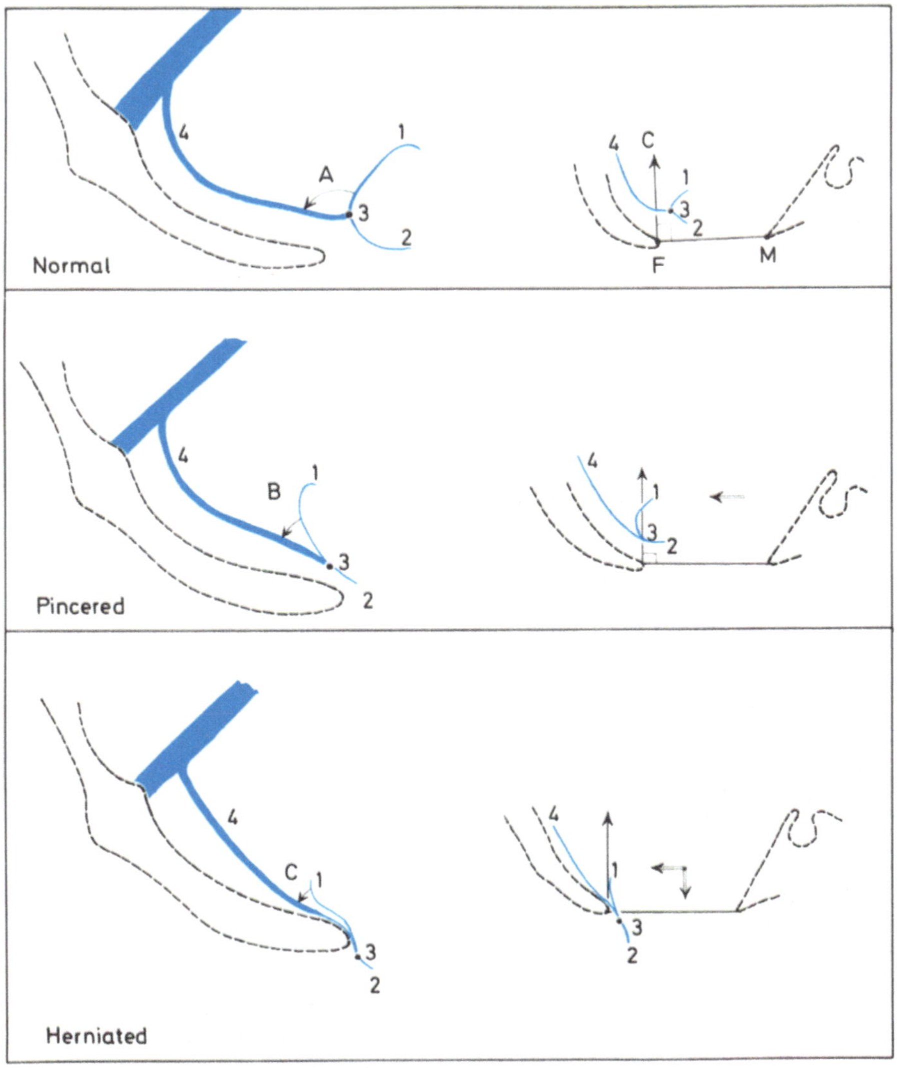

△
Figure 106

Deformation of superior vermian veins in case of beginning hernia of superior vermis. Hernia is asymmetrical and characterized by an elevation and an asymmetrical deformation of superior vermian veins

1 Superior vermian veins
2 Small venous supravermian triangle

Figure 107

Decrease in tonsillovermian angle is a premonitory sign of tonsillar herniation. (See also Figs. 69 and 125)

A Normal relations between superior retrotonsillar vein and inferior vermian vein
B Closure of tonsillovermian angle
C Lowering, backward pushing of the superior retrotonsillar and inferior vermian veins by tonsillar herniation
1 Superior retrotonsillar vein
2 Inferior retrotonsillar vein
3 Copular point
FM Foramen magnum line

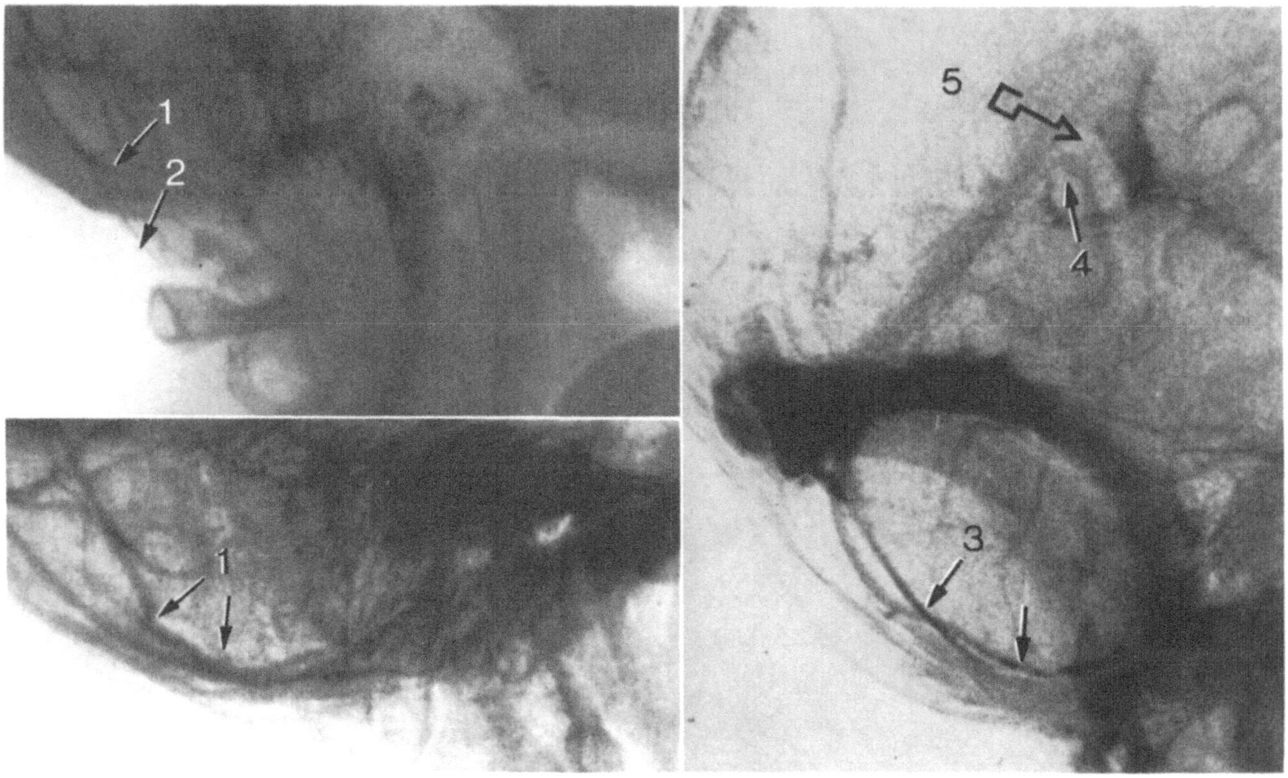

Figure 108

Left: Venous signs of tonsillar herniation in two cases
Right: Double herniation of superior vermis and of tonsils

1 Backward-shifted superior tonsillar vein
2 Backward and downward-shifted inferior tonsillar vein
3 Lowered inferior vermian vein
4 Raised superior vermian vein
5 Reduced supravermian space

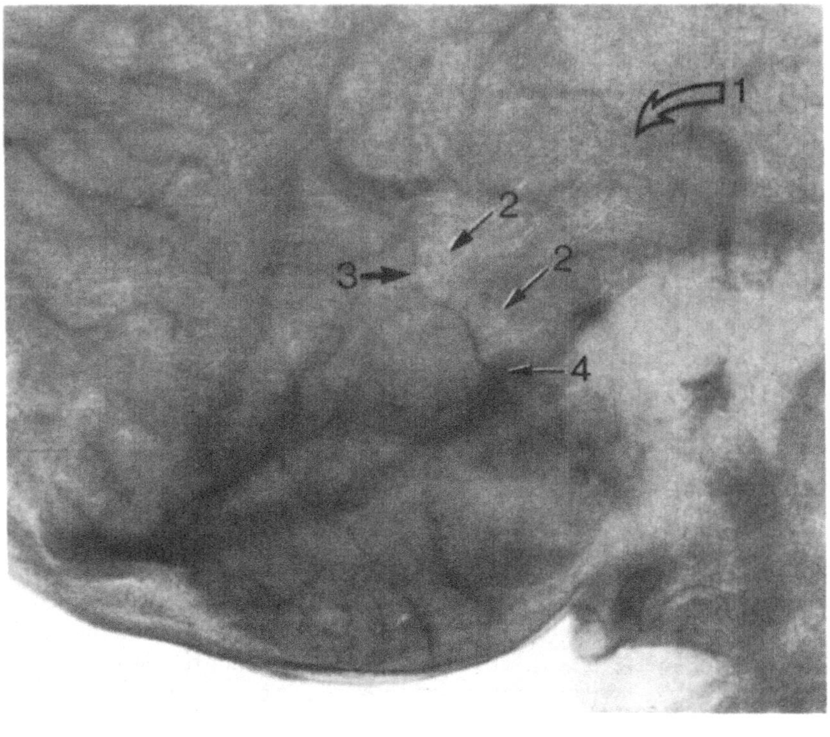

Figure 109

Temporal herniation (*1*) with excluded basal and posterior mesencephalic system due to compression (*2*) and backward-shifted precentral vein (*3*) joining superior petrous vein (*4*)

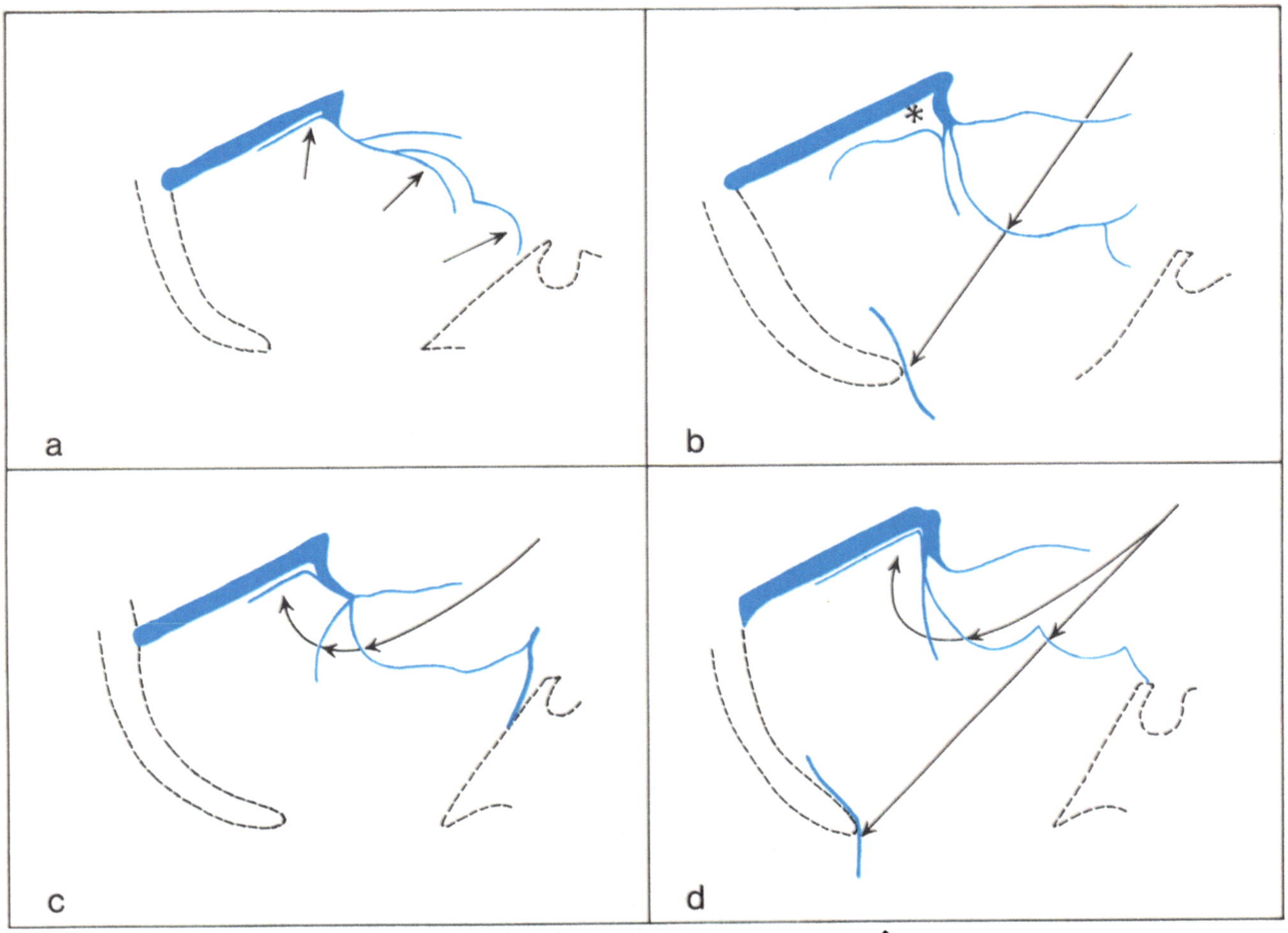

△

Figure 110 a–d

Main varieties of venous signs of associated hernias in posterior fossa

a Cerebellar and superior vermian herniations

b Temporal and tonsillar herniations with normal or magnified area of the supravermian venous triangle (x)

c Temporal and superior vermian herniations

d Temporal, superior vermian, and tonsillar herniations

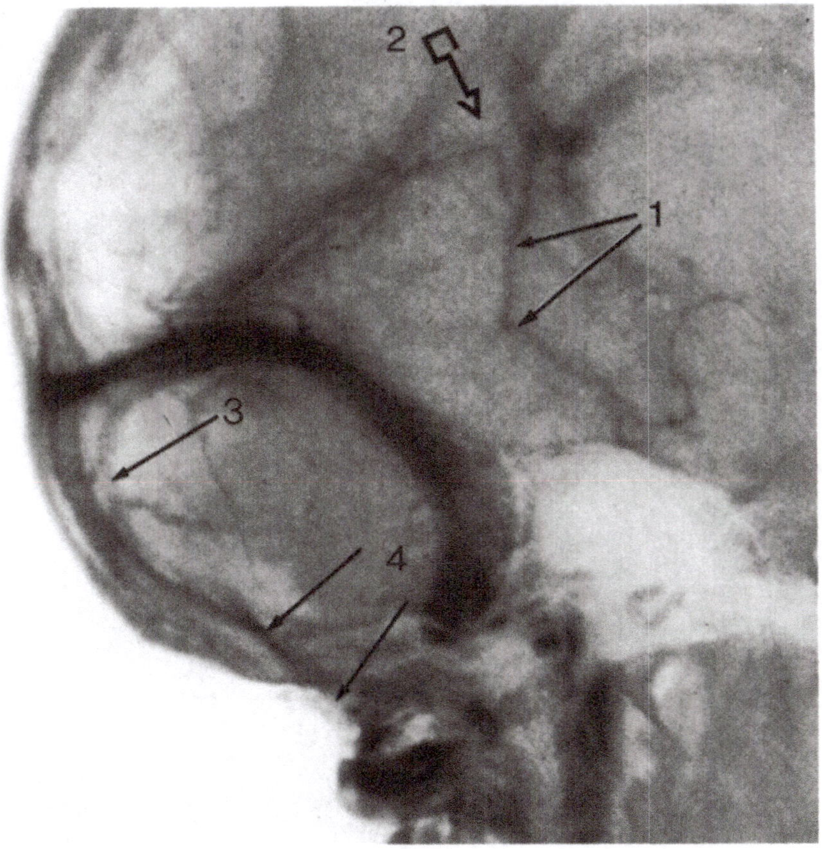

Figure 111

Associated hernias

1 Temporal hernia with considerable compression of posterior mesencephalic and optopeduncular system and a marked backward pushing of precentral vein

2 Absence of superior vermian hernia

3 and *4* Tonsillar herniation

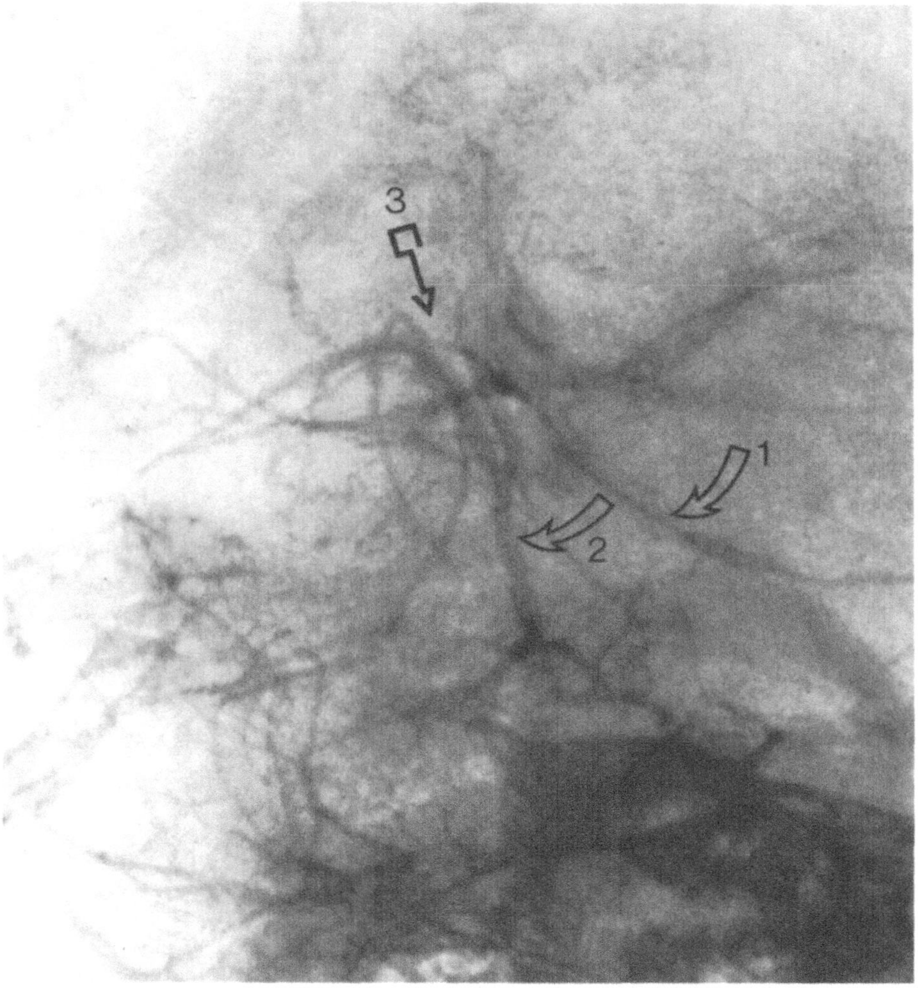

Figure 112

Associated herniations: temporal and superior vermian

1 Hammock-shaped deformation of posterior mesencephalic vein
2 Pushing backward of precentral vein
3 Herniation of superior vermian vein

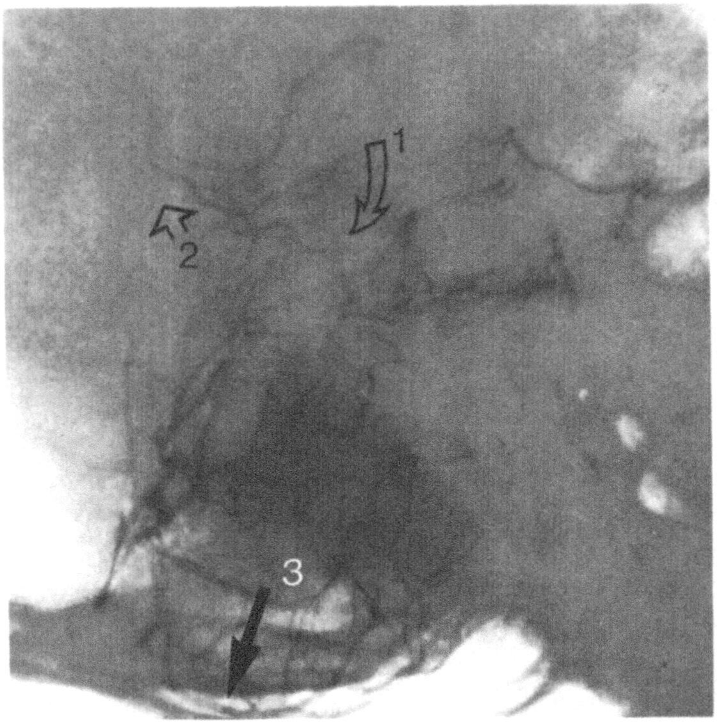

Figure 113

Triple herniation

1 Temporal *2* Superior vermian *3* Tonsillar

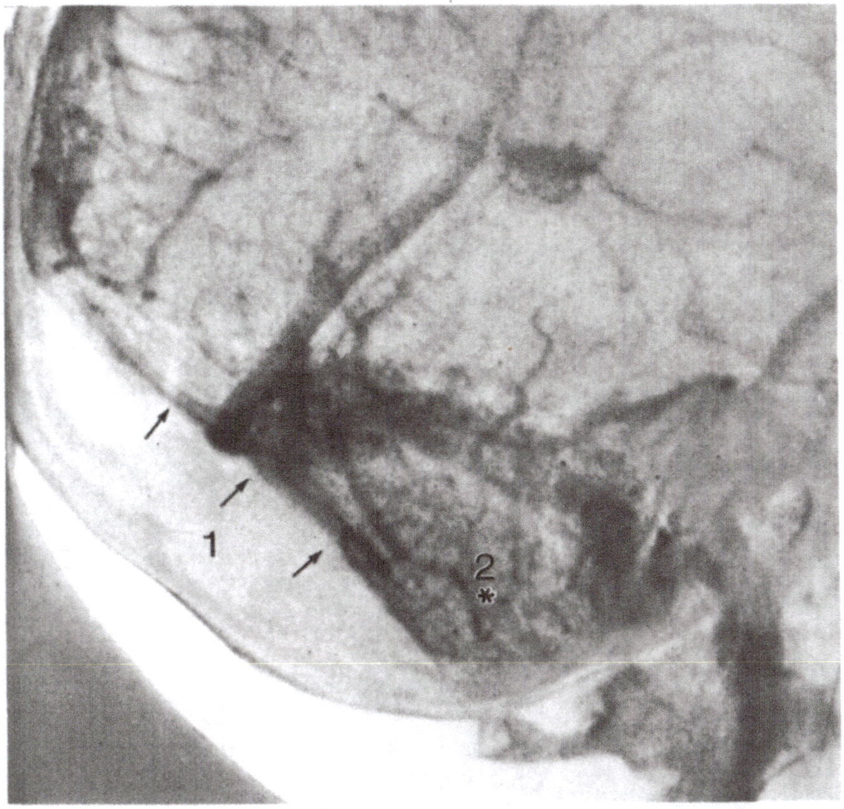

Figure 114

Extradural hematoma of the posterior fossa

1 Avascular area
2 Narrowing of the angle between the vermian vein and the superior tonsillar vein

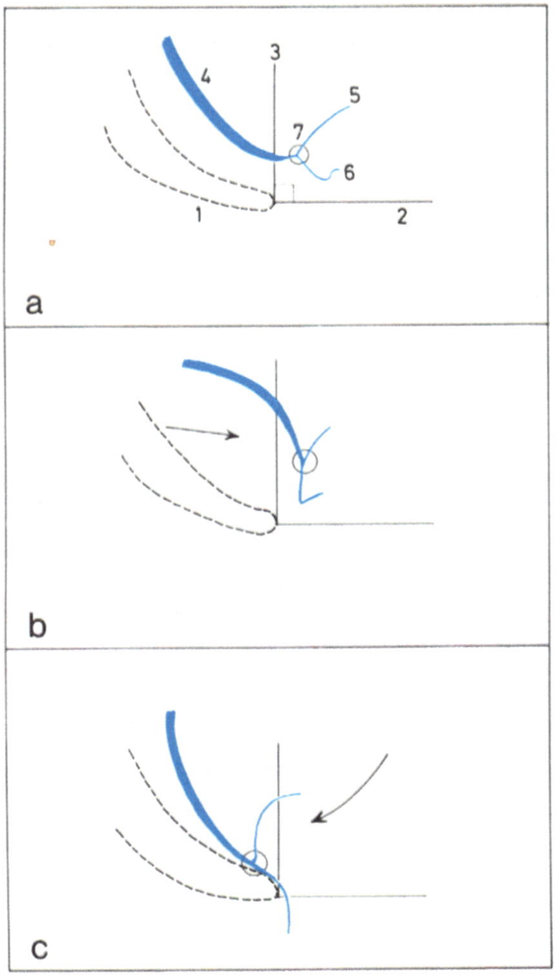

Figure 115 a–c

Tonsillovermian angle (see also Figs. 69, 107, 125)

a Normal conditions. Angle formed by superior retrotonsillar vein and trunk of inferior vermian vein. It measures approximately 90°.

b Posterior expansive growth, causing narrowing of tonsillovermian angle.

c Anterior expansive growth causing narrowing of tonsillovermian angle

1 Occipital squama
2 Foramen magnum line
3 Perpendicular to the foramen magnum line
4 Inferior vermian vein
5 Superior tonsillar vein
6 Inferior tonsillar vein
7 Copular point

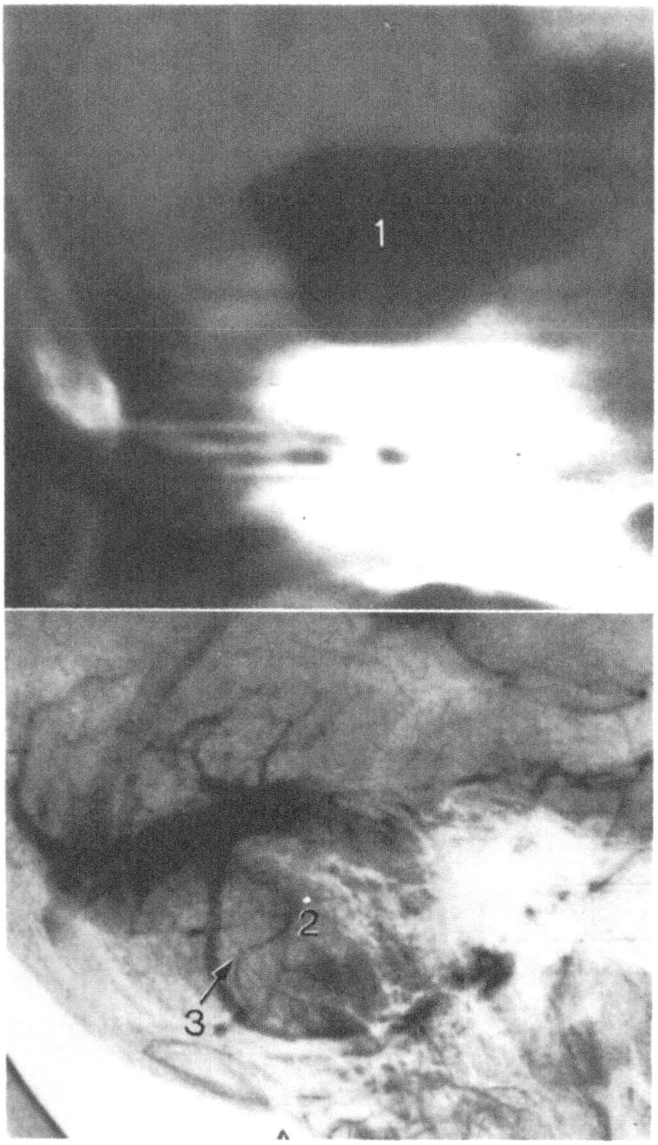

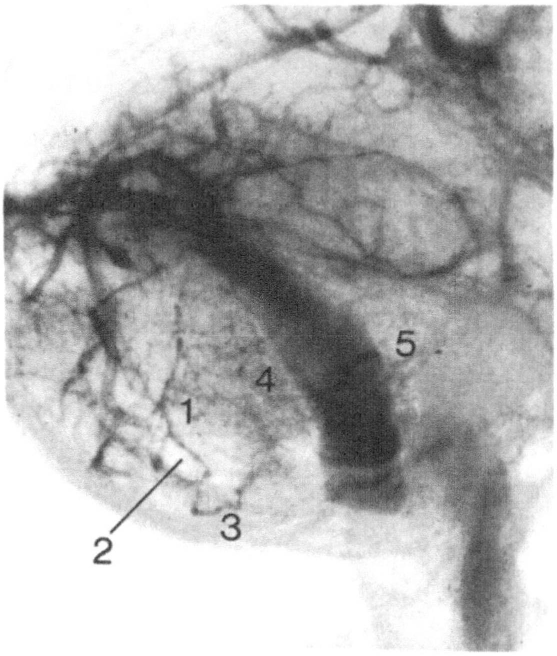

Figure 117

Sarcoma of floor of fourth ventricle

1 Backward-displaced superior retrotonsillar vein
2 Narrowing of tonsillovermian angle
3 Lowering of inferior retrotonsillar vein
4 Abnormal venous network
5 Irregular caliber and stretching of veins of lateral recess

◁ *Figure 116*

Dilated fourth ventricle (*1*), backward-shifted superior retrotonsillar vein (*2*), and narrowed tonsillovermian angle (*3*)

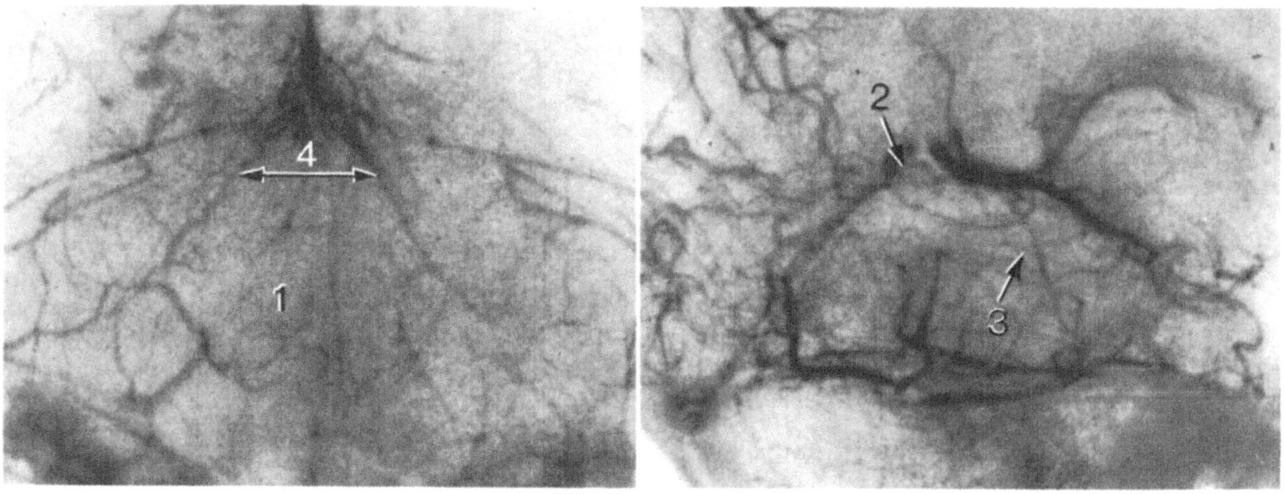

Figure 118 Medulloblastoma of anterior cerebellar vermis

1 Lateral deviation of optopeduncular veins *3* Elevation of precentral vein
2 Herniation in subdural supravermian space *4* Displacement of basal veins by cerebellar herniation

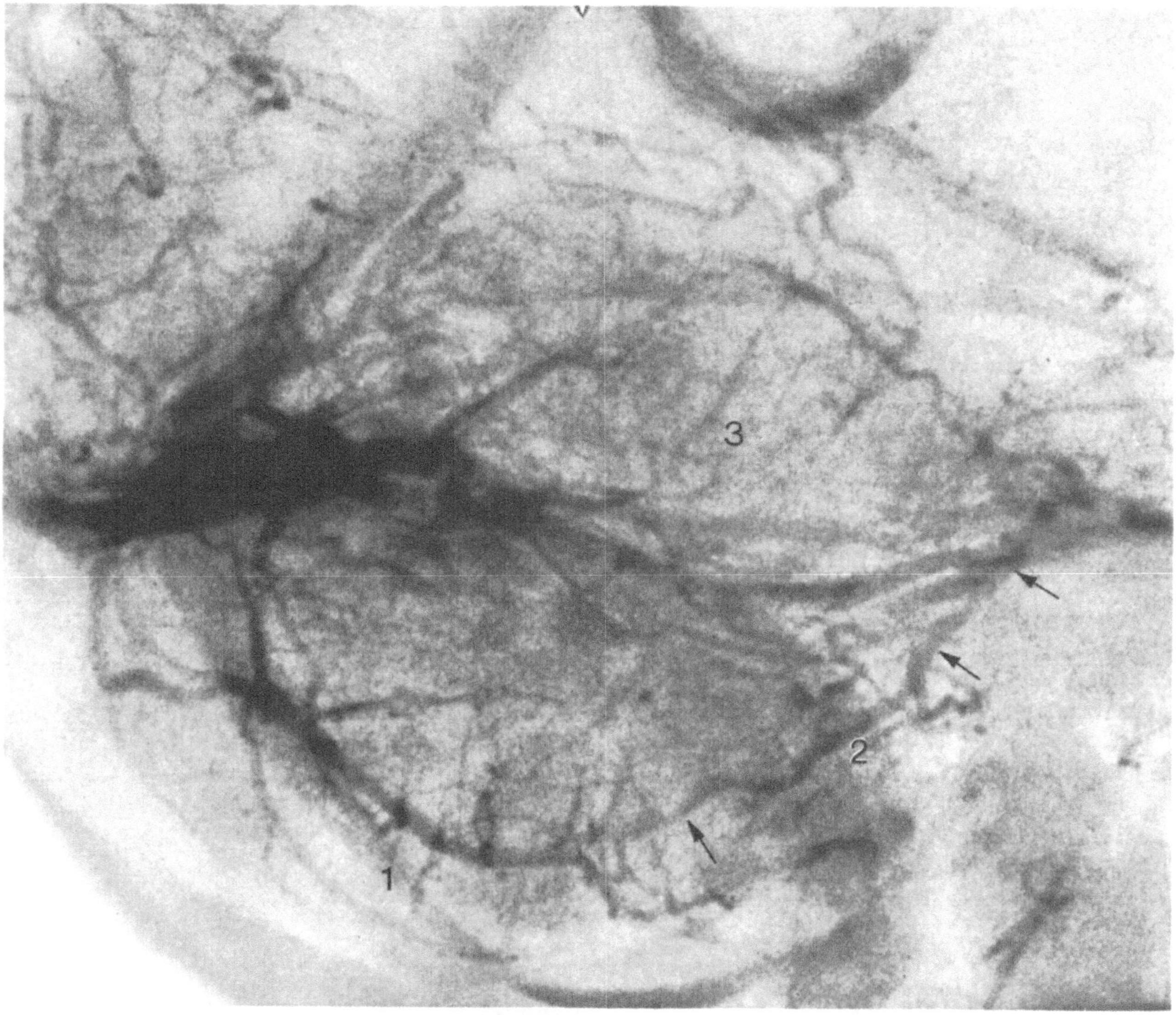

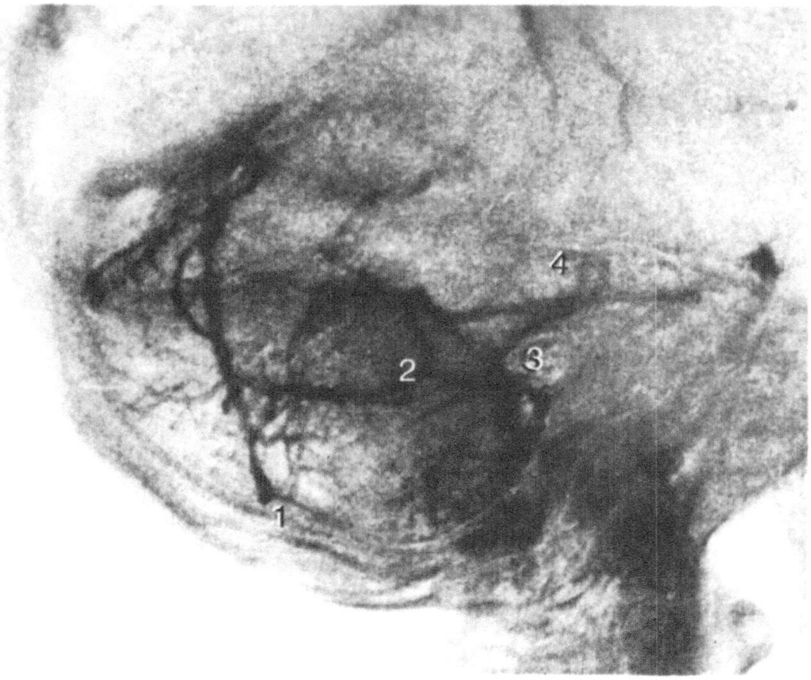

△
Figure 119

Example of the three principal indirect venous signs of tumor in area of fourth ventricle

1 Narrowing on one side of tonsillovermian angle
2 Stretching of vein of lateral recess of fourth ventricle on other side
3 Stretching of various hemispheric veins by an edema

Figure 120

Medulloblastoma of fourth ventricle involving vermis

1 Copular point and narrowing of tonsillovermian angle
2 Stretching of vein of lateral recess of fourth ventricle
3 Lateral medullopontine vein having a stretched concave course
4 Superior petrosal vein

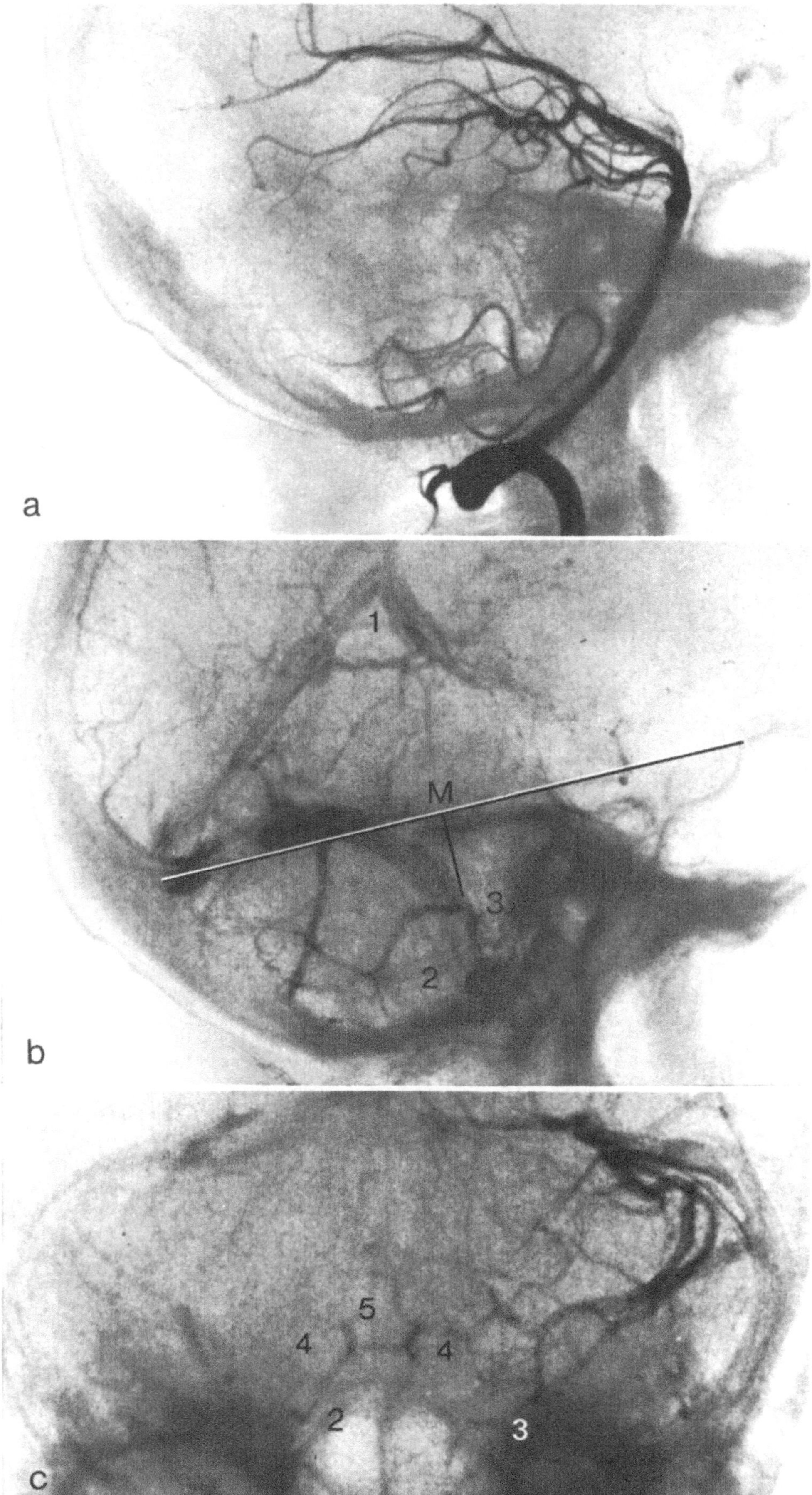

Figure 121 a–c

Medulloblastoma of floor of fourth ventricle developing principally on left

a Arterial signs
b, c Venous signs

1 Herniation in supravermian space
2 Lowered tonsillar veins
3 No displacement of junction point of lateral bulbopontine vein and vein of lateral recess
4 Tonsils
5 Cerebellar vermis

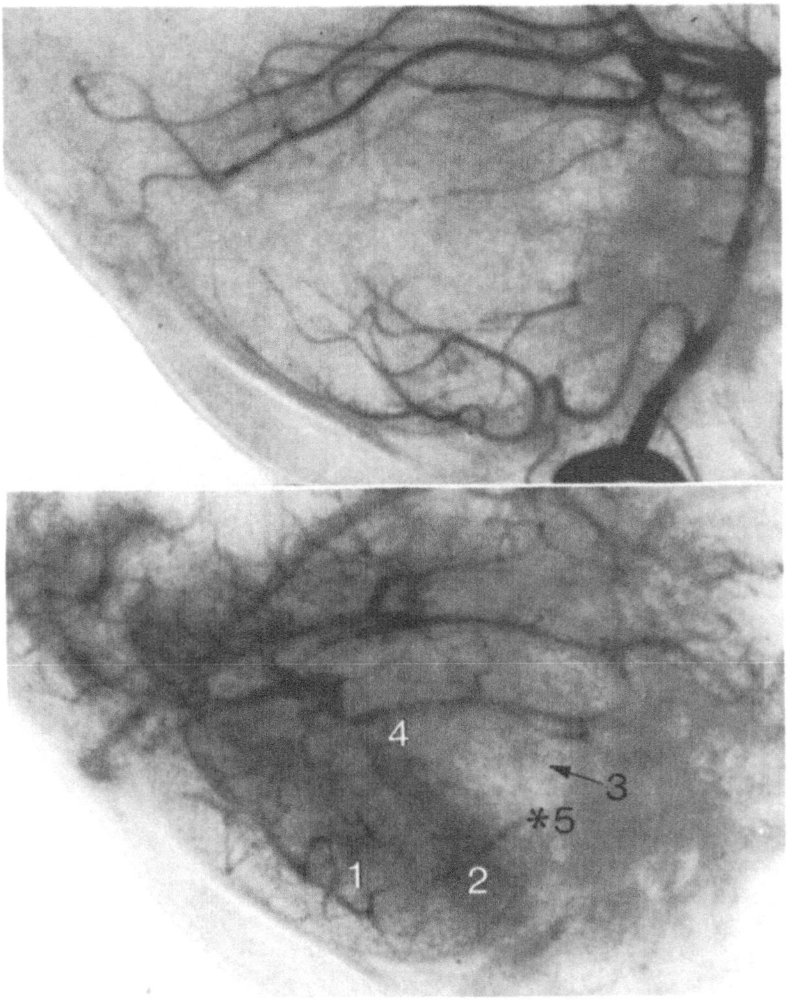

Figure 122

Medulloblastoma of fourth ventricle

1 Narrowing of the tonsillovermian angle
2 Lowering and stretching of vein of lateral recess of fourth ventricle
3 Backward push of lateral medullopontine vein
4 Stretching of cerebellar hemispheric veins
5 Backward displacement of junction point of 2 and 3

Figure 123

Ependymoma of fourth ventricle

1 Posteroinferior cerebellar artery
2 Vascular void corresponding to tumor
3, 3′ Arteriovenous tumoral fistulas in area of vein of lateral recess of fourth ventricle
4 Junction between vein of lateral recess of fourth ventricle and inferior vermian vein
5 Inferior vermian vein

▽

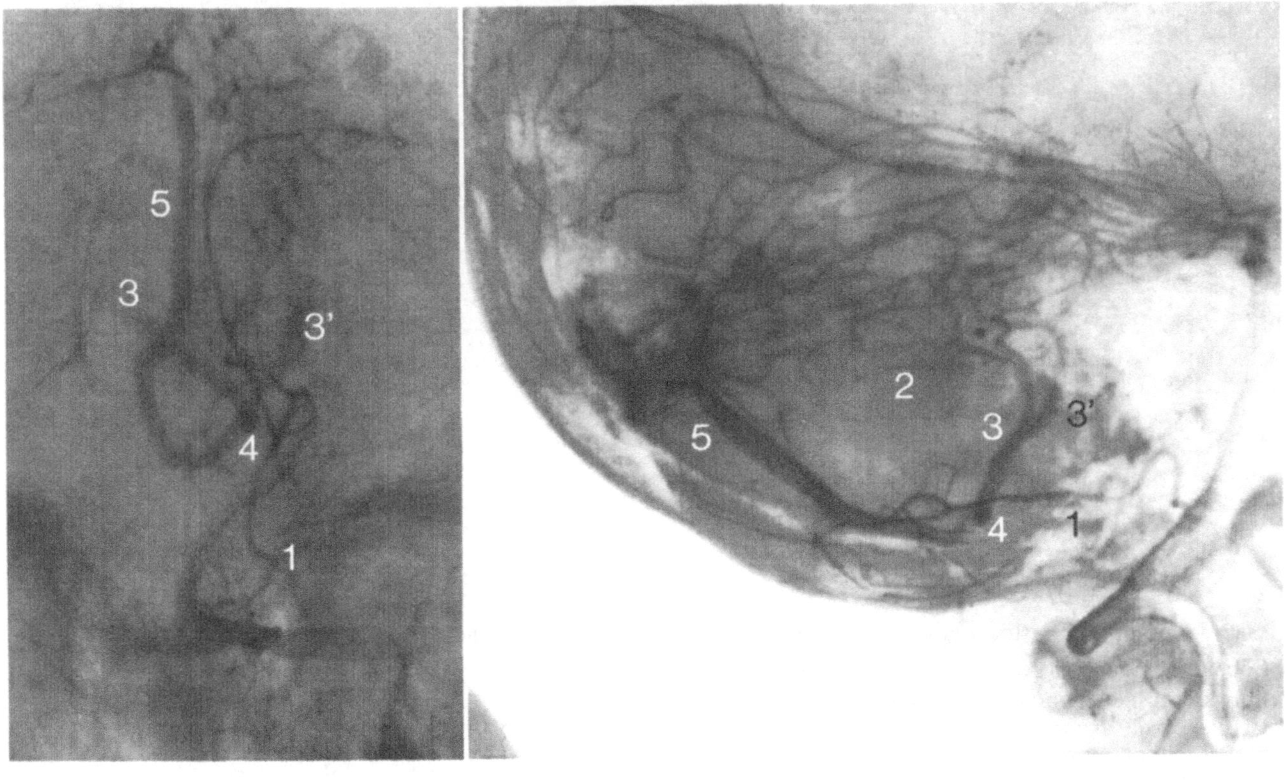

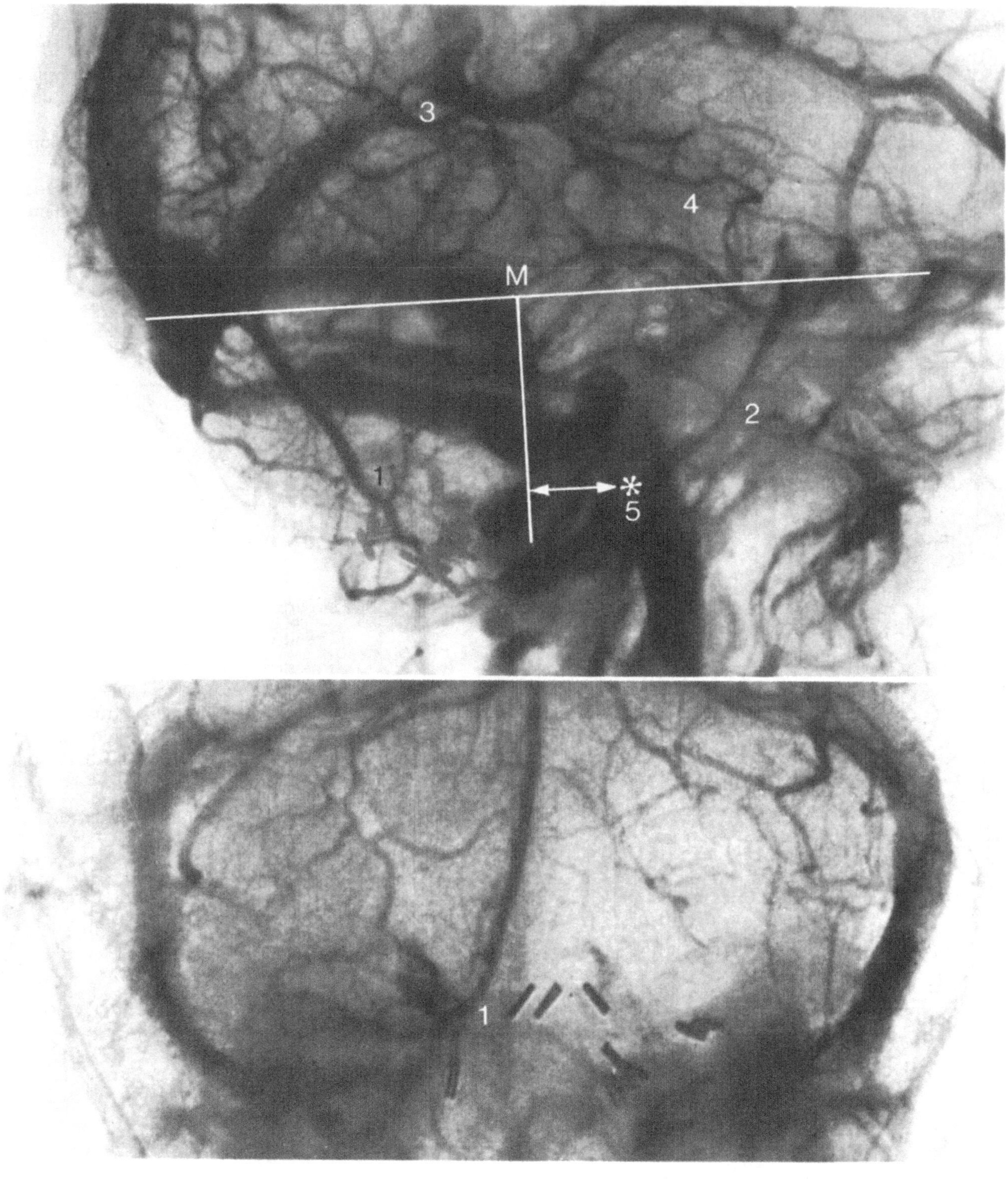

Figure 124

Recurrence of medulloblastoma of fourth ventricle invading cervical medulla and the brain stem

1 Copular point and narrowing of tonsillovermian angle
2 Forward push of lateromedullopontine vein
3 Herniation of superior vermis
4 Dislocation of optopeduncular venous system
5 Forward-displaced junction point of 2 and vein of lateral recess

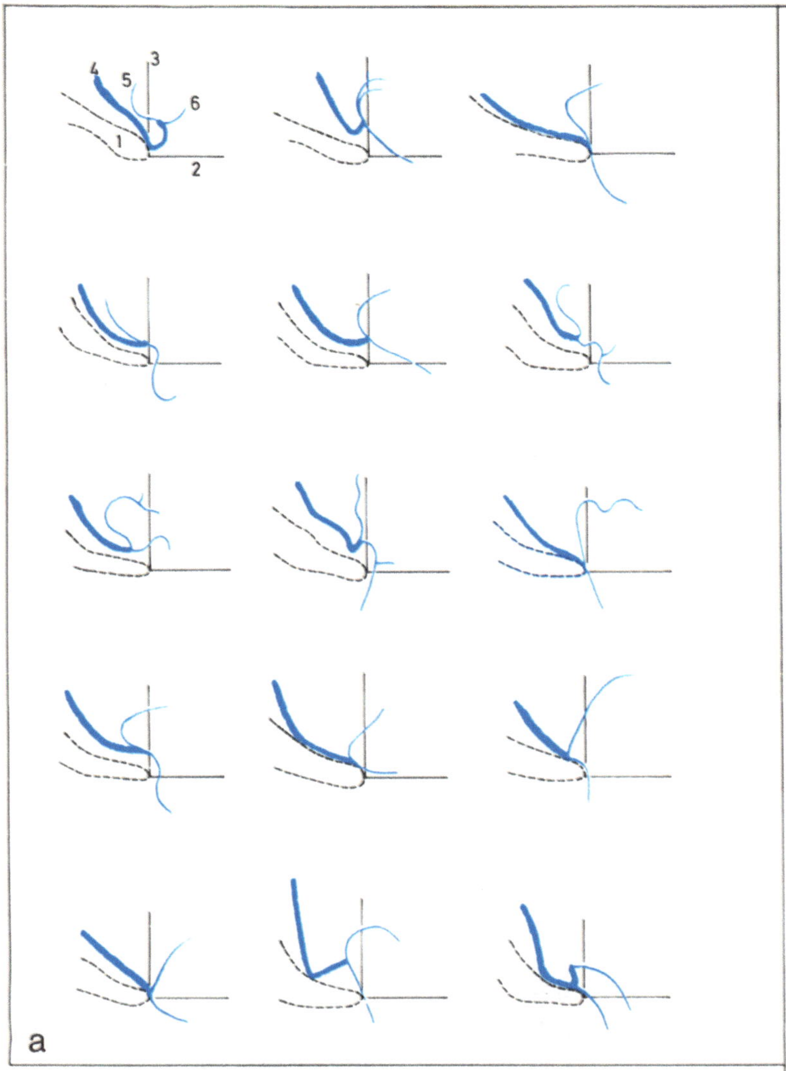

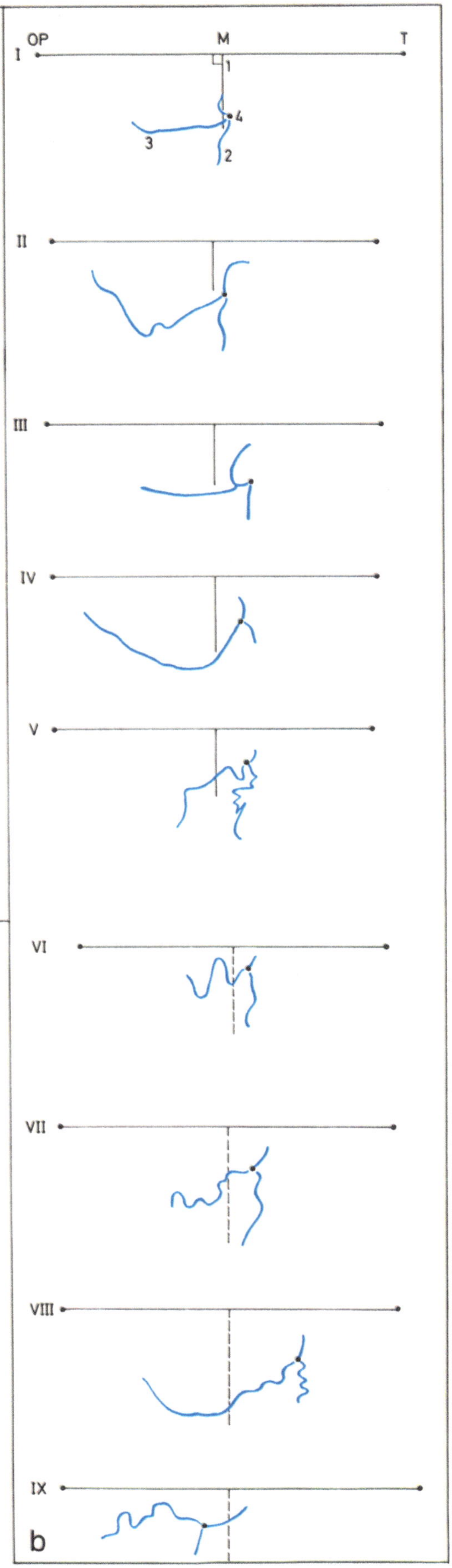

Figure 125a and b

Diagrammatic summary of venous modifications in tumor of fourth ventricle (see also Figs. 107 and 115)

a Different types of changes in tonsillovermian angle in mass expansion of posterior fossa. *1* occipital squama; *2* foramen magnum line; *3* perpendicular to the foramen magnum line; *4* inferior vermian vein; *5* superior retrotonsillar vein; *6* inferior retrotonsillar vein. Normal conditions are found in Figure 69

b Normal forms of vein of lateral recess of fourth ventricle and various modifications by mass expansion of posterior fossa

- I Normal conditions: *OP* occipital protuberance; *T* tuberculum sellae; *OPT* Twining's line; *1* perpendicular dropped on mid-point of Twining's line; *2* lateral medullopontine vein; *3* vein of lateral recess of fourth ventricle; *4* junction of *2* and *3*
- II–VIII Forward-shifted junction point in tumors in area of fourth ventricle
- IX Backward-shifted junction point in a pontohemispheric tumor

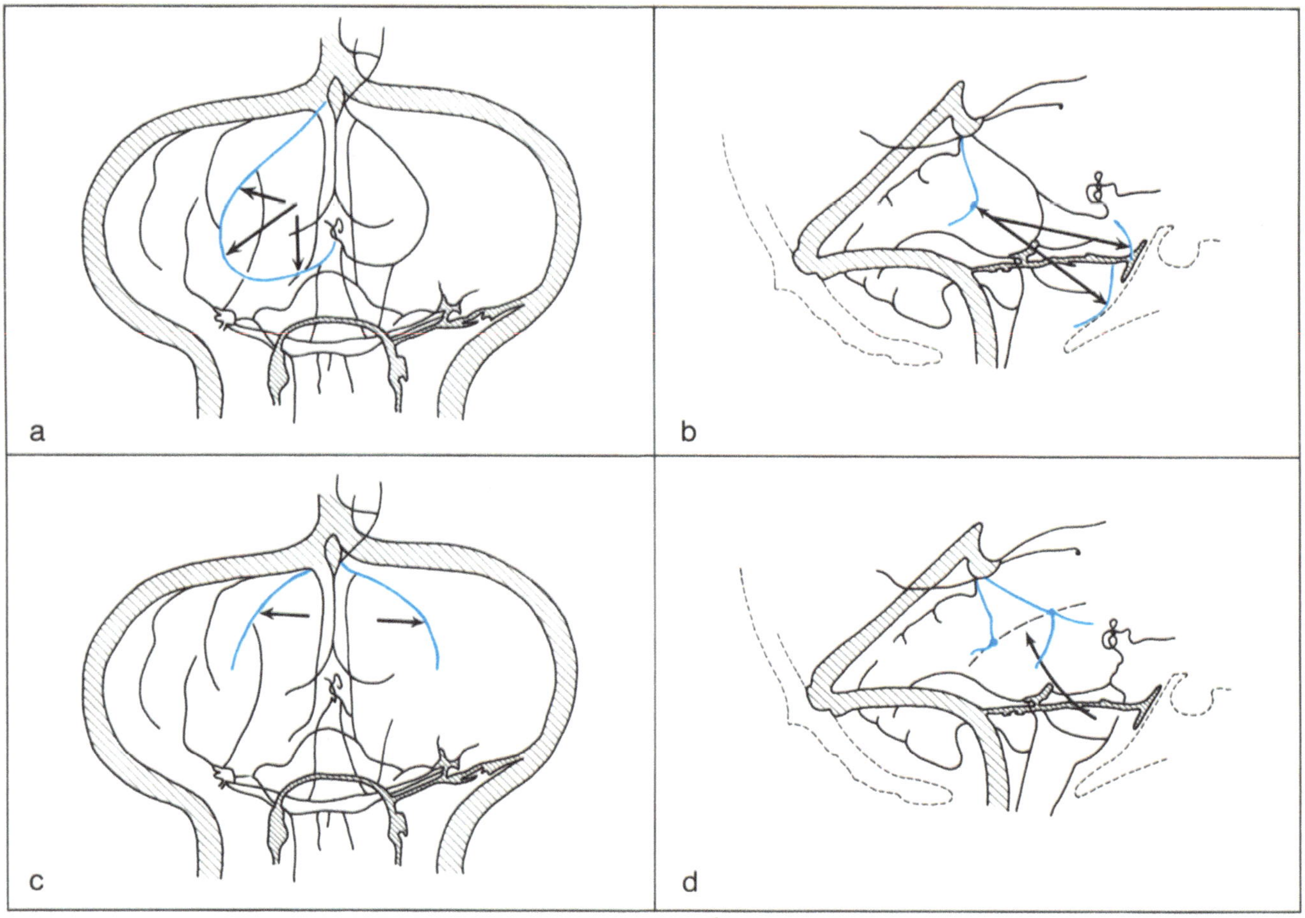

△

Figure 126 a–d Drawing of a few main venous changes in brain stem tumors

a Enlargement of cerebral peduncle
b Enlargement of diameter of pons
c Transversal enlargement of superior brain stem
d Shifted aqueduct of Sylvius demonstrated by its approximate venous landmark: a line joining the collicular point to the confluent point of lateral and posterior mesencephalic veins

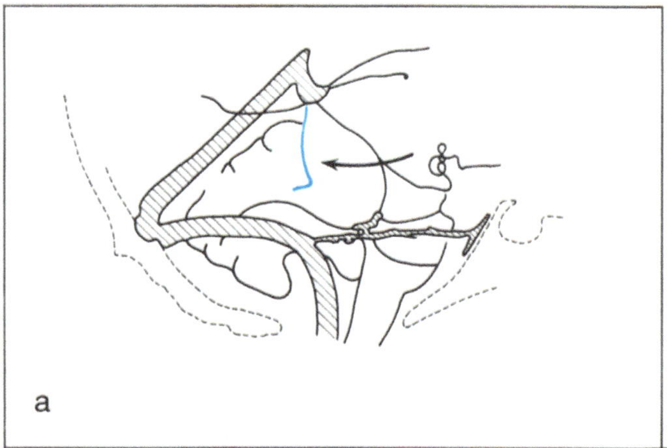

Figure 127 a–c Displacements of precentral vein in brain stem tumors

a Superior brain stem and mesencephalic area
b Pons c Medulla oblongata and fourth ventricle

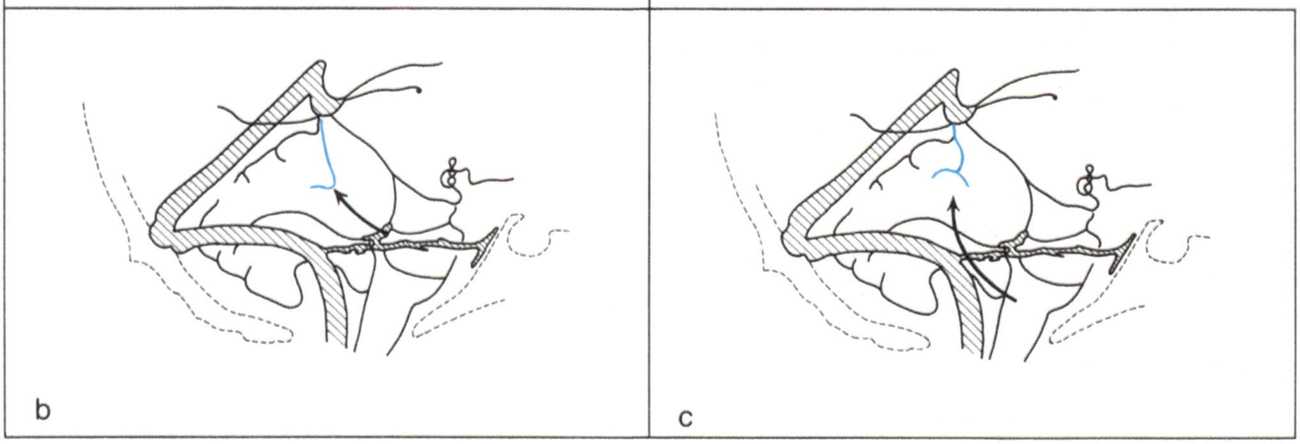

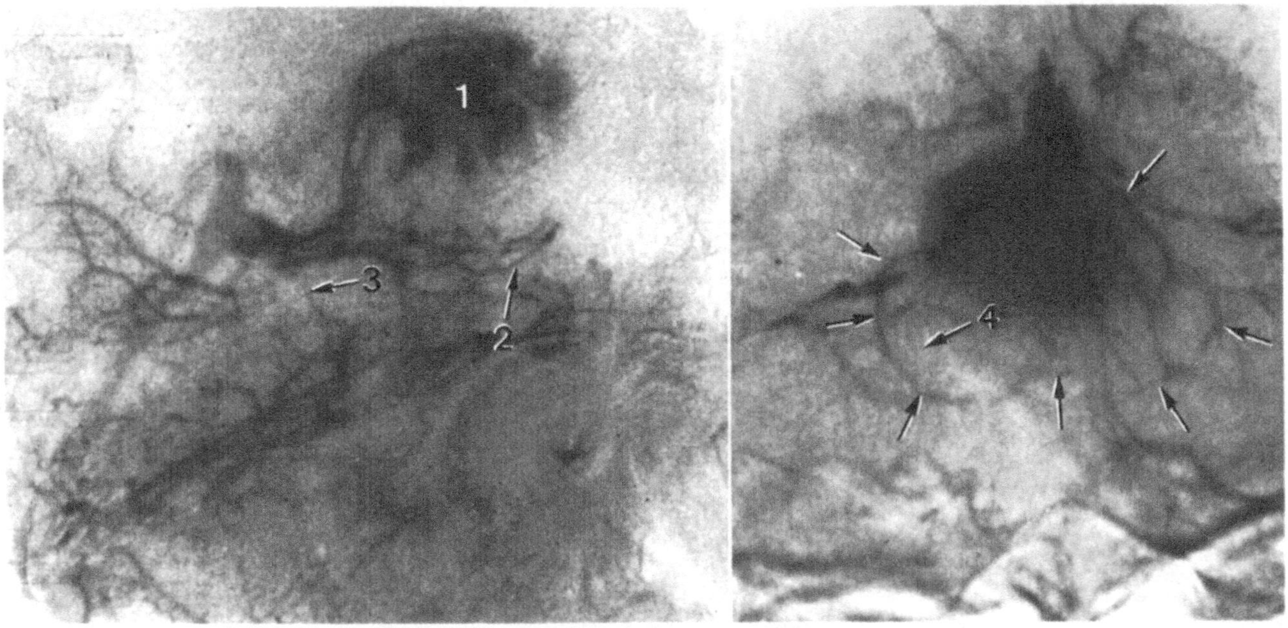

Figure 128 Glioma of thalamus on right side involving right cerebral peduncle

1 blush; *2* lifted posterior mesencephalic vein; *3* anteriorly concave precentral vein; *4* increased surface of right cerebral peduncle

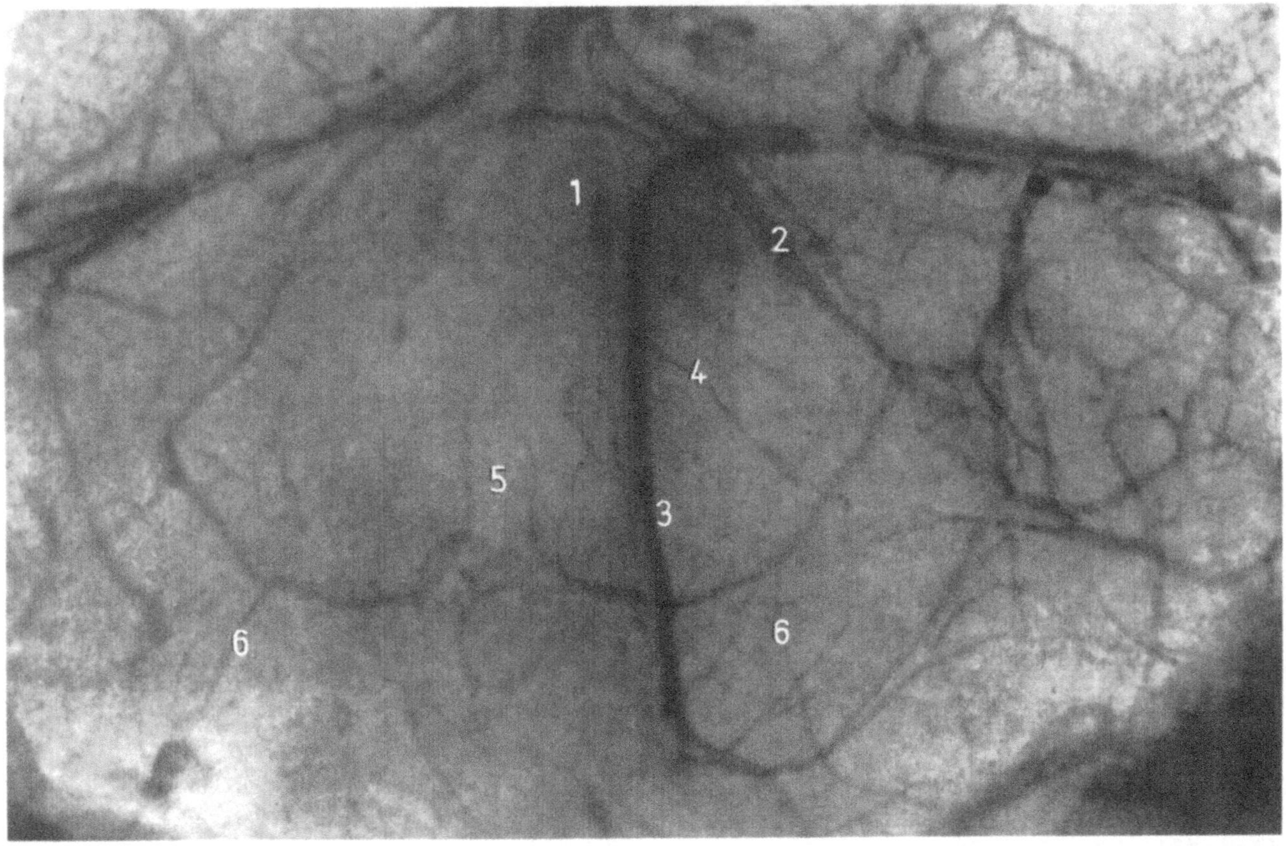

Figure 129 Glioma of pons extending into left cerebral peduncle (Neuroradiology **3**, (2), 75–76 (1971)

1 blush; *2* increased caliber and irregular walls of left posterior mesencephalic vein; *3* stretched tributaries of inferior vermian vein; *4* increased surface of left cerebral peduncle; *5* displacement of interpeduncular space; *6* displacement of inferior segment of left lateral mesencephalic vein

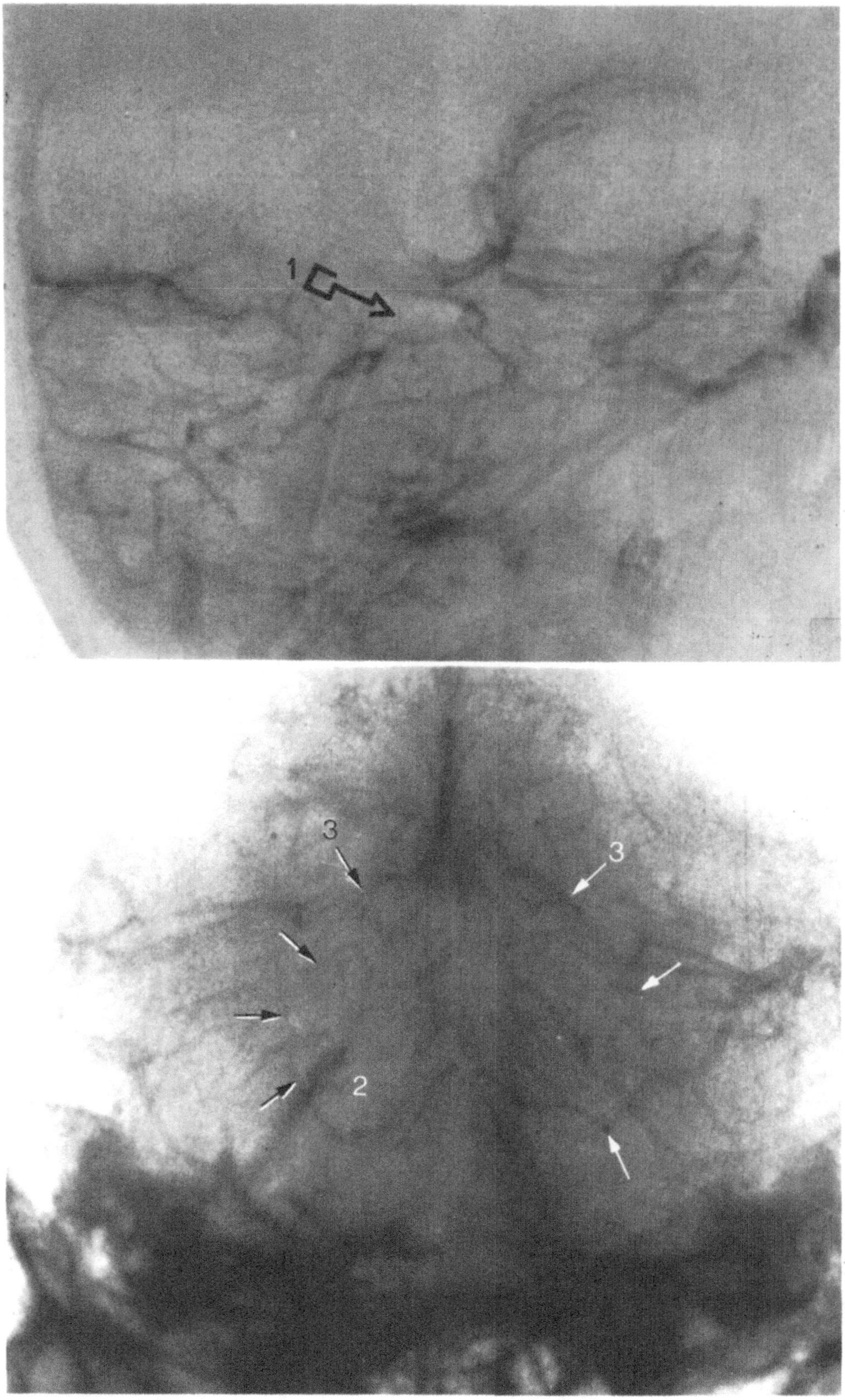

Figure 130

Glioma of pons

1 Absence of herniation of superior vermis
2 Enlarged right cerebral peduncle
3 Compressed and shifted posterior mesencephalic veins

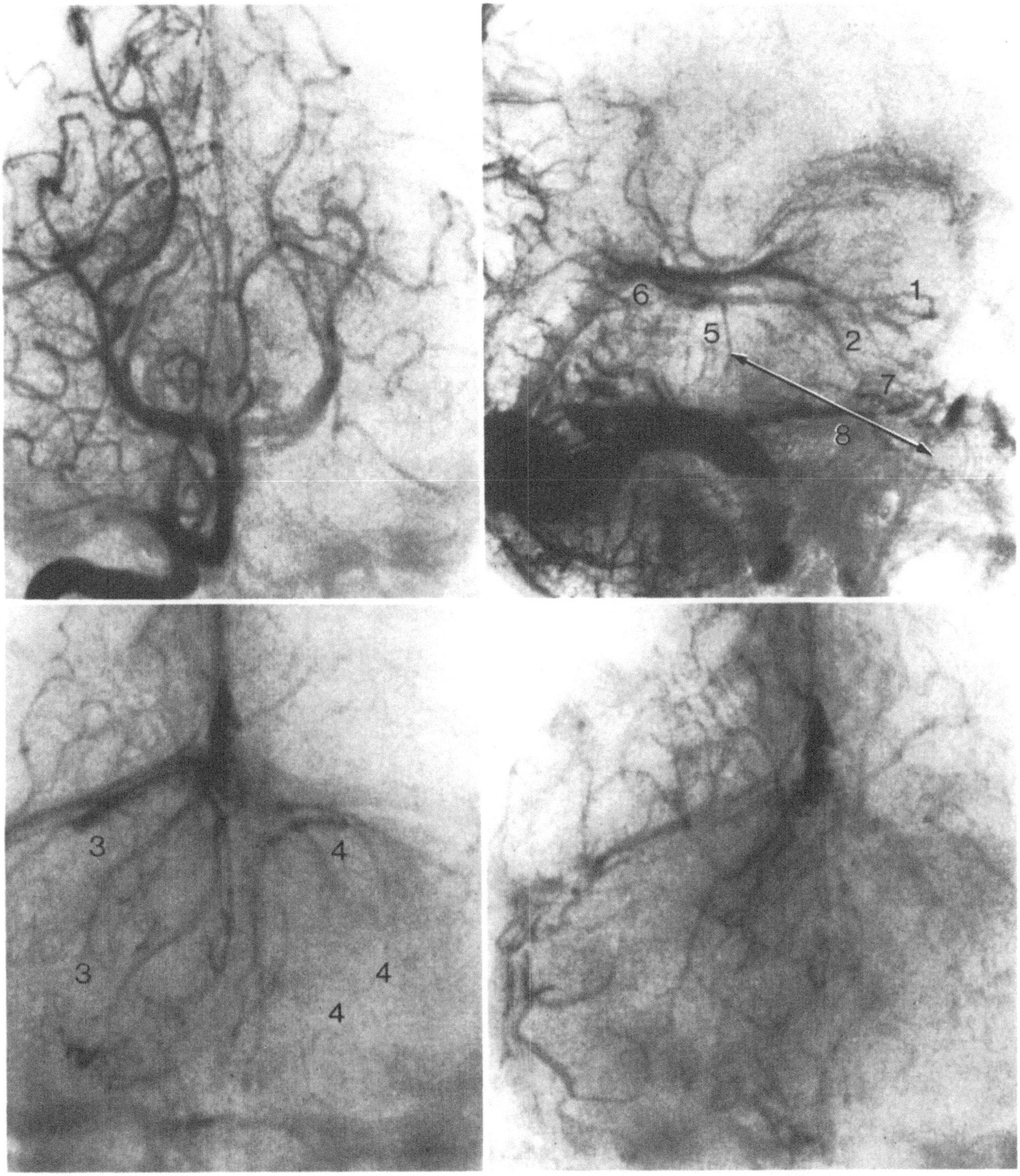

Figure 131

Spongioblastoma of left cerebral peduncle with asymmetry of optopenduncular veins

1 Right optopeduncular veins
2 Left optopeduncular veins
3 Right peripeduncular veins
4 Blurred left peripeduncular veins
5 Stretched precentral vein
6 Herniation of superior vermis
7 Dislocated pontine veins
8 Enlarged diameter of pons

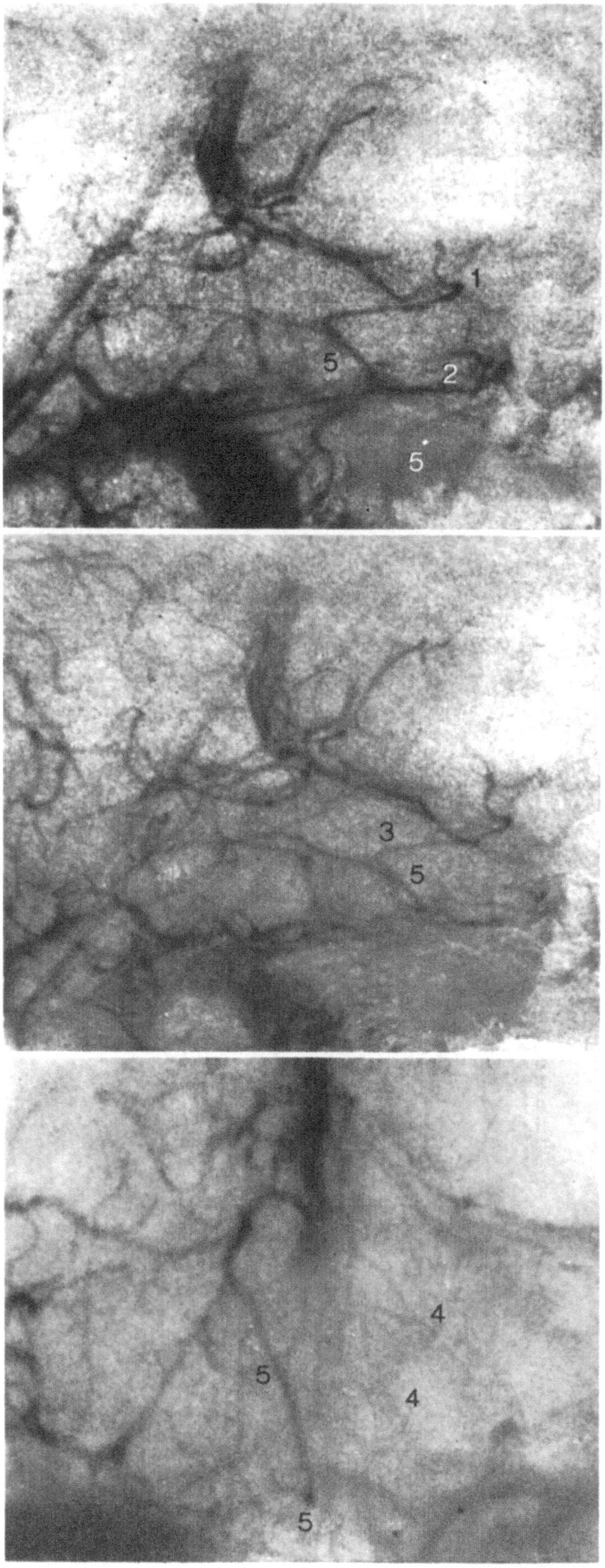

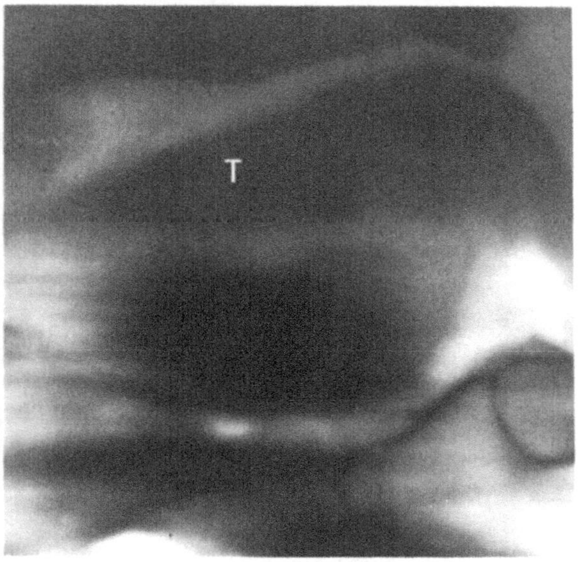

Figure 132

Pontopeduncular tumor (*T*)

1 Dislocated optopeduncular veins
2 Dislocated pontine veins
3 Very important changes in right lateral mesencephalic vein
4 Blurred peripeduncular veins on left side
5 Abnormal dilated pontine vein

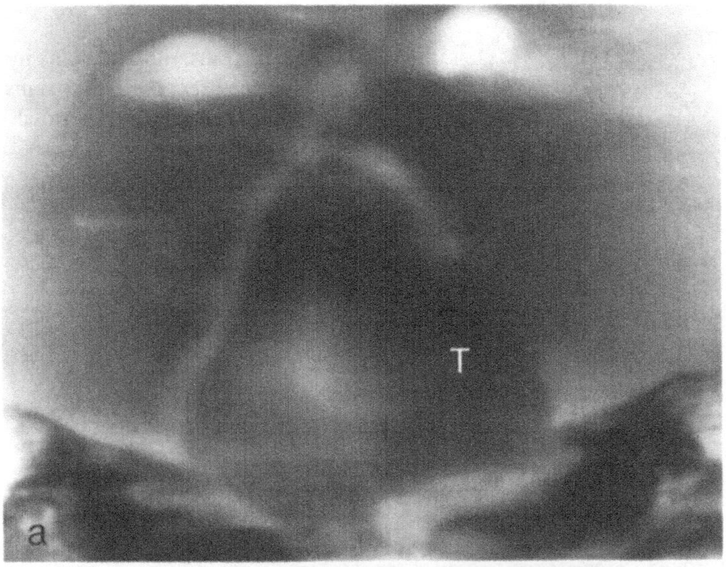

Figure 133 a and b

Lateropontine tumor involving left peduncle and left brachium pontis

a Pneumoencephalography demonstrating location of tumor (*T*)

b Venous phase of vertebral angiography

1 Stretched and elevated optopeduncular veins
2 Enlarged left cerebral peduncle $a < b$, $a' < b'$
3 Normal lateral bulbopontine vein and normal localization of junction point of vein of lateral recess and lateral medullopontine vein (+)
4 Laterally displaced medullopontine vein
5 Hypertrophied and posteriorly shifted lateral mesencephalic vein on left side
6 Normal lateral mesencephalic vein on right side
7 Concave precentral vein
8 Lowered tonsillar vein
9 Landmark of shifted aqueduct of Sylvius
10 Normal vein of lateral recess of fourth ventricle

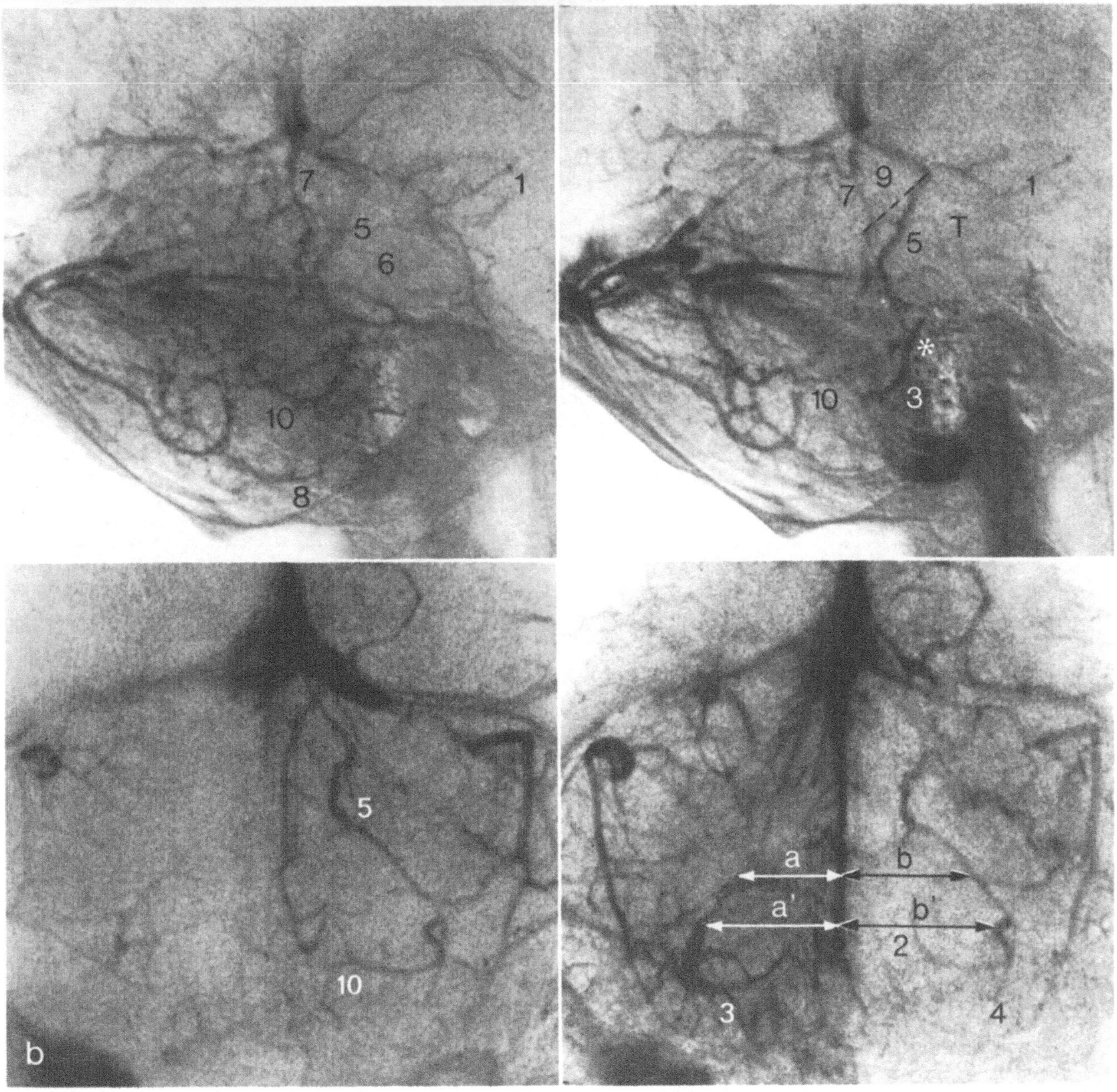

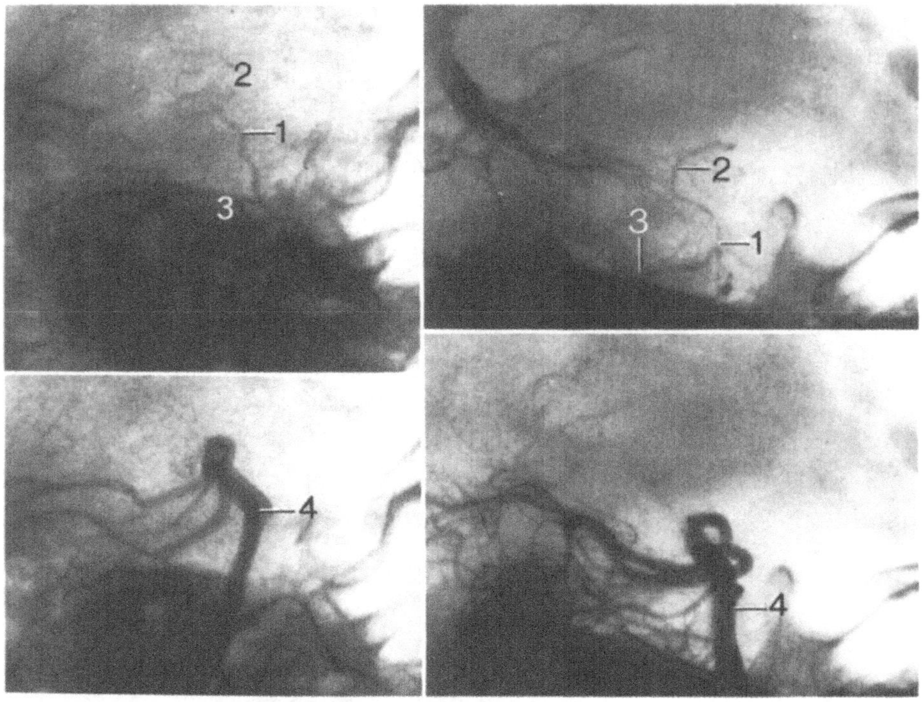

Figure 134

Normal arteries and veins in pontine area. Note that veins are more posteriorly located than arteries

1 Pontine veins
2 Optopeduncular veins
3 Lateral pontine vein
4 Basilar trunk

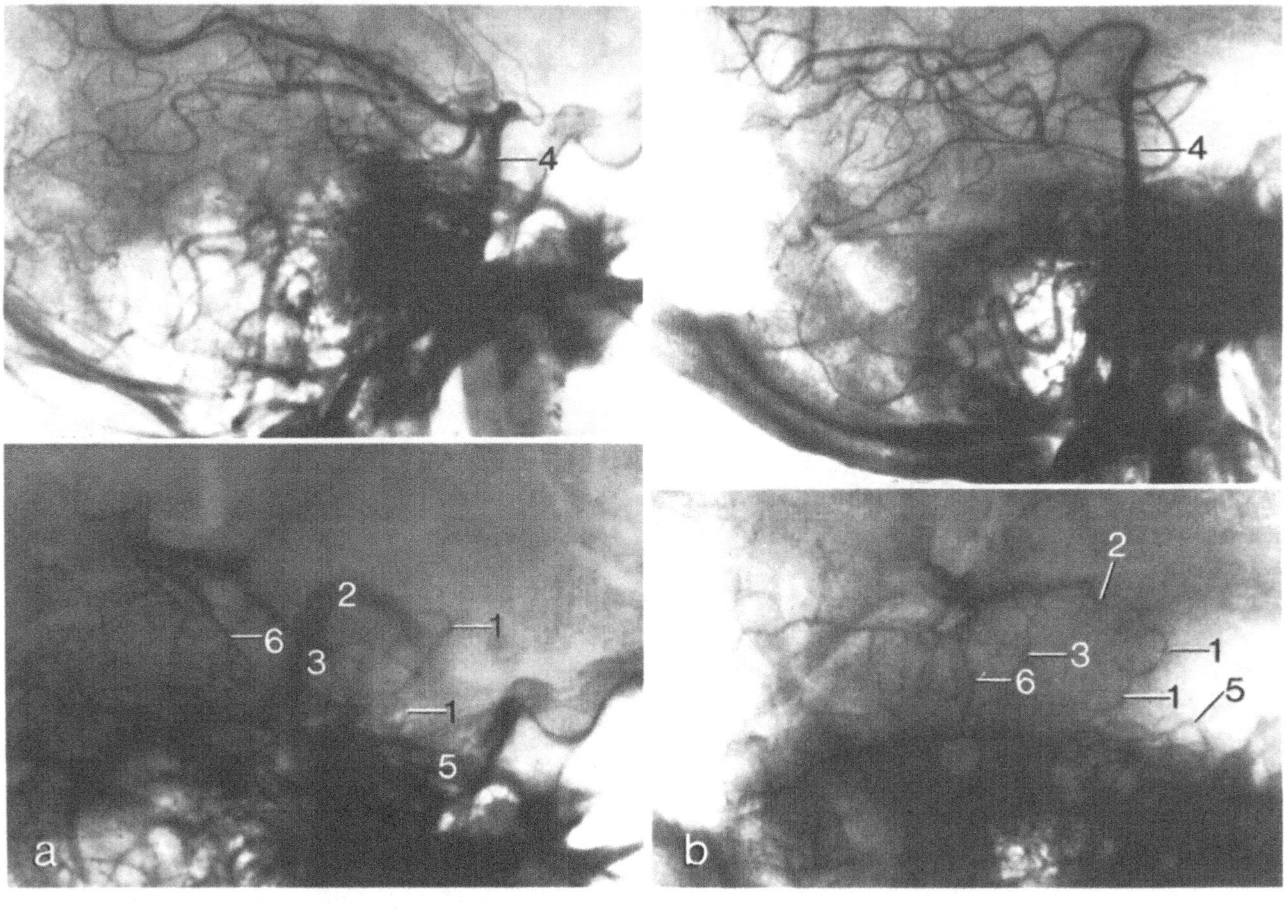

Figure 135 a and b Arteriovenous dissociation in pontine area

a Intrapontine metastases. Note dissociation between arterial and venous displacements
1 prepontine veins; *2* posterior mesencephalic vein; *3* lateral mesencephalic vein; *4* basilar trunk; *5* lateral pontine veins; *6* precentral vein

b Prepontine tumor with harmonious backward-displaced arteries and veins
1 prepontine veins; *2* posterior mesencephalic vein; *3* lateral mesencephalic vein; *4* basilar trunk; *5* lateral pontine veins; *6* precentral vein

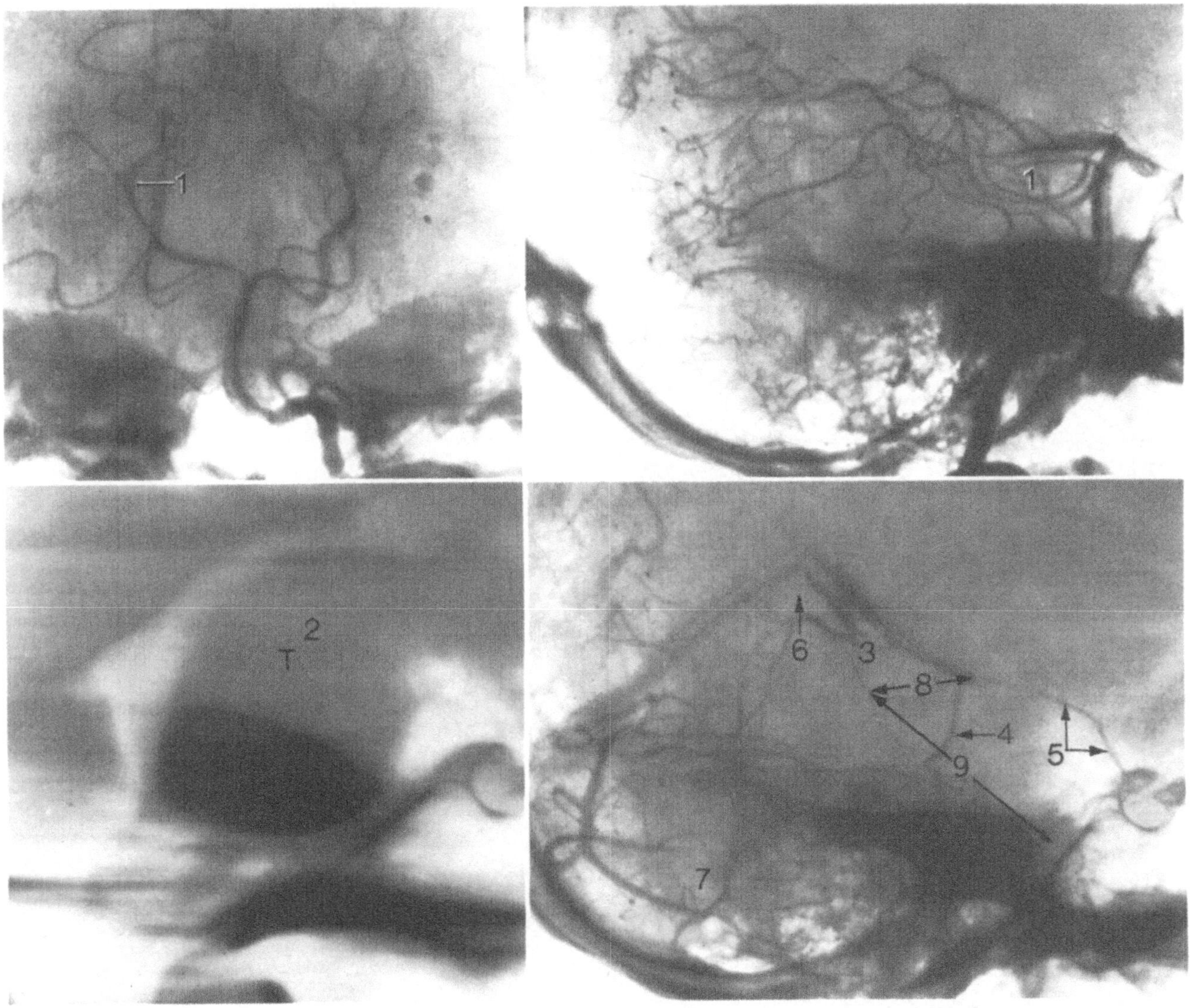

Figure 136 Glioma of pons and cerebral peduncles (*T*)

△

1 Arterial signs
2 Pneumoencephalographic signs
3 Stretched precentral vein
4 Hypertrophic lateral mesencephalic vein
5 Lifted optopeduncular veins
6 Herniation of superior vermis
7 Absence of herniation of tonsils
Dotted lines: landmark of shifted aqueduct of Sylvius (*8*); enlarged a.p. diameter of pons (*9*)

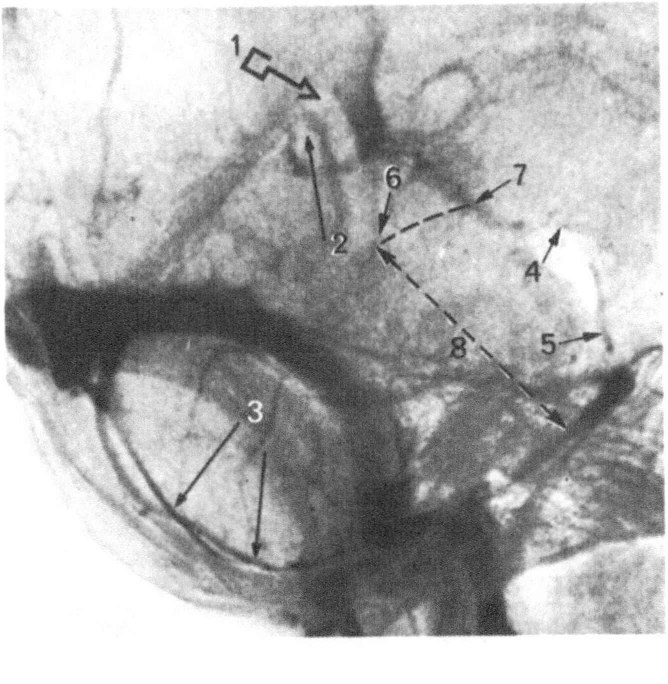

Figure 137

Glioma of pons

1 and *2* Herniation of superior vermis
3 Herniation of tonsils
4 Lifted posterior mesencephalic vein and optopeduncular veins
5 Stretched and lifted pontine veins
6 Lifted collicular point
7 Lifted junction of lateral and posterior mesencephalic veins
Broken lines: landmark of aqueduct of Sylvius (*6–7*) enlarged a.p. diameter of pons (*8*)

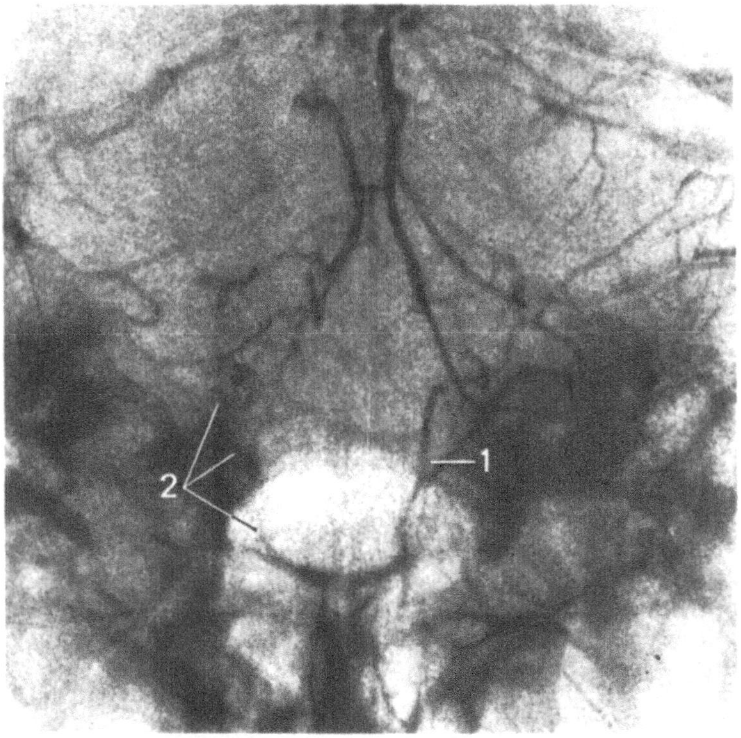

Figure 138

Bulbar involvement of pontine glioma

1 Normal lateral medullopontine vein
2 Compressed and laterally shifted medullopontine vein

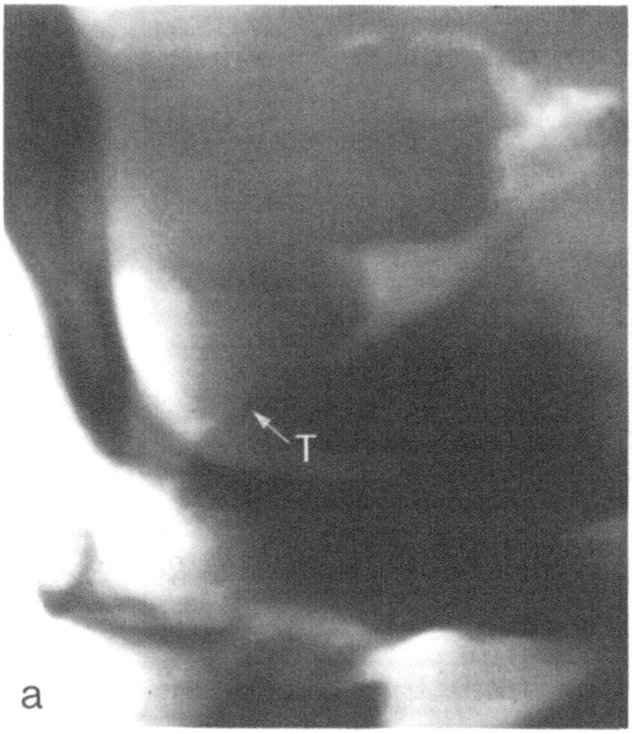

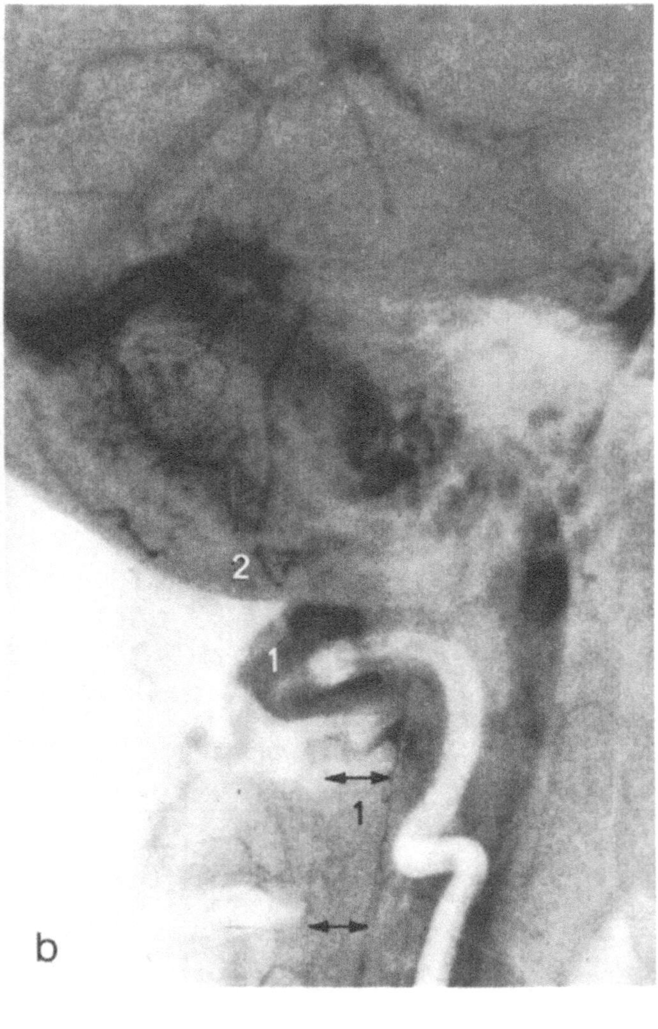

Figure 139a and b

Tumor of medulla oblongata (*T*)

a Pneumoencephalography
b Vertebral venous phase of the vertebral angiography
- ***1*** **Stasis in superior venous system of medulla with delineation of a.p. diameter of medulla**
- ***2*** **Lowered tonsillar veins**

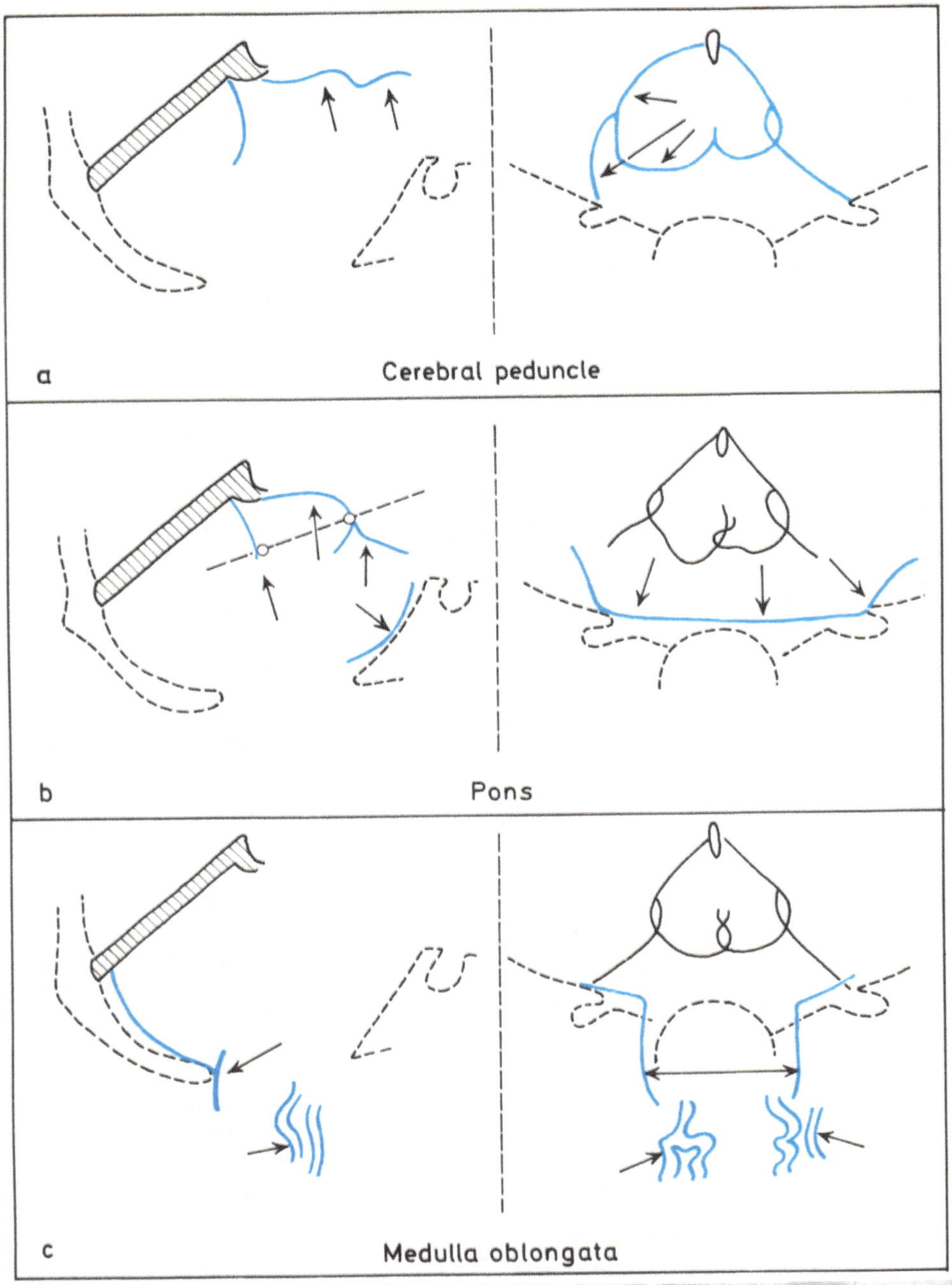

Figure 140 a–c

Drawing of venous signs in the three main topographic varieties of tumors of brain stem

a Tumor of cerebral peduncle
b Tumor of pons
c Tumor of medulla oblongata

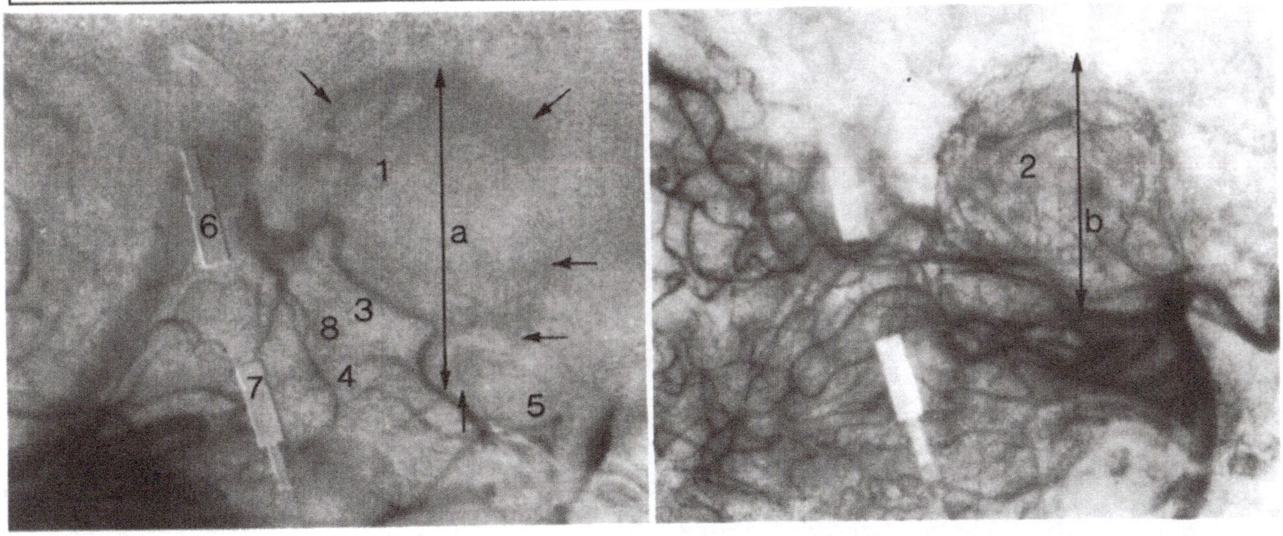

Figure 141

Malignant tumor of thalamus

1 Venous blush
2 Arterial blush
3 Posterior segment of lowered basal vein
4 Backward-pushed precentral vein
5 Lowered optopeduncular veins
6 Empty supravermian triangle
7 Postcentral vein
8 Quadrigeminal veins
a "Venous" diameter of tumor
b "Arterial" diameter of tumor $a > b$

▽

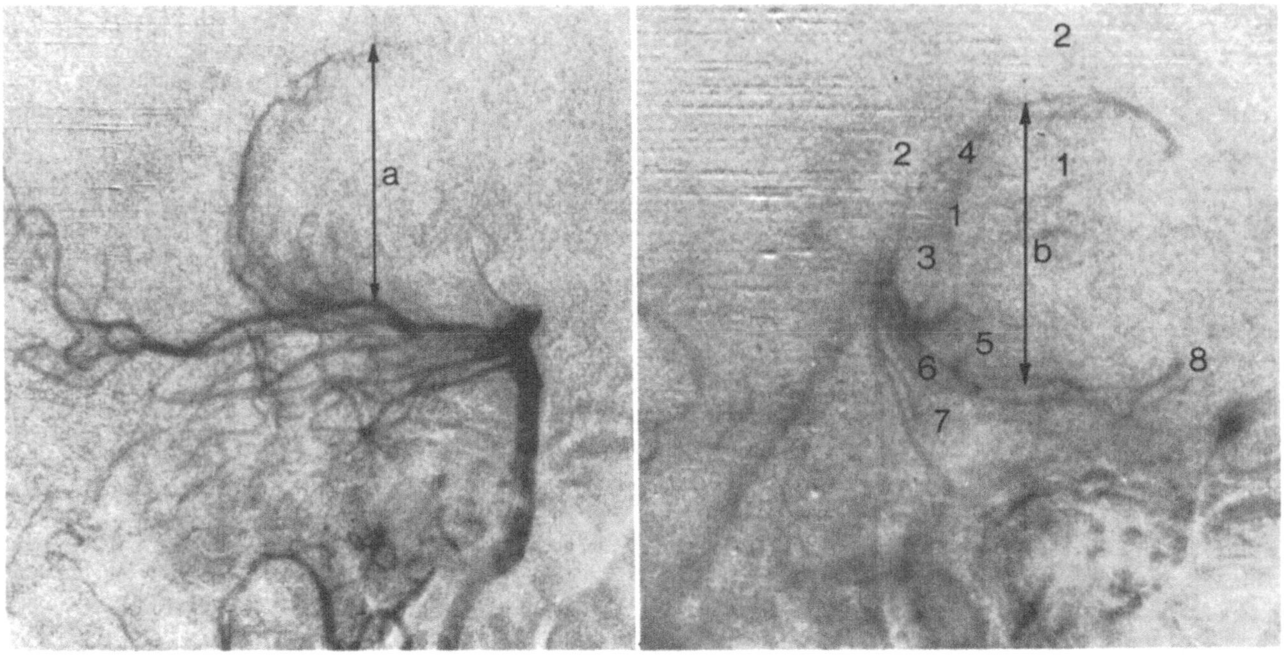

△

Figure 142

Posterior thalamic glioma

1 Superior thalamic vein
2 Superior choroidal vein
3 Pineal veins
4 Blush
5 Posterior mesencephalic vein
6 Quadrigeminal veins
7 Precentral vein
8 Optopeduncular veins
a "Arterial" diameter of tumor
b "Venous" diameter of tumor $a<b$

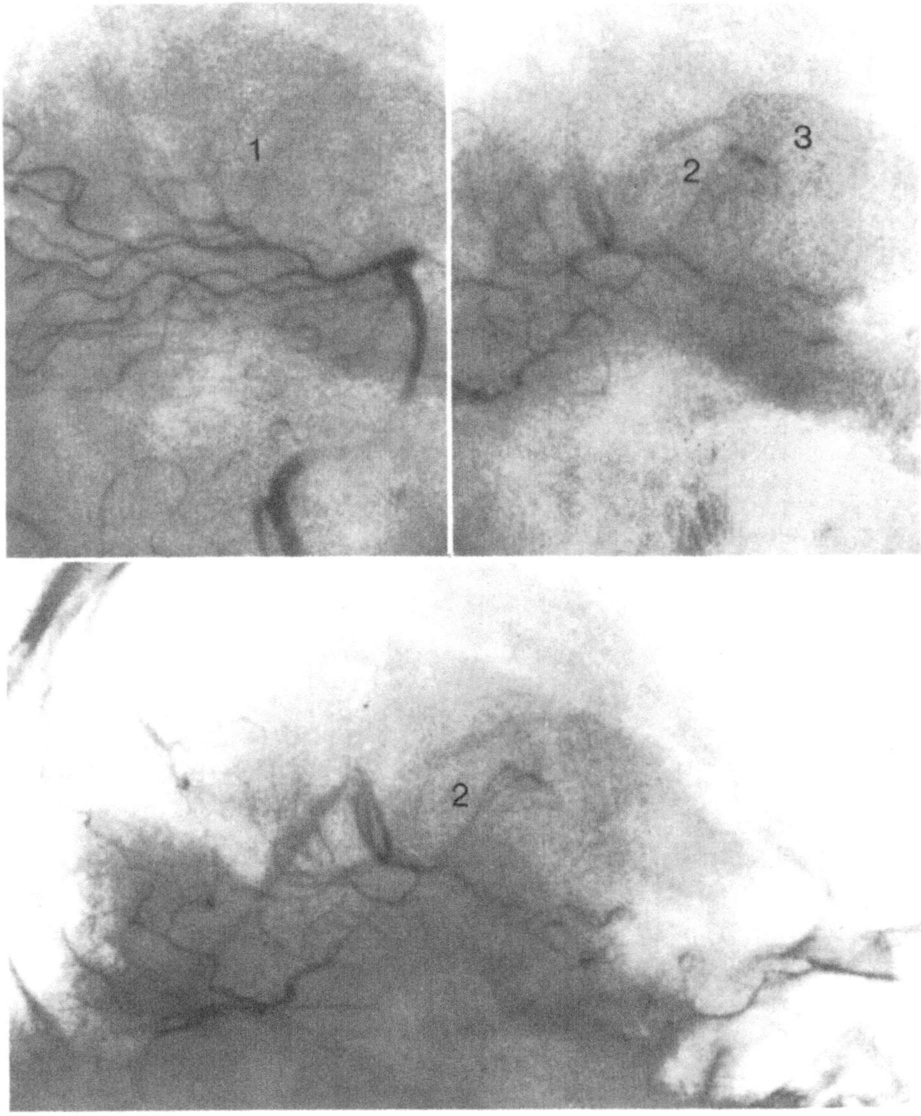

Figure 143

Left thalamic glioma

1 Arterial signs
2 Dissociated left and right thalamic veins
3 Blush

Figure 144

a Thalamic glioma

1 Lifted superior choroidal vein
2 Lifted superior thalamic vein
3 Lowered posterior mesencephalic vein
4 Backward displaced precentral vein
5 Dislocated optopeduncular veins

b Teratoma of pineal gland (pinealoma) in venous (above) and late venous phase of vertebral angiography

1 Avascular area
2 Dislocated pineal veins
3 Straightened precentral vein
4 Tumoral blush

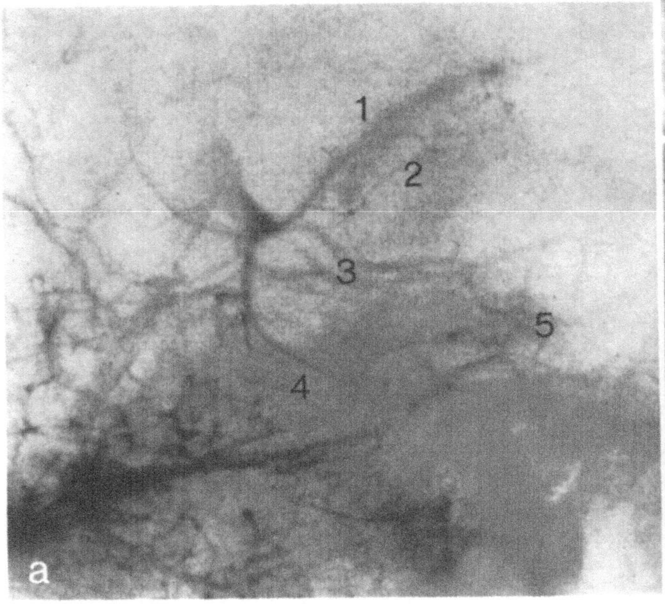

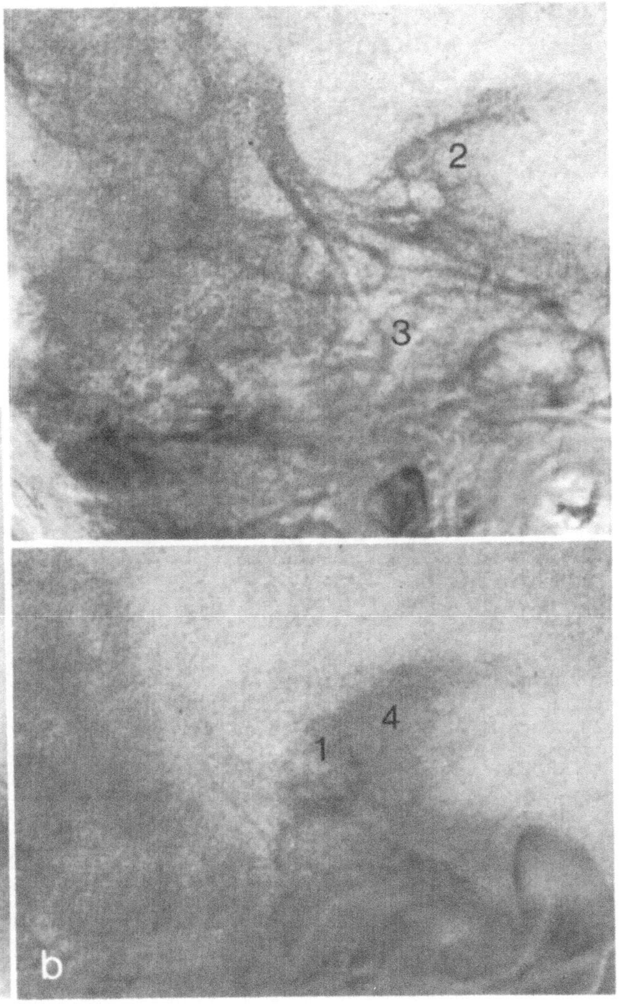

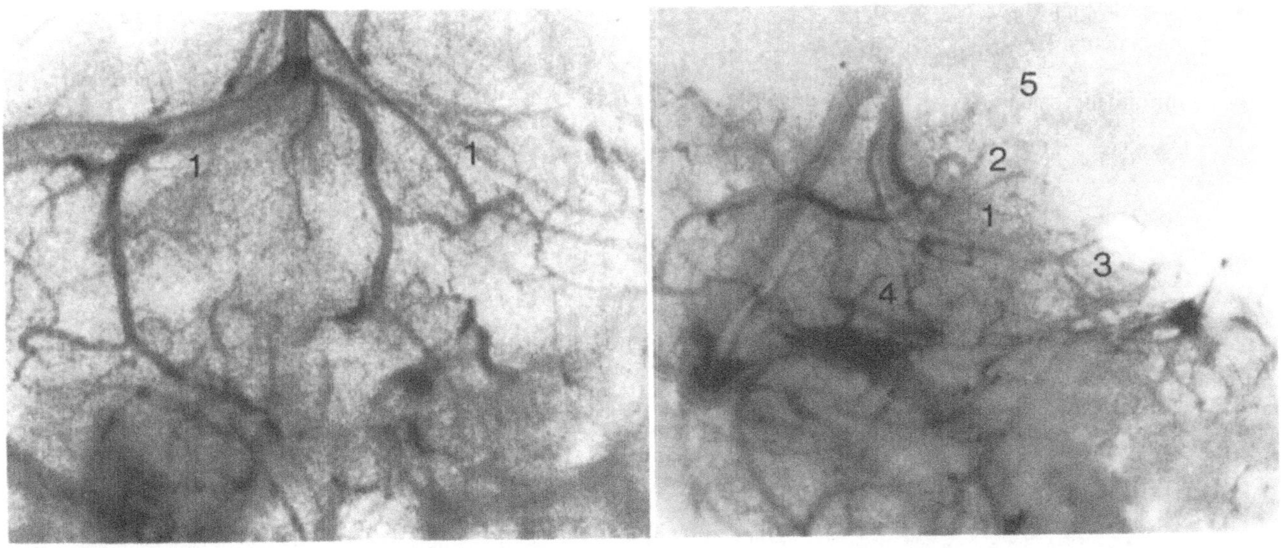

Figure 145

Bilateral predominantly left thalamic glioma

1 Posterior mesencephalic vein: deviated frontally and dislocated laterally
2 Pineal veins: hypertrophied and dislocated
3 Optopeduncular veins: lowered and dislocated
4 Precentral vein: backward displaced
5 Superior thalamic vein: tortuous and lifted

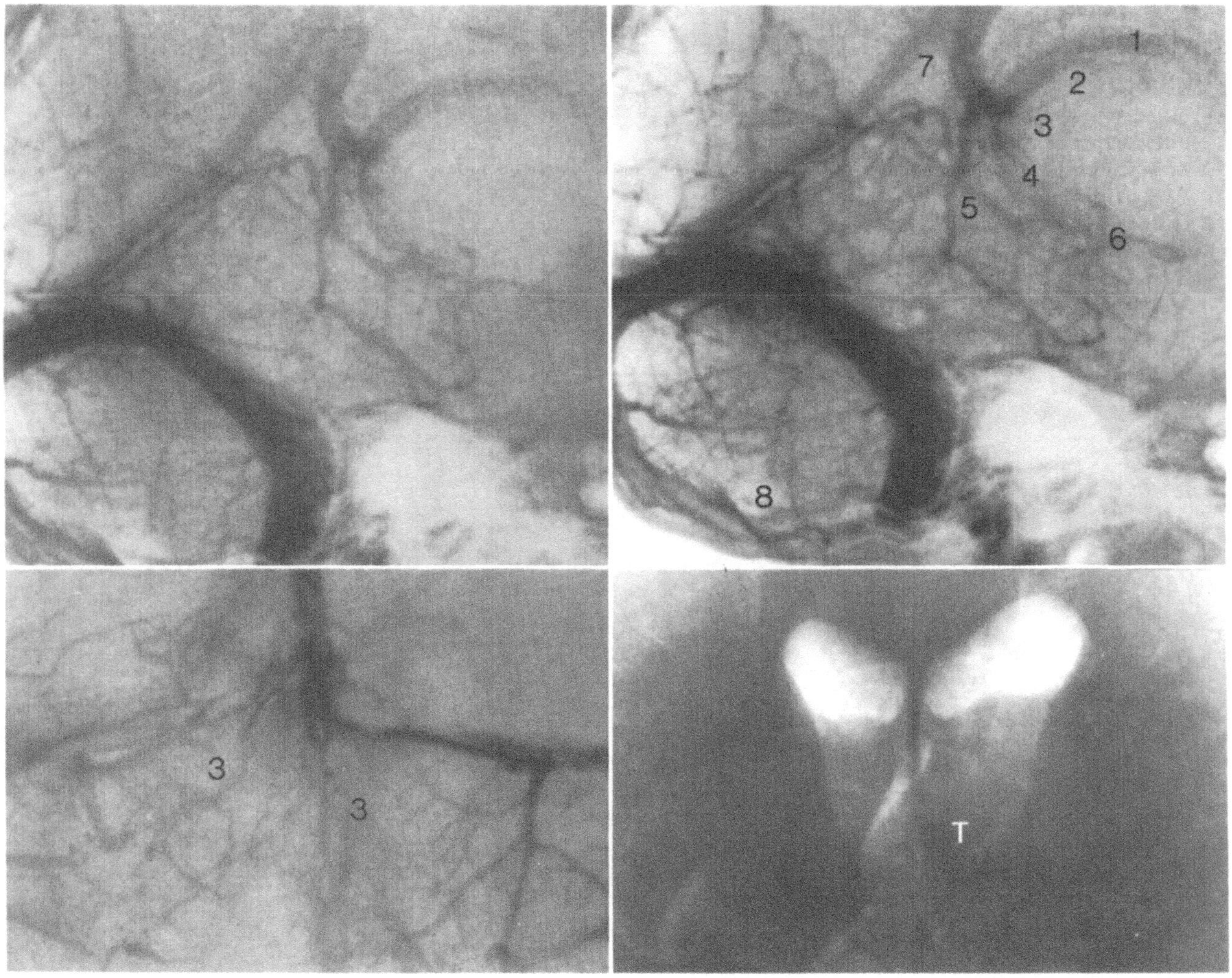

△

Figure 146 Glioma of left thalamus (*T*)

1 Superior choroïdal vein
2 Superior thalamic vein
3 Pineal vein
4 Posterior mesencephalic vein
5 Precentral vein
6 Optopeduncular veins
7 Subdural supravermian space
8 Tonsillar herniation

Figure 147 Cystic tumor of anterior vermis invading two cerebellar hemispheres

1 Pushed-backward precentral vein
2 Stretching of vein of lateral recess of fourth ventricle
3 Backward-displacement of point of junction of vein of the lateral recess and of lateral medullopontine vein
4 Superior petrosal vein

▽

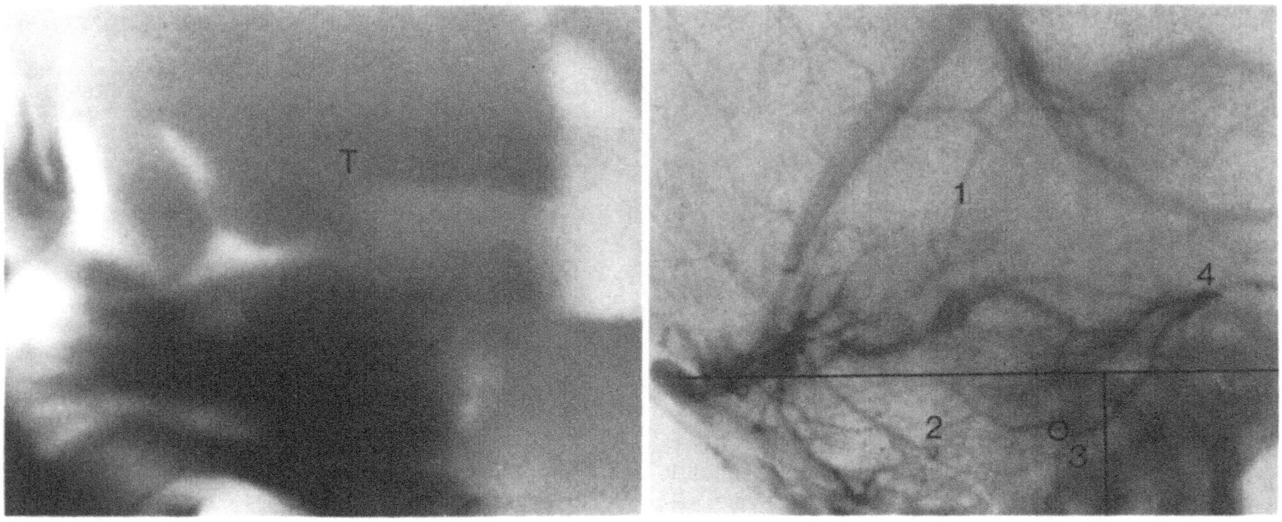

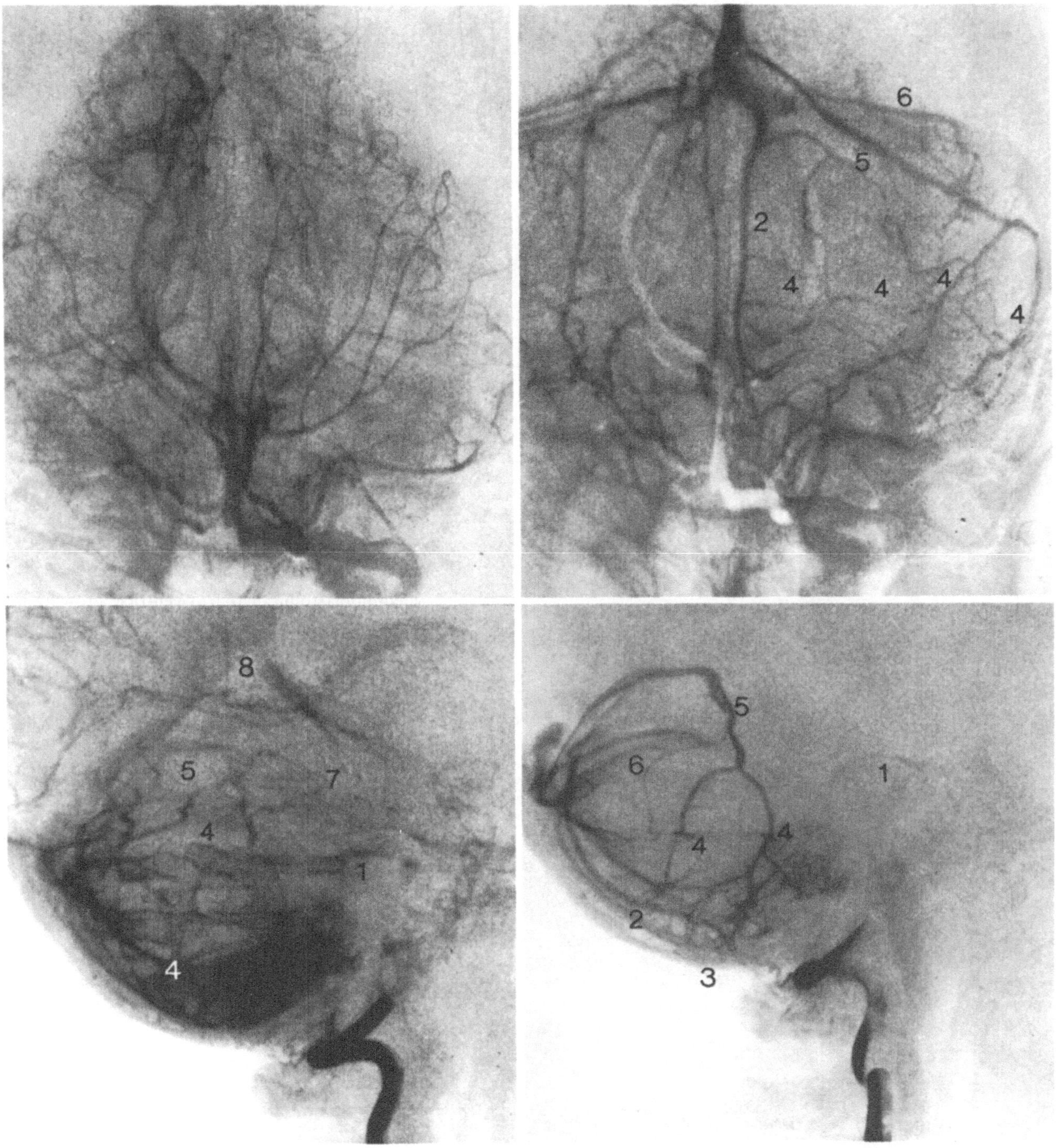

Figure 148

Cystic medulloblastoma of vermis growing into left cerebellar hemisphere

1 Superior petrous sinus
2 Inferior paravermian vein
3 Downward displacement of the inferior retrotonsillar vein
4 Stretched cerebellar hemispheric veins
5 Tentorial vein
6 Lateral sinus
7 Upward-displaced precentral vein
8 Upward-displaced superior vermian vein

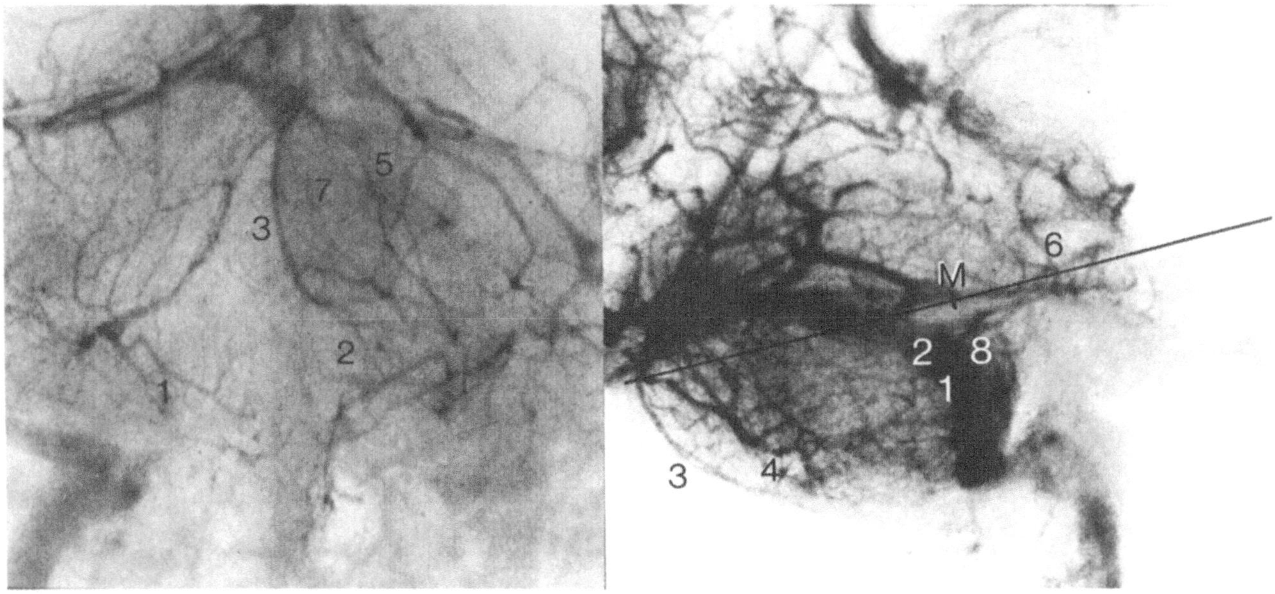

△
Figure 149

Medulloblastoma of fourth ventricle growing into left cerebellar hemisphere

1 Downward displacement and elongation of right branch of vein of lateral recess of fourth ventricle
2 Downward displacement and elongation of left branch of vein of lateral recess of fourth ventricle
3 Bending of inferior paravermian vein
4 Copular point in normal position
5 Narrowness of left lateral mesencephalic vein
6 Forward displacement of lateral mesencephalic vein
7 Tumoral blush
8 Normal junction point of vein of lateral recess and lateral medullopontine vein
M Midpoint of Twining's line

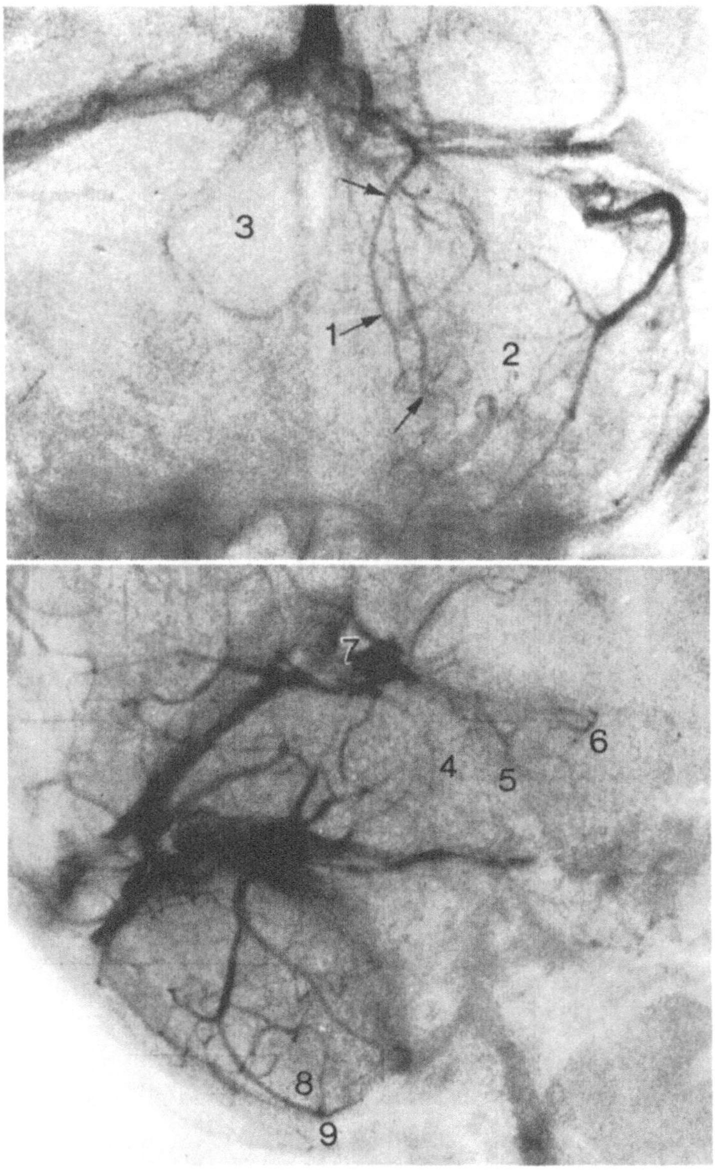

Figure 150

Cystic astrocytoma of left cerebellar hemisphere

1 Hammock of left paravermian vein taking on value of a hemispheric vein
2 Dislocation of hemispheric veins
3 Normal superior cerebral trunk
4 Forward and upward displacement of precentral vein
5 Considerable displacement of lateral mesencephalic vein in same direction as precentral vein
6 Elevation of posterior mesencephalic and of optopeduncular veins
7 Herniation of superior vermis
8 Closure of tonsillovermian angle
9 Lowering of copular point

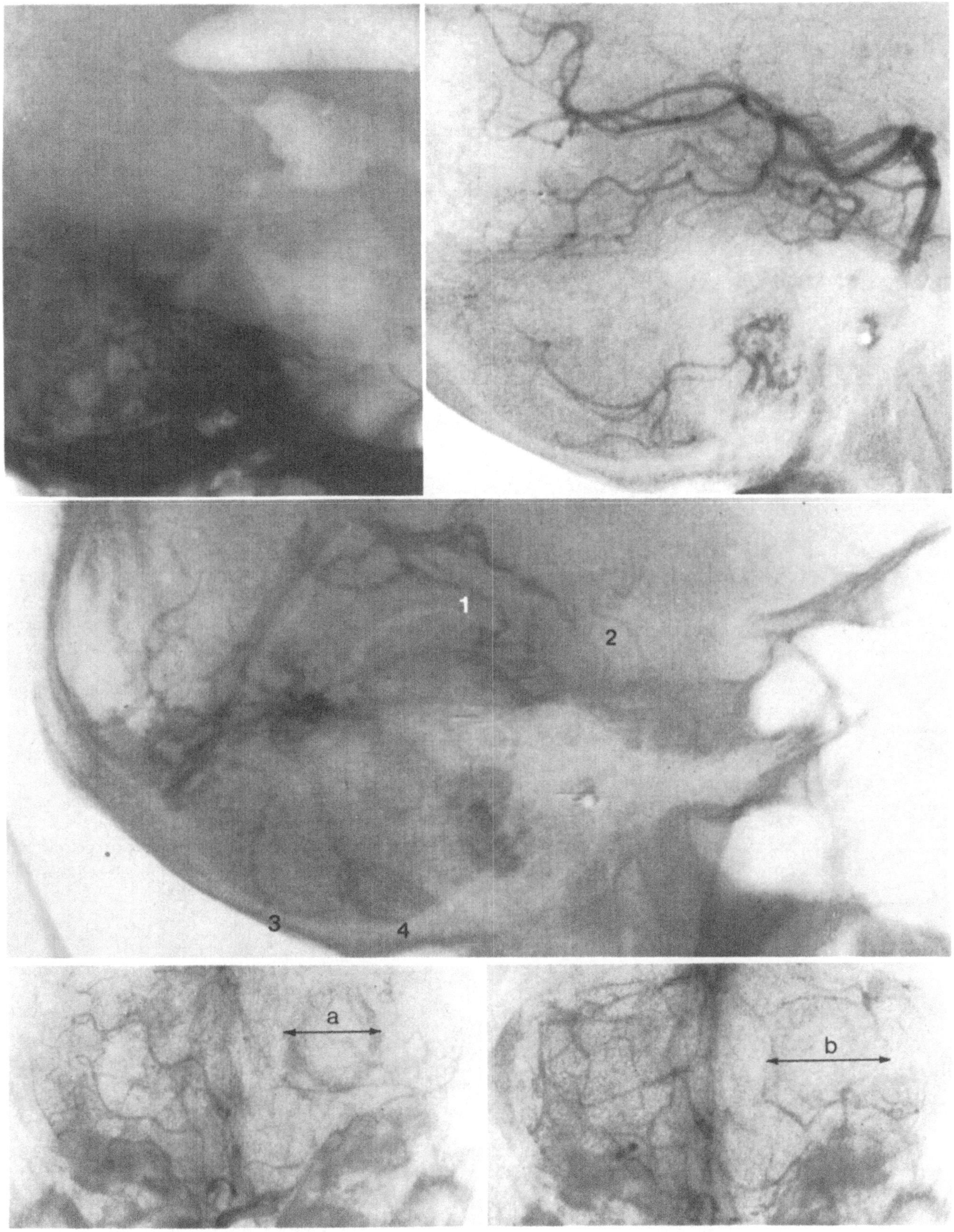

Figure 151 Metastasis of left cerebellar hemisphere

1 Marked elevation of precentral vein (parallel to posterior mesencephalic vein)
2 Lifted prepontine and optopeduncular veins
3 Lowered inferior vermian vein
4 Lowered hemispheric vein
a Tumoral area (arterial signs)
b Tumoral area (venous signs)
$a<b$

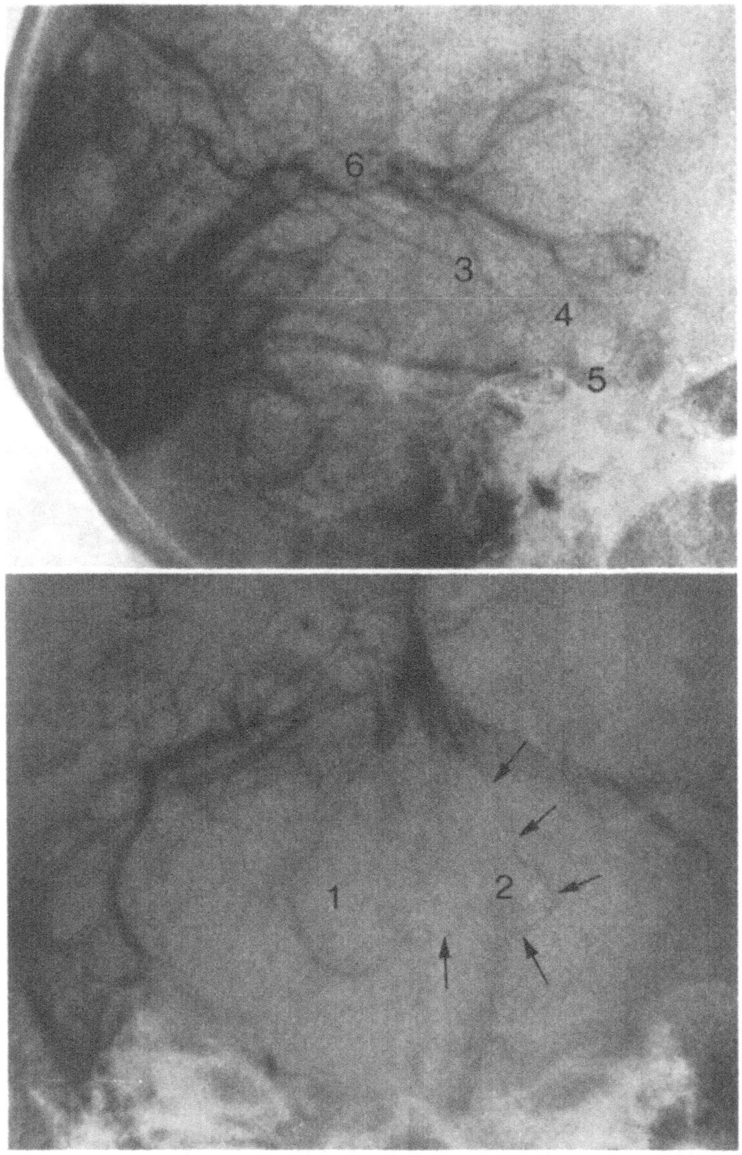

Figure 152

Metastasis of left cerebellar hemisphere (frontal and lateral view, venous phase)

1 Regular contours of right cerebral peduncle
2 Irregular contours of left cerebral peduncle with signs of elevation and rotation of superior cerebral trunk
3 Elevation of precentral vein
4 Elevation of lateral mesencephalic vein
5 Forward displacement of prepontine veins
6 Herniation of superior vermis

Figure 153a and b

Neurinoma of left acoustic nerve in case of RECKLINGHAUSEN's disease also extending in posterior fossa and in supratentorial space

a *1* Typical arterial signs of supratentorial extension characterized by flattened angle of posterior cerebral artery
2 Marginal arteries of tumor belonging to posteroinferior cerebellar artery
3 Displaced basal artery

b Typical venous signs of both infra- and supratentorial extension of tumor.
1 Flattened angle of left peduncular vein, a sign of supratentorial extension
2 Oblique position of superior petrous vein (inverted obliquity, as in case of Fig. 158). See also Figure 165

▽

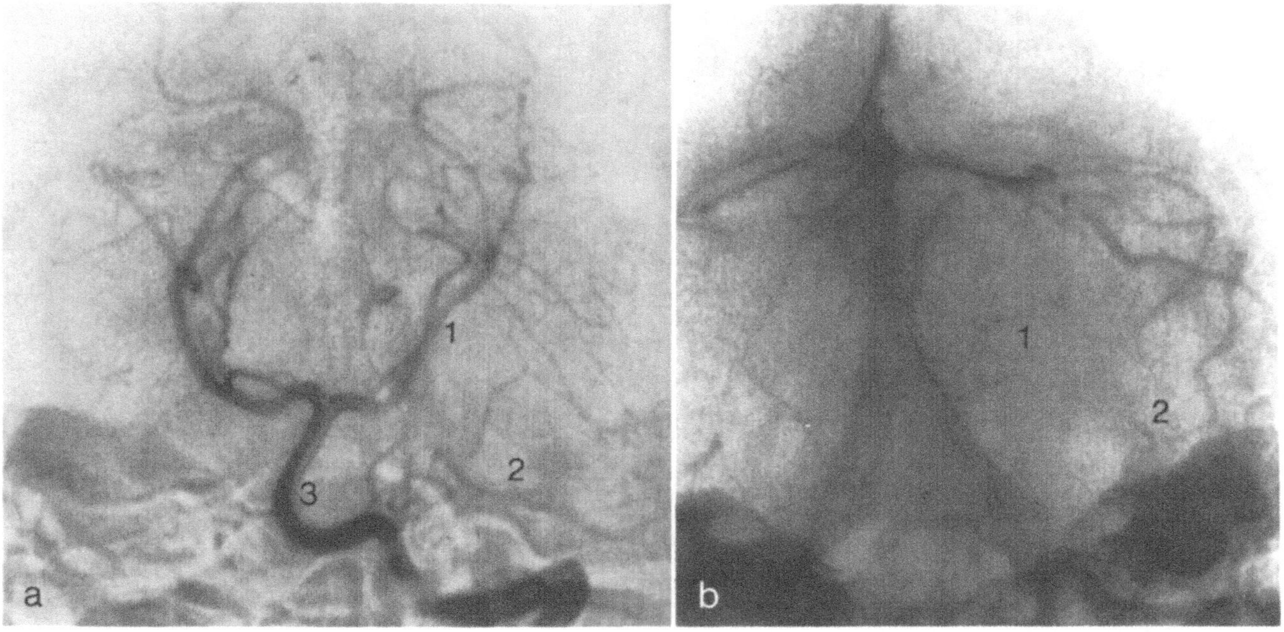

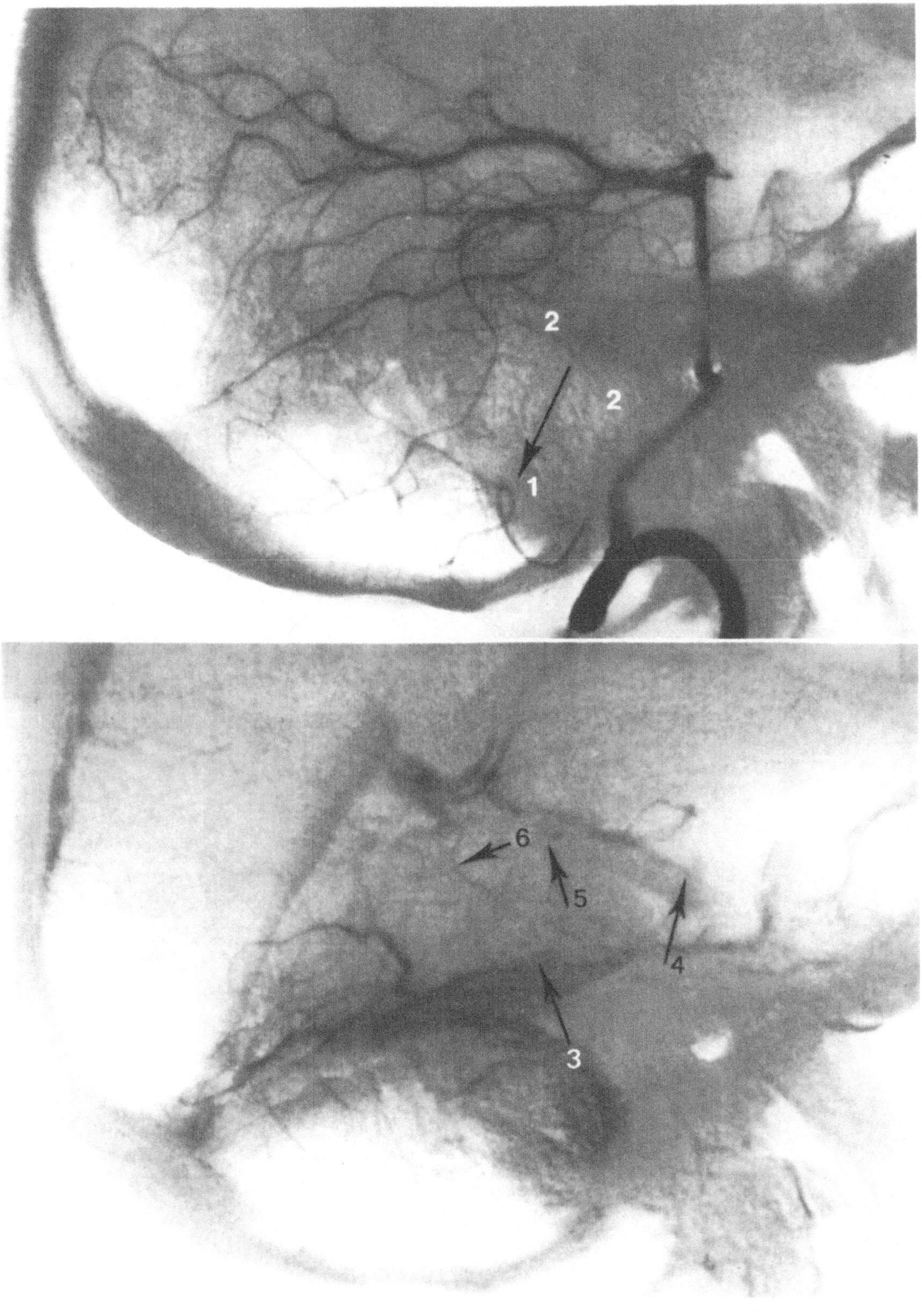

Figure 154

Acoustic neurinoma in lateral view

1 Lowered posteroinferior cerebellar artery
2 Stretched curve of middle cerebellar artery
3 Lifted veins of cerebellopontine angle
4 Compressed and shifted optopeduncular veins
5 Lifted lateral mesencephalic veins
6 Posteriorly displaced precentral vein with anterior concavity (paramedian development of tumor)

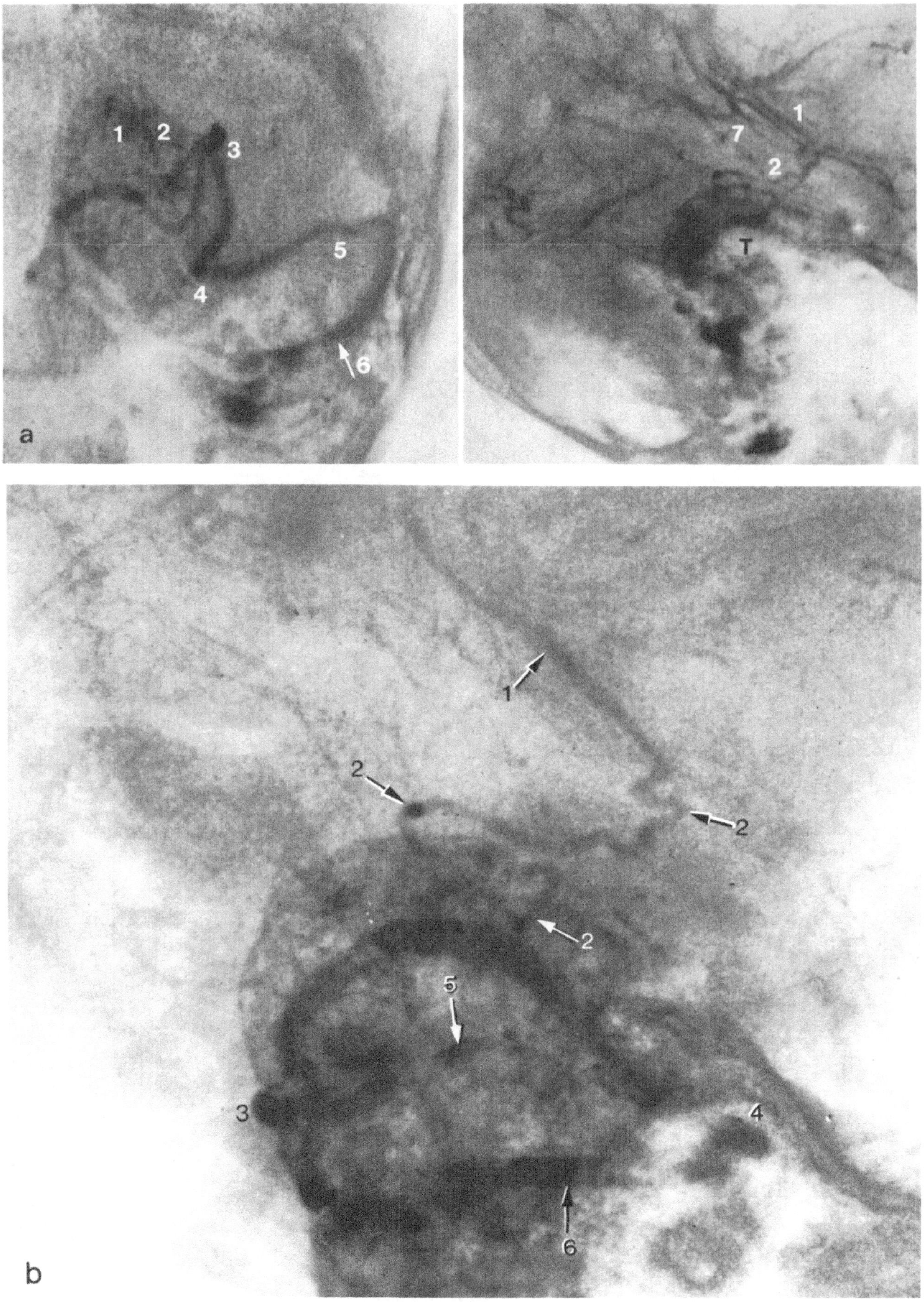
1
2
3
4
5
6
a
1
7
2
T
1
2
2
2
5
3
4
6
b

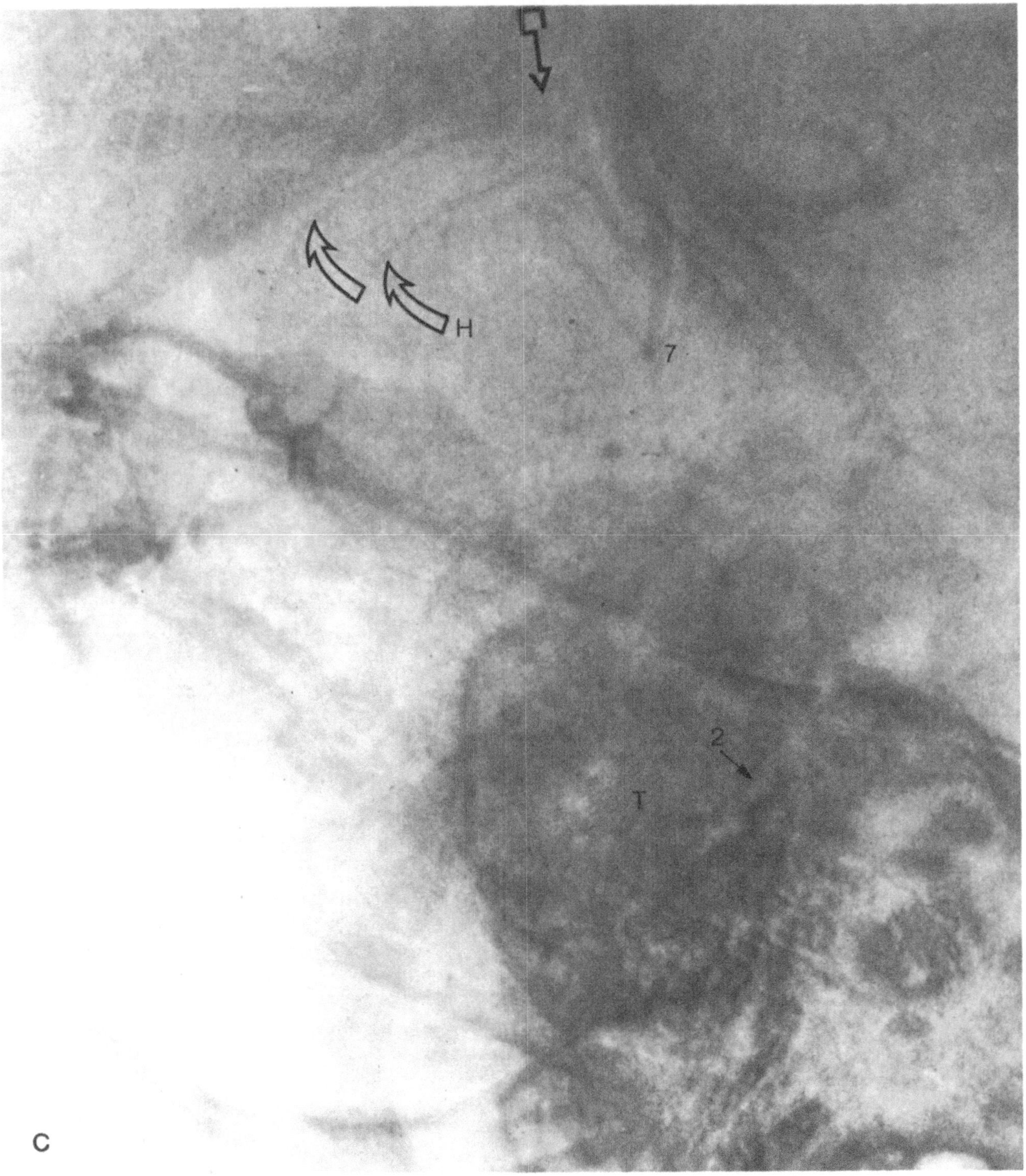

Figure 155 a–c

Neurinoma of left acoustic nerve with important venous signs

T Irregular tumoral blush
H Beginning herniation of superior cerebellar veins
1 Drainage through posterior mesencephalic vein
2 Drainage through lateral mesencephalic vein
3 Displaced superior petrous vein (parasol sign) with its affluents (see also Figs. 157 and 162)
4 Venous drainage through connecting vein with superior petrous sinus
5 Superior petrous sinus
6 Inferior petrous sinus
7 Elevated precentral vein

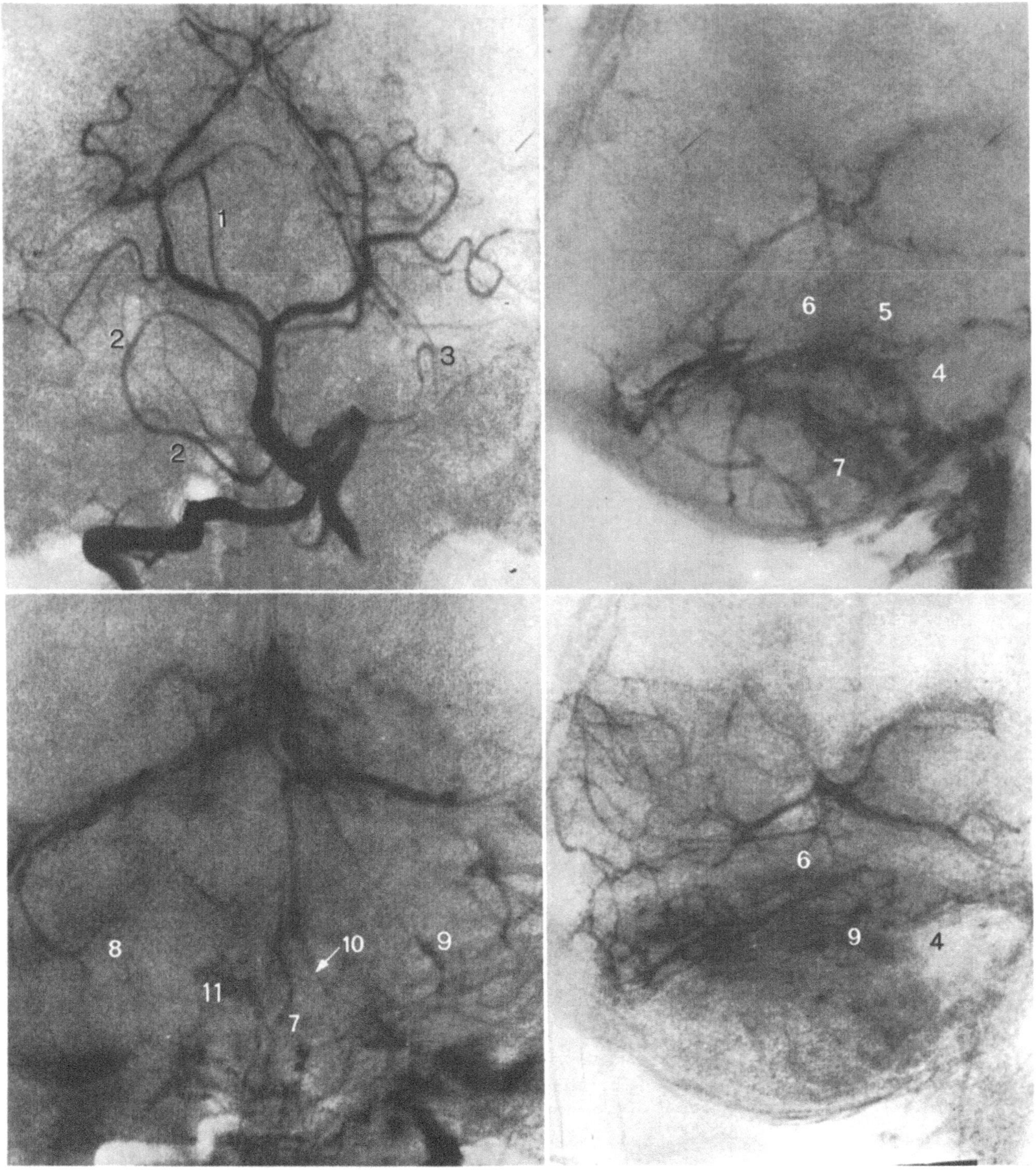

Figure 156

Neurinoma of right cerebellopontine angle

1 Very markedly displaced superior cerebellar artery
2 Dislocated branches of middle cerebellar artery describing an abnormally large loop in right cerebellopontine angle
3 Normal left middle cerebellar artery with normal loop at level of internal auditory canal
4 Avascular area corresponding to tumor
5 Anteriorly and superiorly shifted lateral mesencephalic vein
6 Posteriorly shifted precentral vein
7 Vein of lateral recess of fourth ventricle delineating posteroinferior and medial pole of tumor
8 Nonopacified superior petrous vein, normally found on left side (*9*)
9 Normal left superior petrous vein
10 Normal left vein of lateral recess of fourth ventricle
11 Tumoral blush

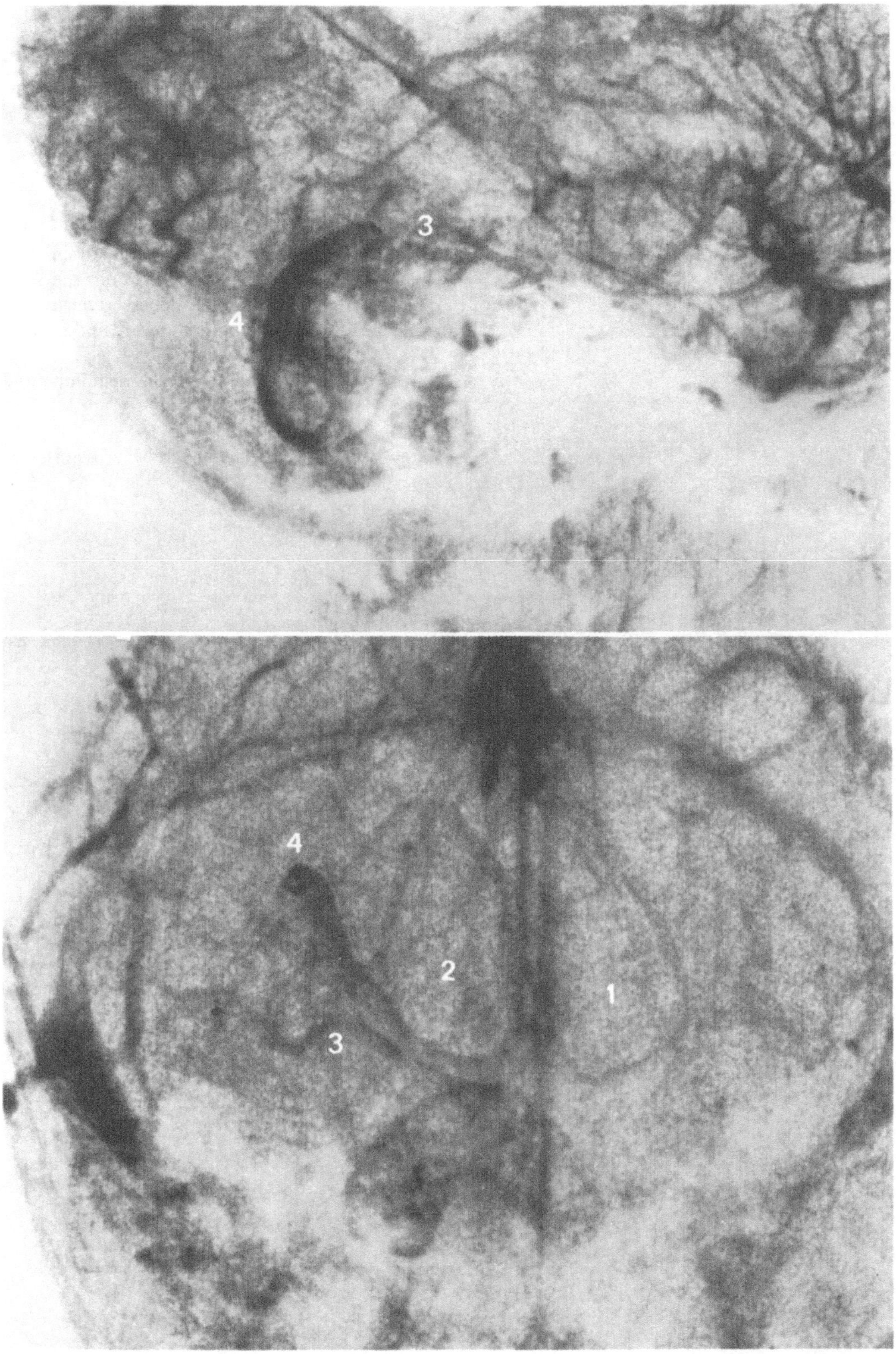

Figure 157 Neurinoma of right acoustic nerve

1 Normal left cerebral peduncle
2 Slightly deformed right cerebral peduncle
3 Hypertrophied lifted and laterally shifted right superior petrous vein (parasol sign: cf. Figs. 155 and 162)
4 Very hypertrophied and posteriorly shifted vein of lateral recess of fourth ventricle

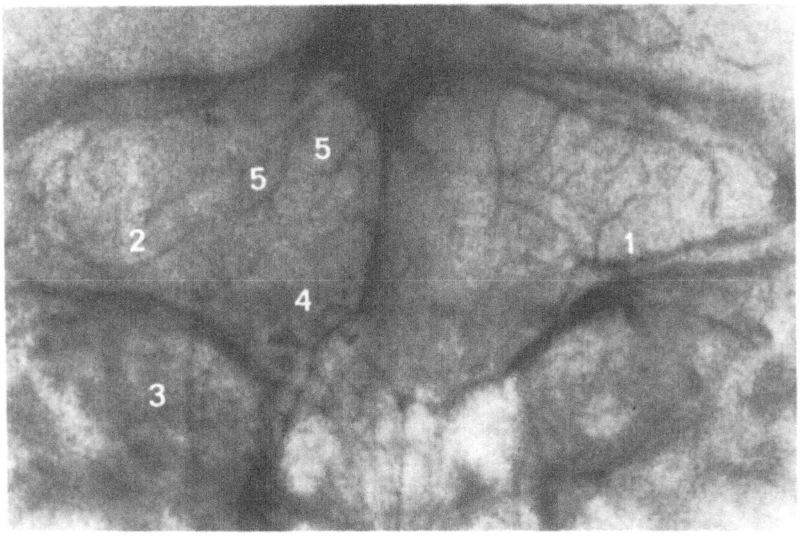

Figure 158

Neurinoma of right cerebellopontine angle

1 Normal left superior petrous vein
2 Dislocated venous system of right superior petrous vein having an inverted oblique orientation upward and laterally (cf. Fig. 153)
3 Laterally shifted lateral medullopontine vein
4 Tumoral stain
5 Deviated veins of cerebellar hemisphere

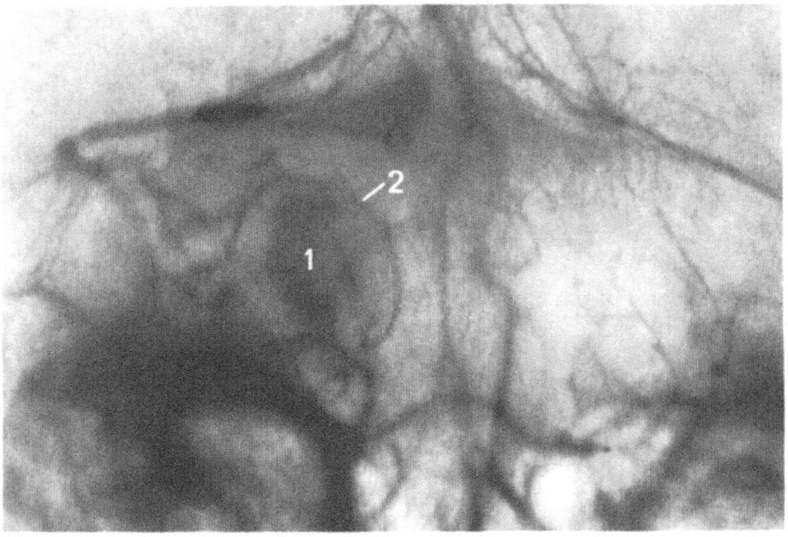

Figure 159

Meningioma of right cerebellopontine angle. Tumoral blush (*1*) and bordering vein (*2*) of meningioma extended to midline

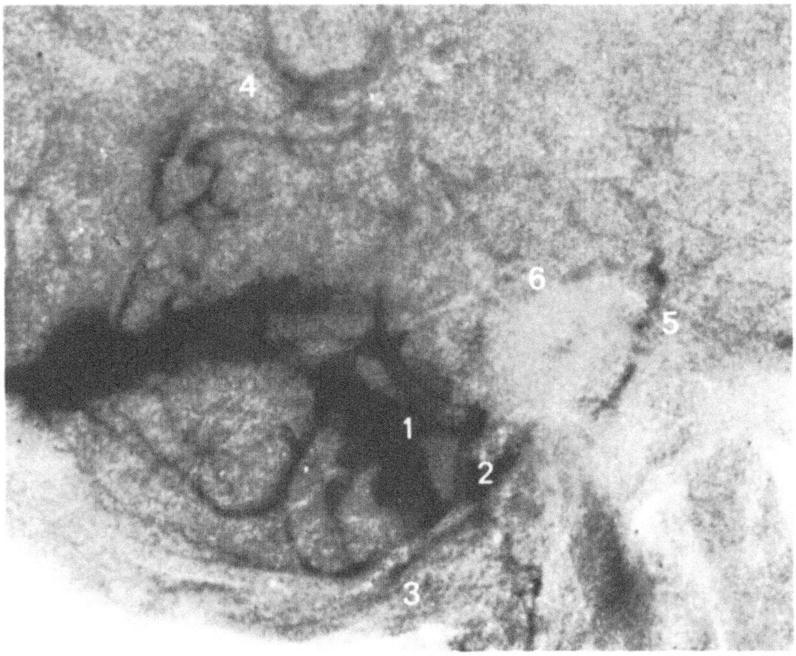

Figure 160

Meningioma of cerebellopontine angle with prepontine extension

1 Venous hammock of vein of lateral recess of fourth ventricle
2 Lateral medullopontine vein
3 Lowered tonsillar vein
4 Empty subdural space above the superior cerebellar vermis
5 Anteriorly displaced pontine veins
6 Spheric avascular area of the tumor

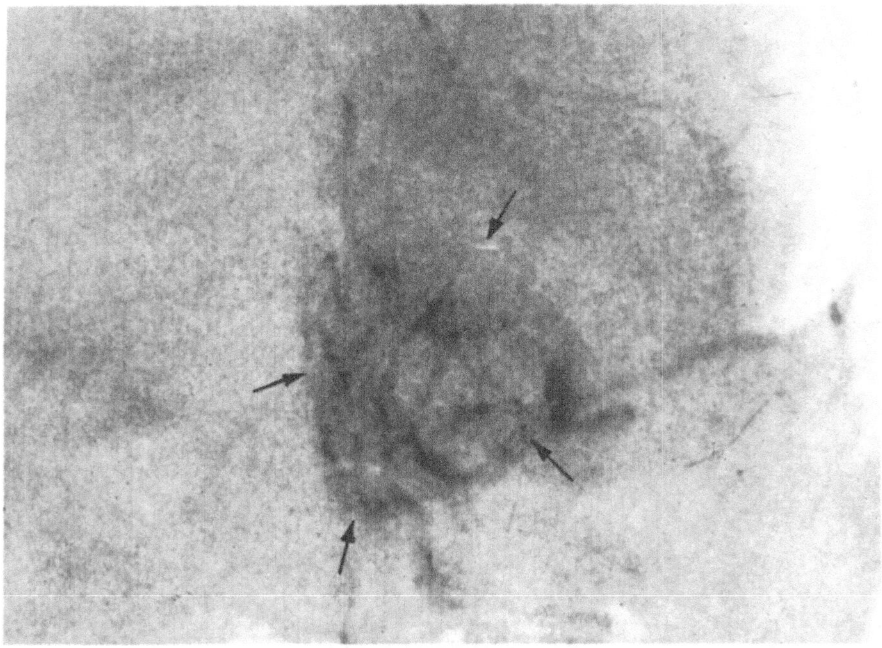

Figure 161

Meningioma of left cerebellopontine angle. Heterogeneous blush with numerous venous vessels

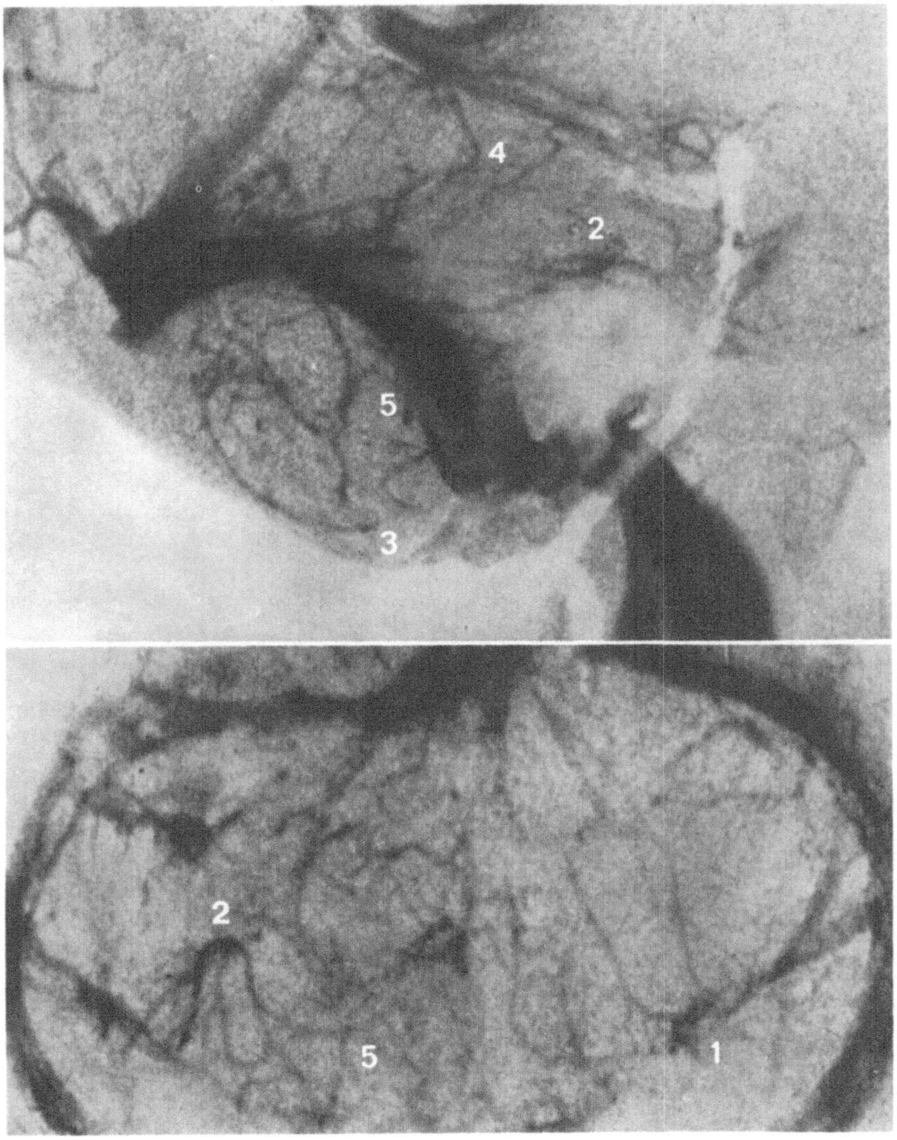

Figure 162

Meningioma of right cerebellopontine angle

1 Normal superior petrous vein
2 Upward and laterally shifted superior petrous vein (parasol sign). Cf. Figs. 155 and 157
3 Lowered tonsillar veins
4 Upward-displaced precentral vein and lateral mesencephalic vein
5 Lowered and posteriorly displaced vein of lateral recess of fourth ventricle

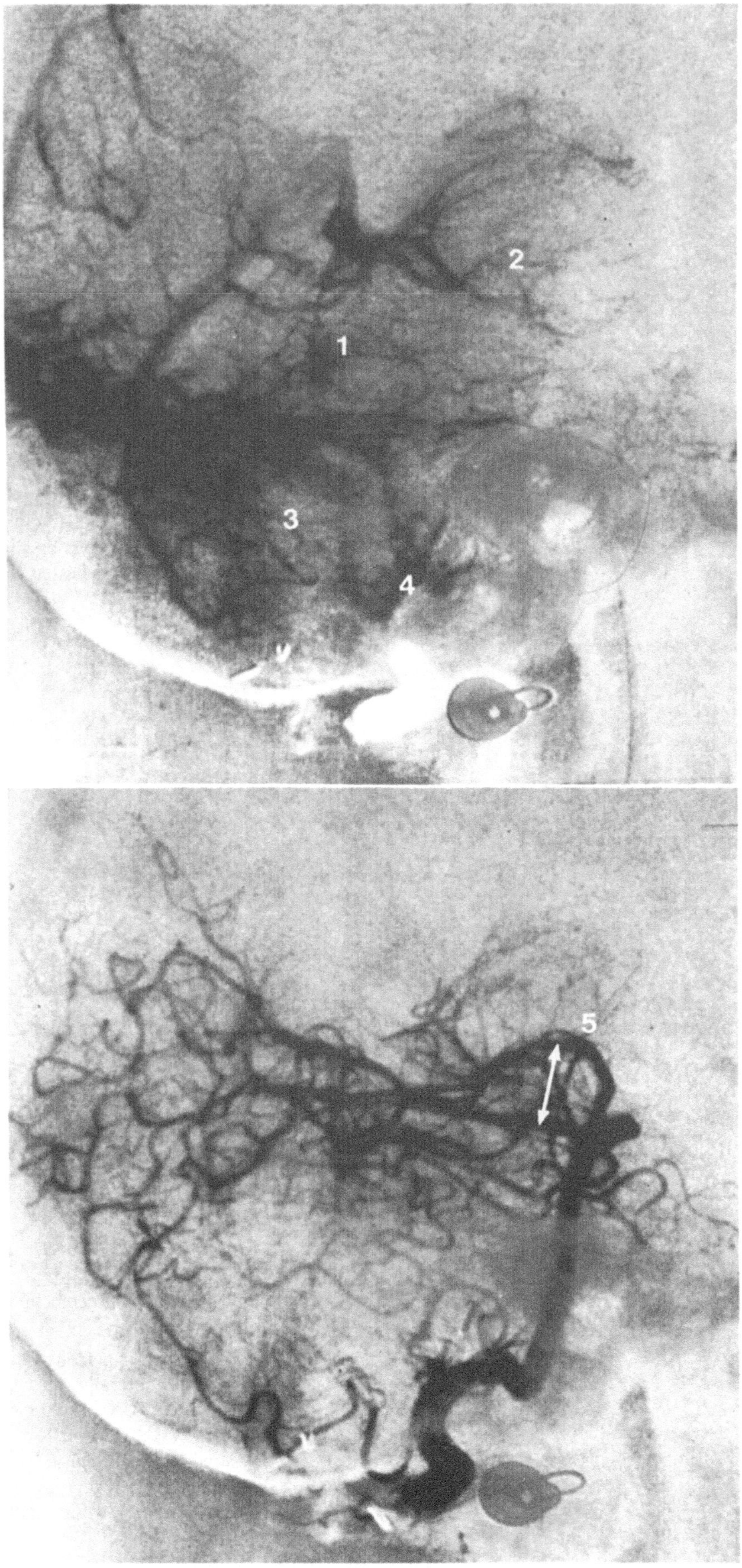

Figure 163

Meningioma of right cerebellopontine angle (recurrence with supratentorial extension)

1 Posteriorly shifted and anteriorly concave precentral vein
2 Upward-shifted right posterior mesencephalic vein, a sign of supratentorial extension
3 Posteriorly shifted vein of lateral recess of fourth ventricle
4 Venous stasis at inferior pole of tumor
5 Shifted right posterior cerebral artery, arterial sign of supratentorial extension (arrow)

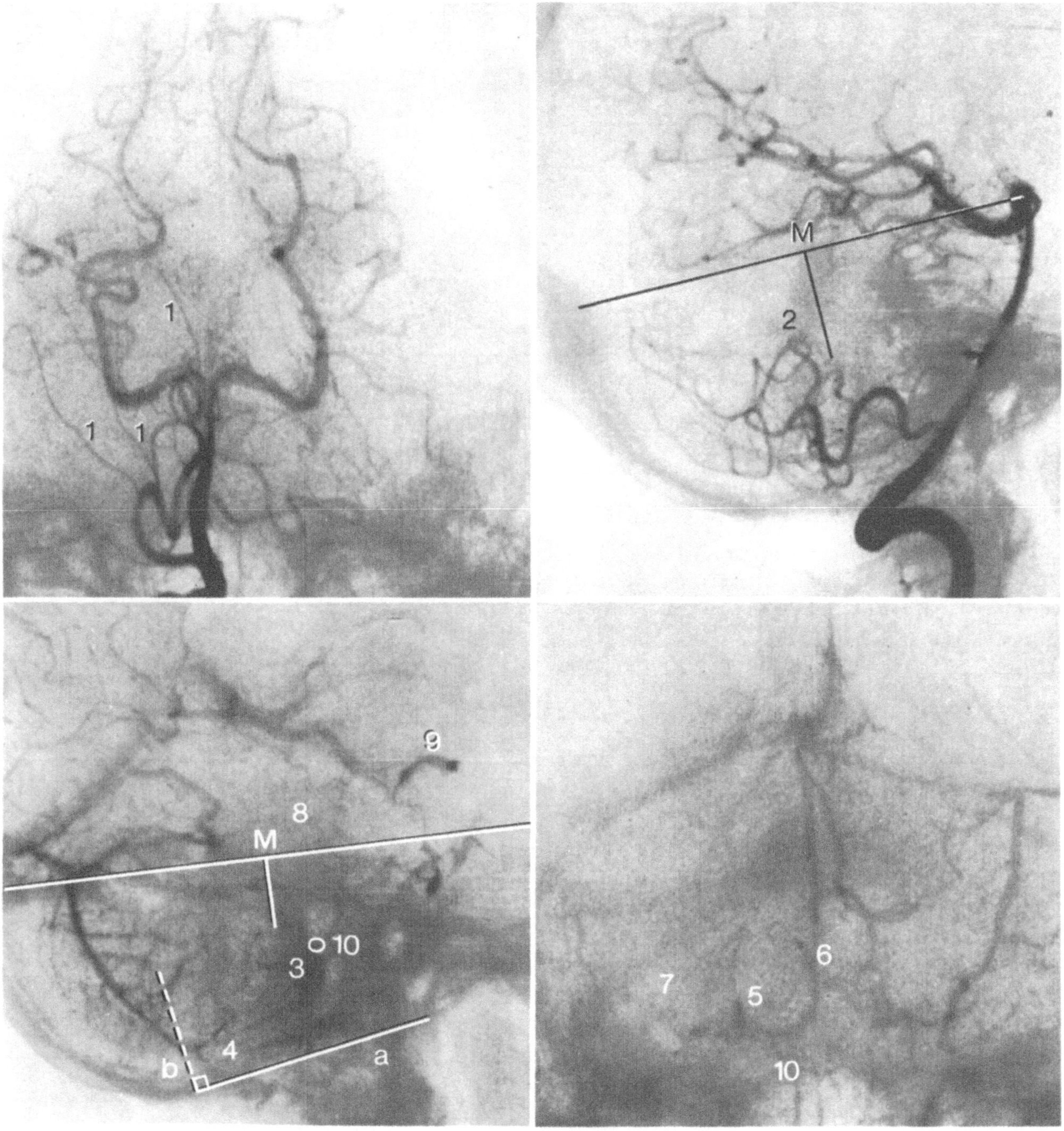

Figure 164

Epidermoid tumor of right cerebellopontine angle

1 Stretched branches of middle cerebellar artery
2 Posteriorly shifted choroidal arteries of fourth ventricle. *M* (midpoint of line joining internal occipital protuberance and extremity of basilar artery), according to MEGRET
3 Venous hammock at inferior pole of the tumor
4 Nondisplaced tonsillar veins and copular point located above line of foramen magnum (*a*) and in front of perpendicular through posterior margin of foramen magnum
5 Medial displacement of superior petrosal vein
6 Medial displacement of vein of lateral recess of fourth ventricle
7 Tumoral area without veins
8 Upward displacements of precentral vein
9 Dilated and displaced optopeduncular veins
10 Junction of lateral recess of fourth ventricle and lateral medullopontine vein shifted forward (1 cm in front of perpendicular dropped on midpoint of Twining's line

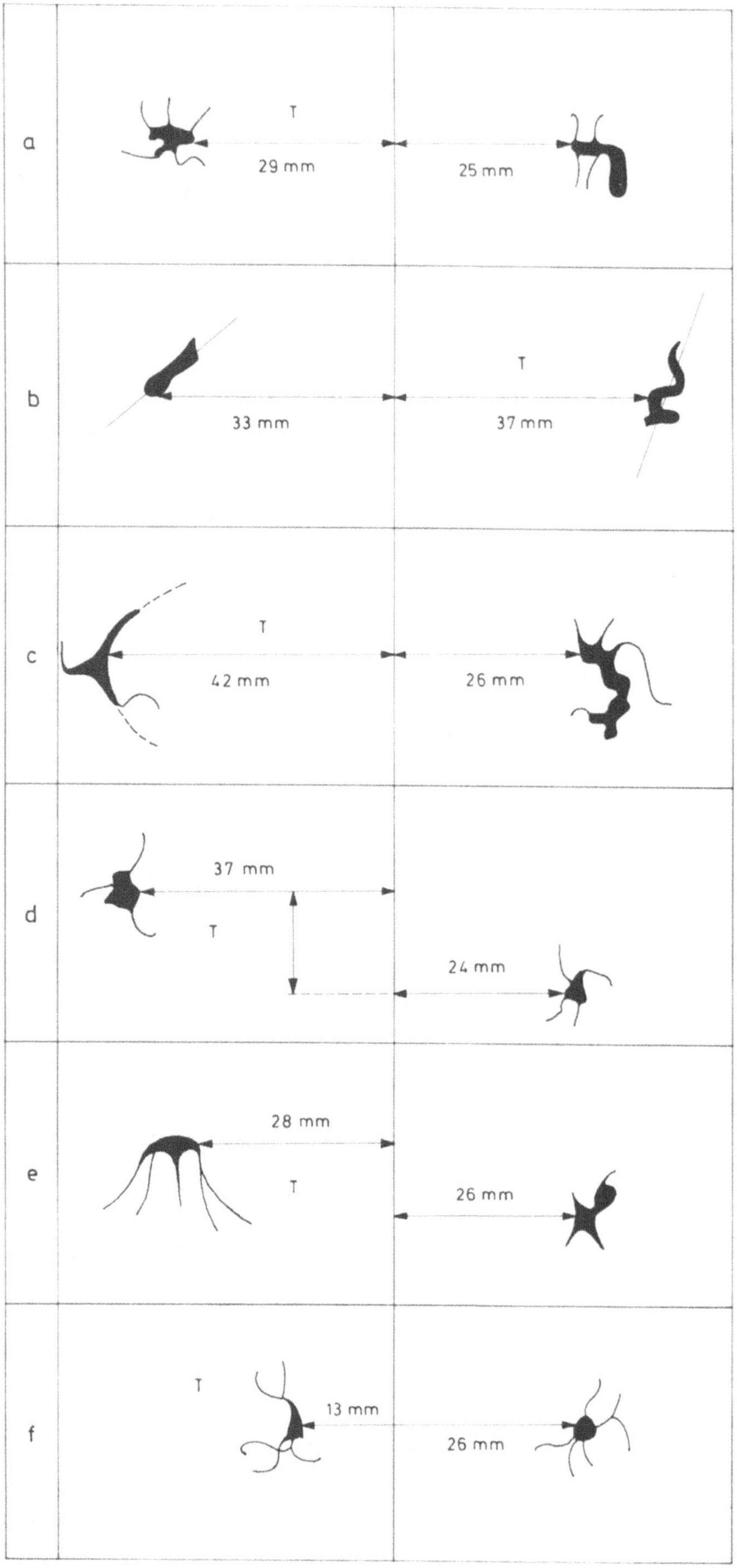

Figure 165 a–f

Displacements of superior petrosal vein in six cases of tumor (*T*) in the cerebellopontine angle. Displacement toward exterior is constant, averaging 10 mm. Uplift is considerable in two cases, measuring 1 cm. One is to note as well the deformation of superior petrosal vein which faces upward and outward on pathologic side with a concavity toward midline (b, c), a stellar or parasol arrangement of the affluents (e), and the possibility of a displacement of superior petrous vein toward midline (f)

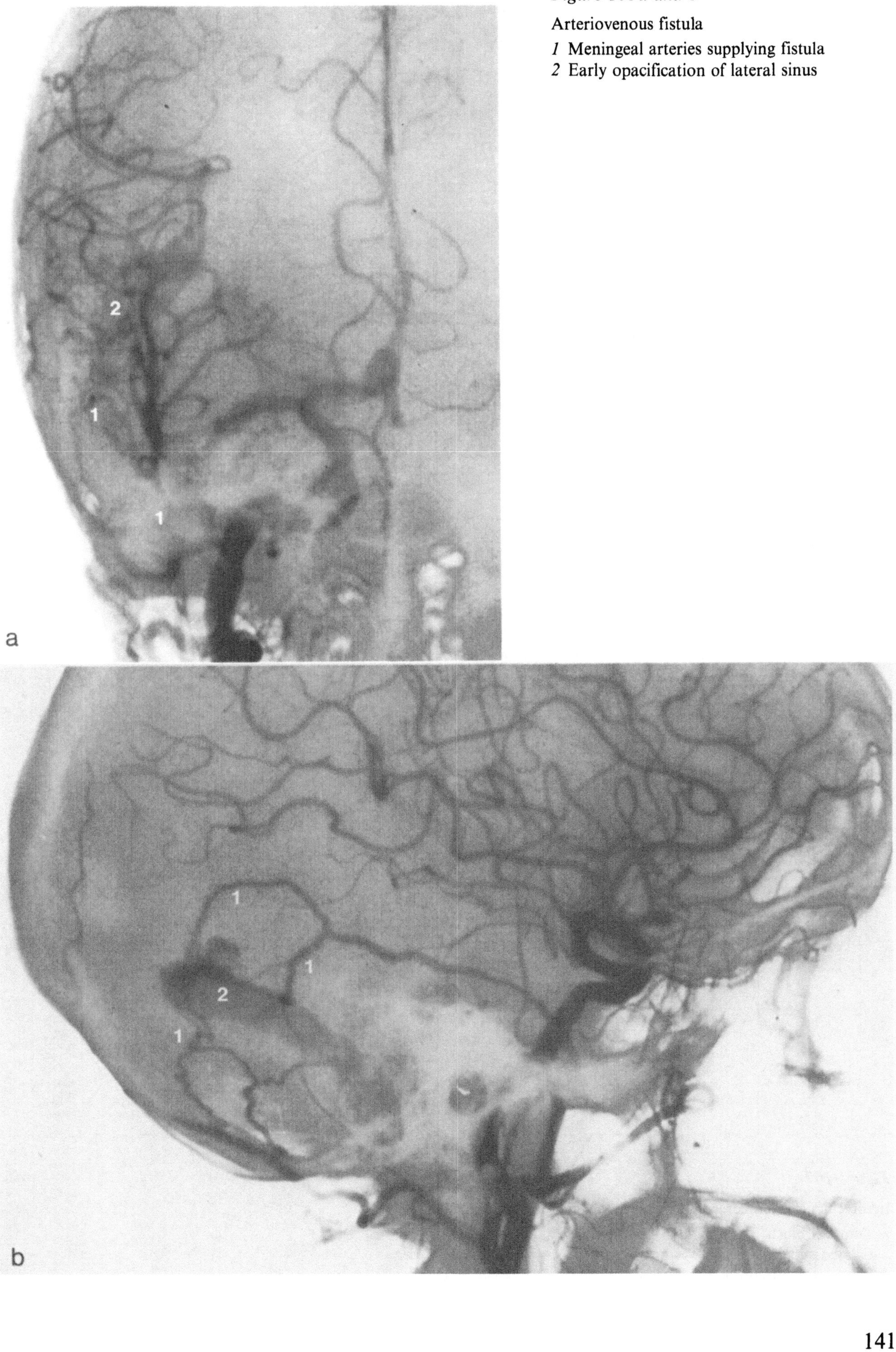

Figure 166 a and b

Arteriovenous fistula

1 Meningeal arteries supplying fistula

2 Early opacification of lateral sinus

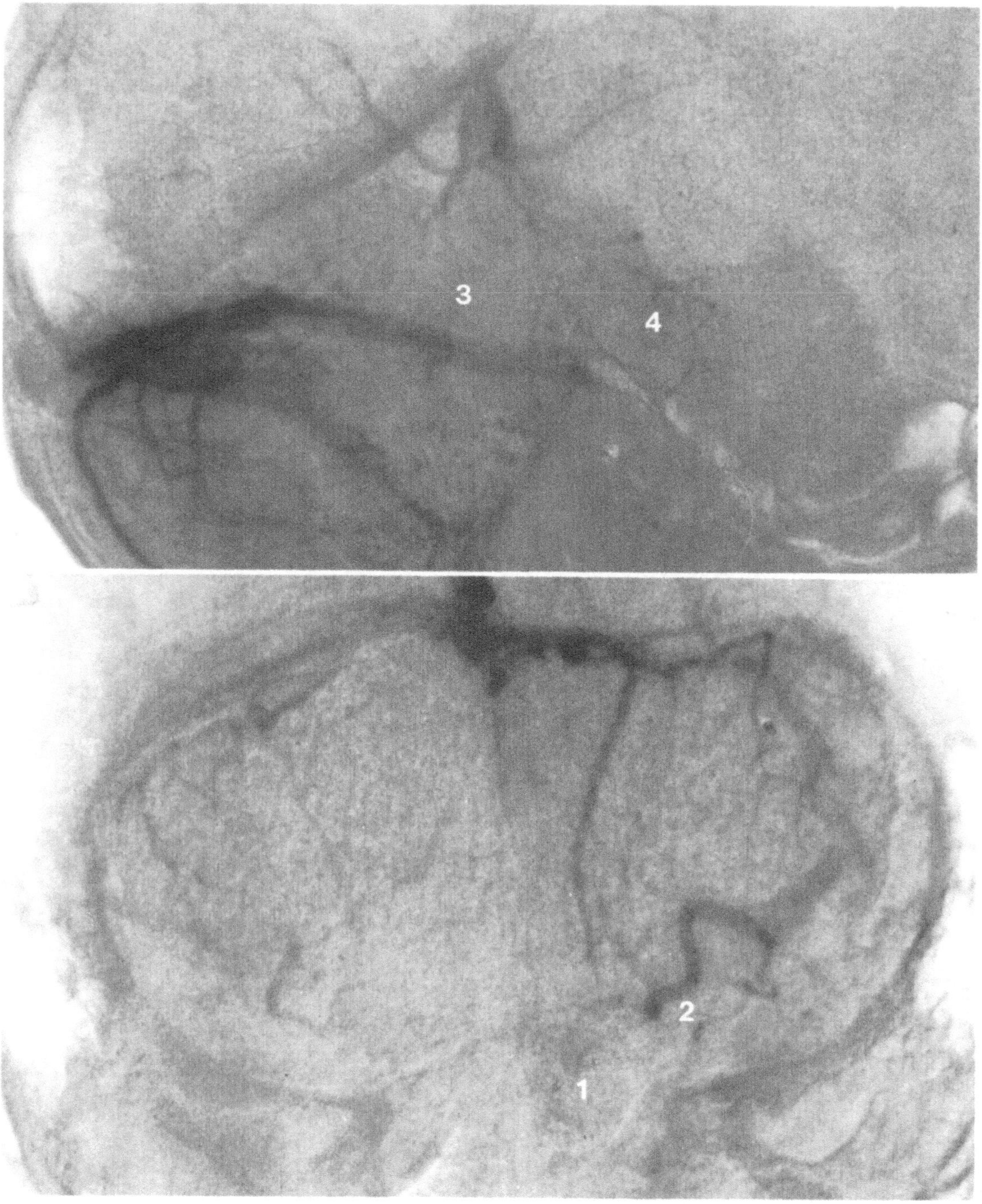

Figure 167

Involvement of left cerebellar angle by malignant tumor of sellar region

1 Tumoral blush
2 Dilated superior petrosal vein
3 Backward-shifted precentral vein
4 Dislocated optopeduncular veins

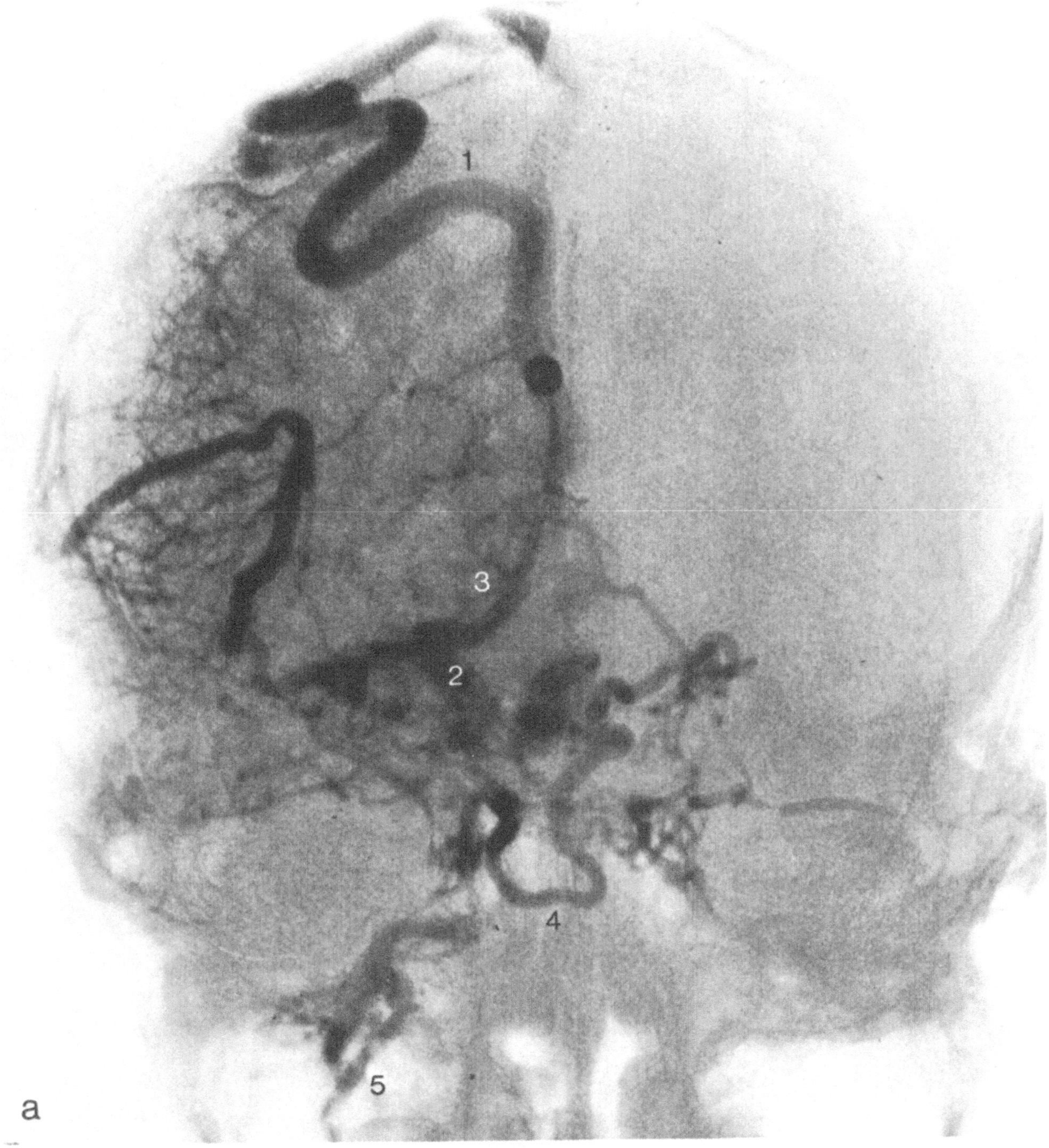

Figure 168 a and b

Arteriovenous aneurysm supplied by internal carotid artery and having multiple drainage channels both supratentorial, subtentorial, and spinal

1 Drainage toward superior longitudinal sinus
2 Optopeduncular vein
3 Basal vein drainage
4 Drainage by prepontine venous system toward venous spinal plexuses (5)

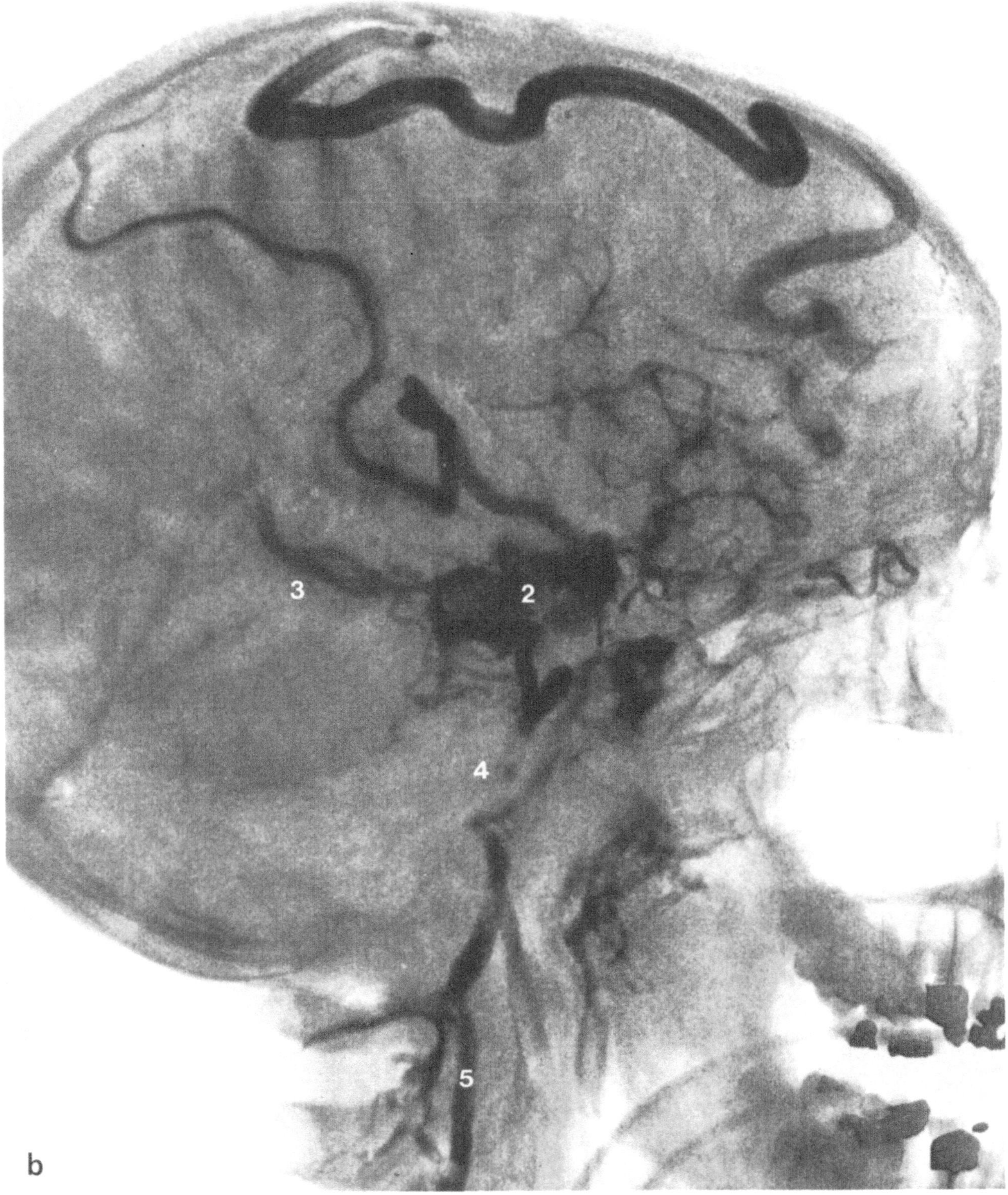

Figure 168b (legend see p. 143)

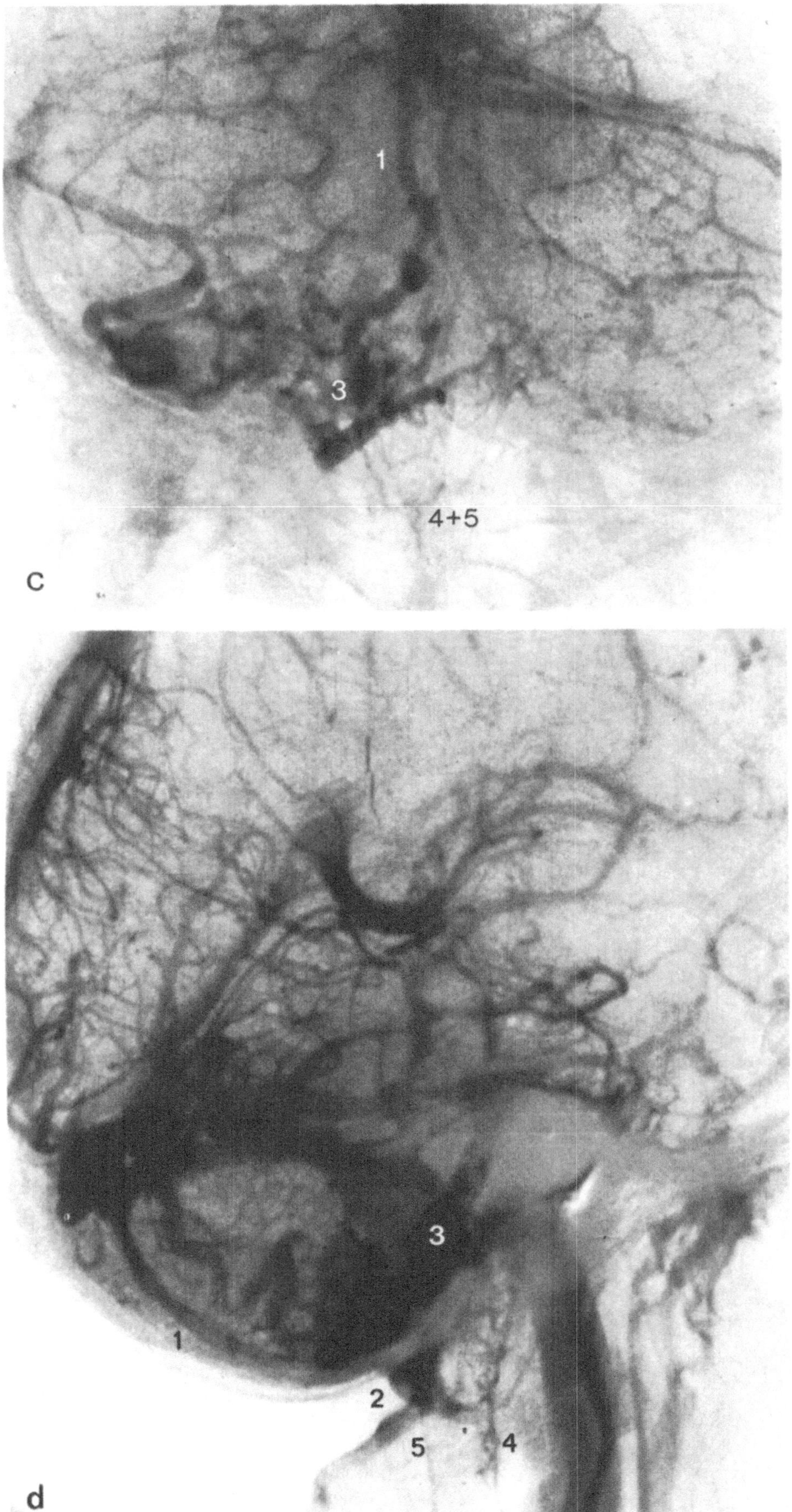

Figure 168 c and d

Arteriovenous malformation with multiple drainage veins

1 Posterior vermian vein
2 Vertebral vein
3 Lateral bulbopontine vein
4 Anterior veins of bulb
5 Posterior veins of the bulb

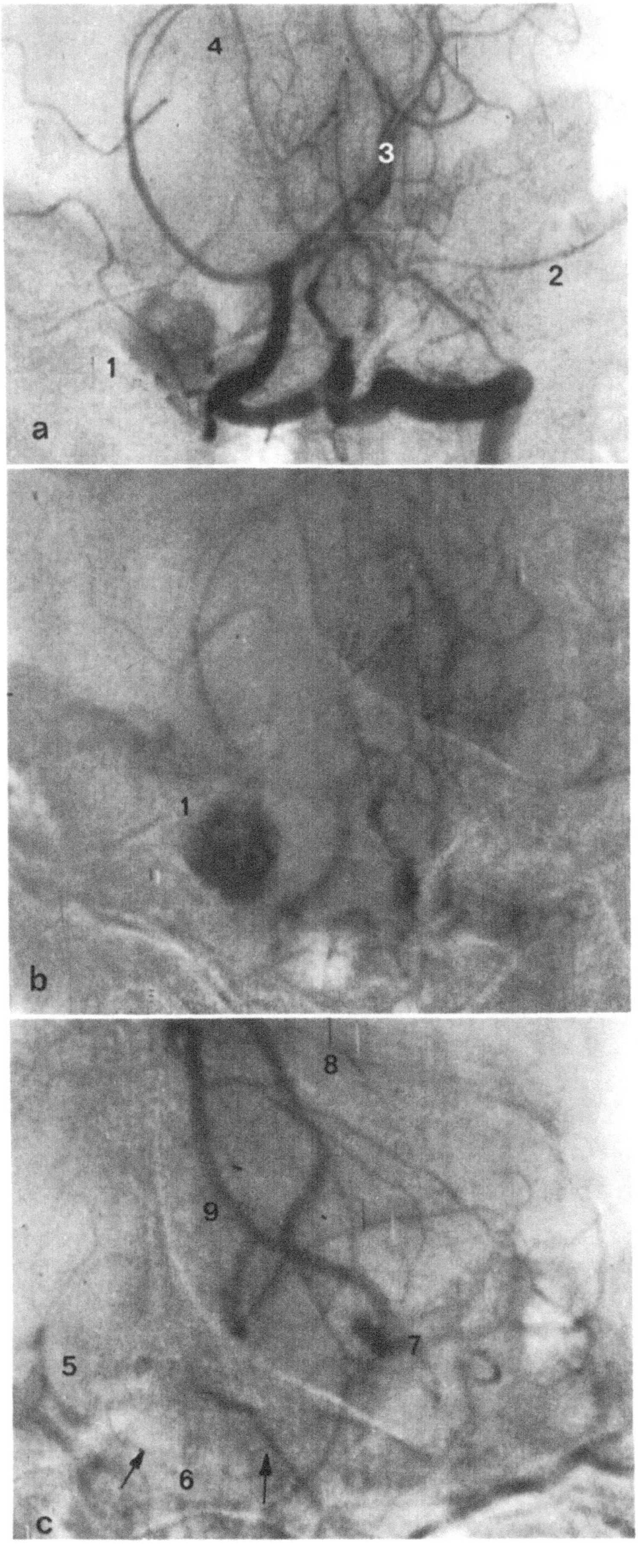

Figure 169 a–d

Hemangioblastoma of posterior fossa

a Left vertebral angiography.
  1 Opacification of tumoral nucleus during arterial phase
  2 and *3* Hammock of middle cerebellar artery and of posterior cerebral and cerebellar arteries
  4 Controlateral deviation of inferior vermian vein. Whole of arterial anomalies prove existence of an important associated cyst

b The fleshy area (*1*) is particularly well observed during intermediary phases.

c The venous drainage is important and is a factor in establishment of diagnosis.
  5 Right superior petrosal vein delimits rim of tumor
  6 Prebulbar vein and prepontine vein
  7 Dilated left superior petrosal vein
  8, *9* Large drainage veins

d Right vertebral angiography.
  10 Large drainage veins characteristic of angioreticuloma
  11 Association of aneurysm of anterior communicating artery

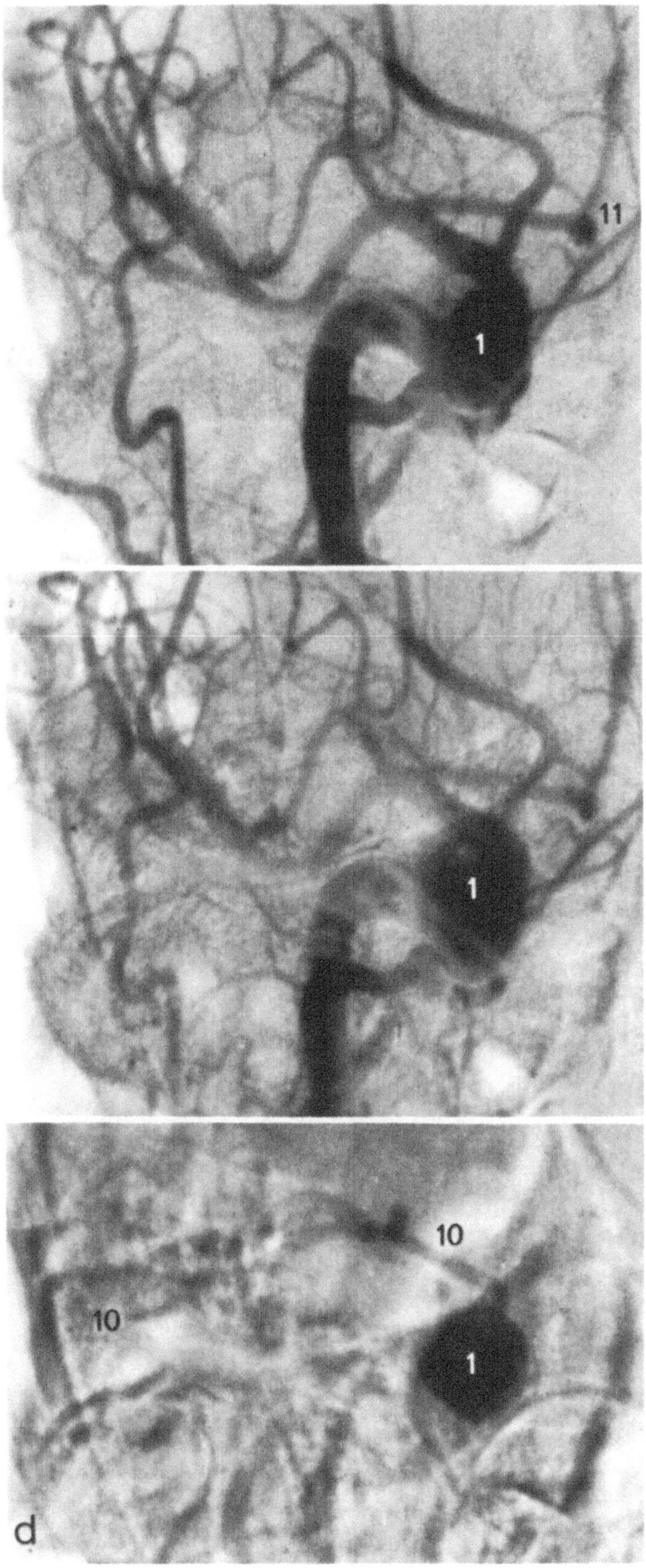

Figure 169d (legend see p. 146)

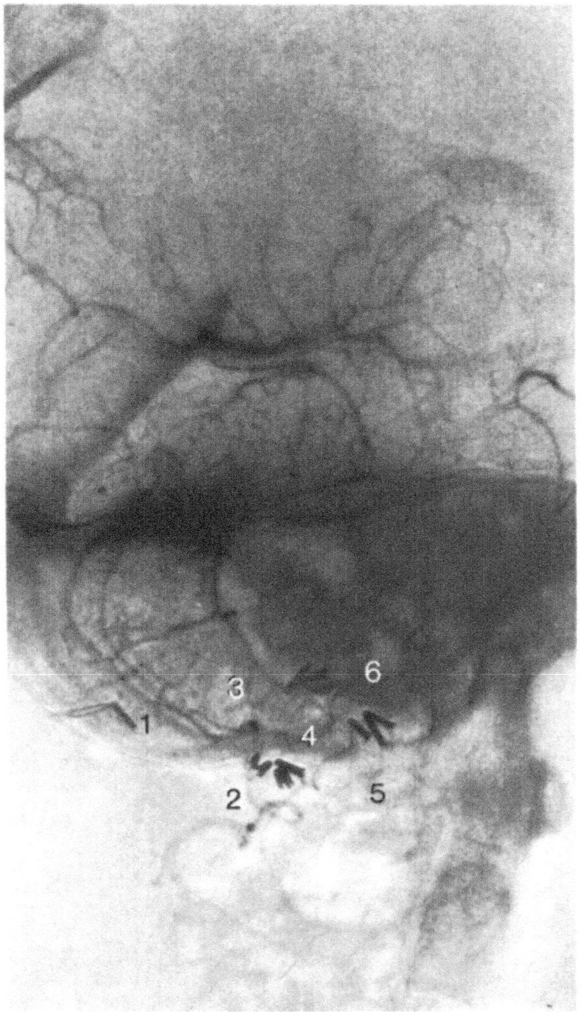

△
Figure 170

Chiari malformation (surgical verification)

1 Inferior vermian vein
2 Inferior retrotonsillar vein
3 Superior retrotonsillar vein
4 Vein of lateral recess of fourth ventricle
5 Lateral bulbopontine vein
6 Junction of lateral bulbopontine vein and vein of lateral recess of fourth ventricle

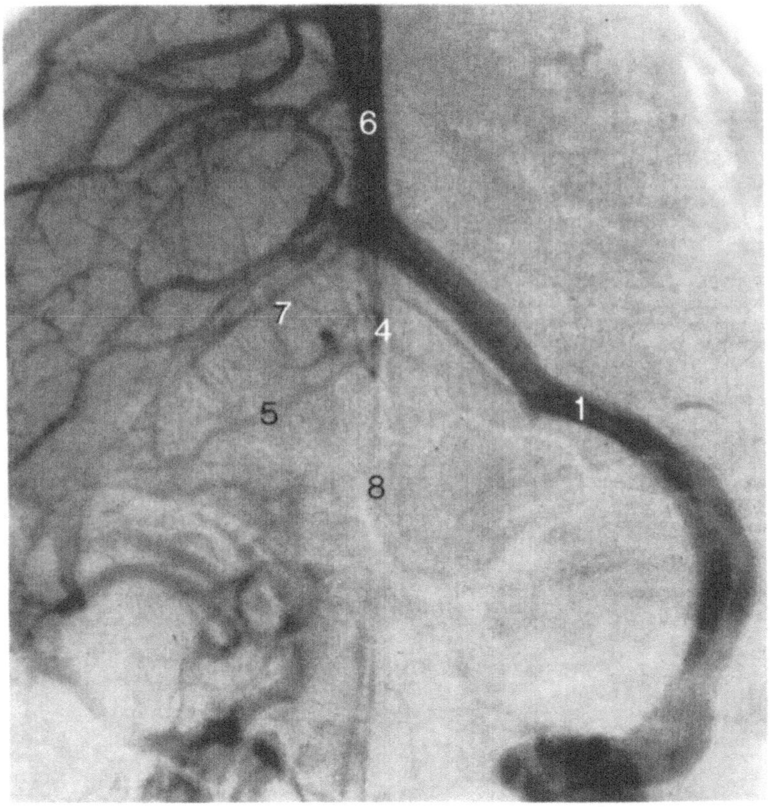

Figure 171

Dandy-Walker malformation (Prof. PICARD, Nancy). Carotid angiography demonstrating topographic anomaly of sinuses and increased volume of posterior fossa

1 Lateral sinus
2 Straight sinus
3 Great vein of Galen
4 Internal cerebral vein
5 Basal vein
6 Inferior longitudinal sinus
7 Tentorial vein
8 Increased volume of posterior fossa

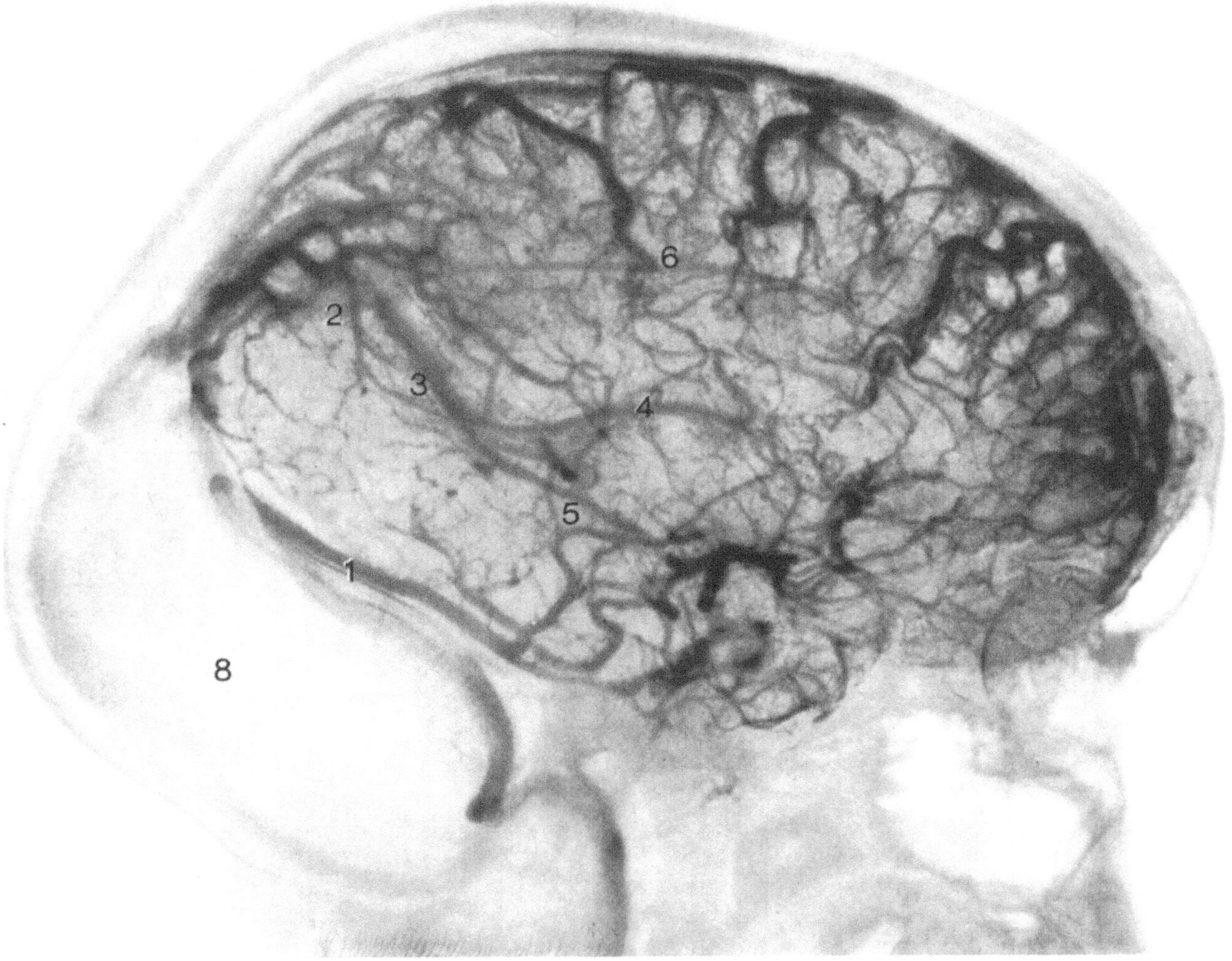

Bibliography

1. Abrams, H.L.: Angiography. Boston: Little, Brown & Co. 1961
2. Aubin, M.L., Metzger, J.: Visibilité du tronc cérébral. Neurochirurgie **15**, 7, 509–518 (1969)
3. Aubin, M.L., Philippon, J., Viano, J., Metzger, J.: Phlébogramme de la fosse postérieure. Son intérêt dans le diagnostic des processus expansifs sous-tentoriels. J. Radiol. Électrol. **53**, 1, 55–60 (1972)
4. Atkinson, W.J.: Anterior inferior cerebellar artery: its variation, pontine distribution and significance in surgery of cerebellopontine angle tumors. J. Neurol. Neurosurg. Psychiat. **12**, 137–151 (1949)
5. Babin, E., Capesius, P., Megret, M., Braun, J.P., Ludwiczak, R.: A propos de quelques particularités topographiques du phlébogramme profond de l'enfant. Ann. Radiol. (Paris) **19**, 5, 525–529 (1976)
6. Babin, E., Megret, M.: Variations in the drainage of the basal vein. Neuroradiology **6**, 154–161 (1973)
7. Banna, M., Foster, J.B., Tomlinson, B.E. (Newcastle upon-Tyne): Non-filling of the intracranial circulation in association with carotid hypoplasia and dural sinus thrombosis. Angiology **23**, 9, 536–548 (1972)
8. Belloni, G., du Boulay, G.: The choroidal and the copular point. Brit. J. Radiol. **47**, 557, 261–264 (1974) 10 réf.
9. Ben Amor, M.: Radio-anatomie normale et pathologique des veines précentrale et brachiale. Ann. Radiol. **15**, 105–112 (1972)
10. Ben Amor, M., Marion, Ch., Heldt, N.: Normal and pathological radioanatomy of the superior choroid vein. Neuroradiology **3**, 16–19 (1971)
11. Ben Amor, M., Wackenheim, A.: Les veines cérébrales profondes sus-tentorielles. Radioanatomie normale (I). Ann. Radiol. **16**, 731–757 (1973)
12. Ben Amor, M., Wackenheim, A.: Les veines cérébrales profondes sus-tentorielles. Radioanatomie II. Pathologie tumorale. Ann. Radiol. **17**, 71–99 (1974)
13. Ben Amor, M., Wackenheim, A.: Les veines cérébrales profondes sus-tentorielles. Radioanatomie III. Pathologie non tumorale. Ann. Radiol. **17**, 161–182 (1974)
14. Billewicz, O., Ben Amor, M.: The posterior ventricular branches of the internal cerebral and basilar veins in cases of deep cerebral tumours infiltrating the basal ganglia. Neuroradiology **1**, 223–227 (1970)
15. (a) Billewicz, O., Ben Amor, M.: The posterior ventricular branches of the internal cerebral and basilar vein. Neuroradiology **2**, 37–45 (1971)
    (b) Billewicz, O.: The normal and pathological radioanatomy of the lateral mesencephalic vein. Neuroradiology **8**, 295–299 (1975)
16. Bradac, G.B.: The ponto-mesencephalic veins (radioanatomical study). Neuroradiology **1**, 52–57 (1970)
17. Bradac, G.B., Holdorff, B., Simon, R.S.: Aspects of the venous drainage of the pons and the mesencephalon. Neuroradiology **3**, 102–108 (1971)
18. Bradac, G.B., Martins, L.F.: Remarks about angiographic findings in cases of pontine tumor. Neuroradiology Germ. **7**, 2, 105–108 (1974)
19. Bradac, G.B., Simon, R.S., Holtz, U.: Zur Auswertung der venösen Phase des Vertebralisangiogramms. Fortschr. Röntgenstr. **117**, 6, 630–637 (1972)
20. Bradac, G.B., Wackenheim, A., Braun, J.P.: Contribution à l'étude du phlébogramme de l'angiographie vertébrale. Neurochirurgia **12**, 1–16 (1969)
21. Braun, J.P.: Les veines cérébrales. France méd. **7**, 317 (1967)
22. Braun, J.P., Tournade, A.: The veins of the lateral recess of the IVth ventricle. Neuroradiology Germ. **7**, 1, 9–13 (1974)
23. Braun, J.P., Tournade, A.: Venous Drainage in the Craniocervical Region. Neuroradiology **13**, 155–158 (1977)
24. Braun, J.P., Tournade, A., Wackenheim, A.: The importance of normal venous landmarks in the topography of the mesencephalon, pons and cerebellum. Neuroradiology **10**, 43–46 (1975)
25. Braun, J.P., Wackenheim, A.: Capillarographie cérébrale. J. Radiol. Électrol. **50**, 532 (1969)
26. Braun, J.P., Wackenheim, A.: Capillarographie cérébrale. Neurochirurgia **12**, 94 (1969)
27. Bray, P.F., Carter, S., Taveras, J.M.: Brainstem tumors in children. Neurology (Minneap.) **1**, 8 (1958)
28. Browning, W.: The Veins of the Brain and Its Envelopes. Brooklyn, New York: F.B. O'Connor, Jr. 1884
29. Bull, J., Kozlowski, P.: The angiographic pattern of the petrosal veins in the normal and pathological. Neuroradiology **1**, 20–26 (1970)
30. (a) Castan, P.: Tumeurs glomiques tympaniques et jugulaires. Montpellier 1973
    (b) Castan, P., Bouzige, J.P., Castan-Tarbouriech, E.: Les hydrocephalies de l'enfant. Expansion Sci. 1975, 1 vol.
31. Chase, N.E., Taveras, J.M.: Cerebral angiography in the diagnosis of suprasellar tumors. Amer. J. Roentgenol. **84**, 154 (1961)
32. Columella, F., Papo, J.: Vertebral angiography in supra tentorial expansive processes. Acta radiol. (Stockh.) **46**, 178 (1956)
33. Constant, P., Montagnac, M., Piton, J., Caille, J.M.: Intérêt de l'angiographie vertébrale dans les tumeurs

intra-axiales de la fosse postérieure chez l'enfant. Bordeaux méd. Fr. 7, 8, 1223–1232 (1974)
34. Crosby, E.C.: Correlative Anatomy of the Nervous System. New York: MacMillan Co. 1962
35. (a) David, M., Bernard-Weil, E., Dilenge, D.: Les tumeurs de la glande pinéale. Ann. Endocr. Paris **24**, 287 (1963)
(b) Debrun, G., Chartres, A.: Infra and supratentorial arteriovenous malformations. Neuroradiology **3–4**, 184–192 (1972)
36. Decker, K.: Klinische Neuroradiologie. Stuttgart: G. Thieme 1960
37. Decker, K., Backmund, H.: Angiographie des Hirnkreislaufs. Stuttgart: G. Thieme 1968
38. Di Chiro, G.: Angiographic patterns of cerebral convexity veins and superficial dural sinuses. Amer. J. Roentgenol. **87**, 308–321 (1962)
39. Di Chiro, G., Doppmann, J.L.: Endocranial drainage of spinal cord veins. Radiology **95**, 555 (1970)
40. Djindjian, R., Bories, J.: Principes généraux du diagnostic des processus expansifs intracrâniens par l'angiographie vertébrale. Diagnostico neuroradiologico. Edité par Sole-Llenas et A. Wackenheim. Barcelona: Toray, 1967, p. 343
41. Djindjian, R., Hurth, M., Thurel, C.: Cervico-cranial phlebography of angiomas of the spinal cord. Neuroradiology **1**, 42–46 (1970)
42. Duvernoy, H.M.: The Superficial Veins of the Human Brain. Berlin-Heidelberg-New York: Springer 1975
43. Ebel, K.D.: Craniostenosis Roentgenological and craniometric features. Pediat. Radiol. Germ. **2**, 1, 1–13 (1974). (36 ref.)
44. Gabrielsen, T.O., Amundsen, P.: Pontine arteries in vertebral angiography. Amer. J. Roentgenol. **106**, 296–302 (1969)
45. Gabrielsen, T.O., Amundsen, P.: The pontomesencephalic veins. Radiology **92**, 889–896 (1969)
46. Gabrielsen, T.O., Seeger, J.F., Crane, J.D.: Veins of the upper cervical spinal cord in vertebral angiography. Acta radiol. Diagnosis **13**, 2, 801–807 (1972)
47. Galligioni, F., Bernardi, R., Pellone, M., Iraci, G.: Superficial sylvian vein in normal and pathologic cerebral angiography. Amer. J. Roentgenol. **107**, 565–578 (1969)
48. Galligioni, E., Bernardi, R., Pellone, M., Iraci, G.: Some observations on deep cerebral phlebogram in hydrocephalus due to supra or subtentorial space-occupying lesions. Amer. J. Roentgenol. **107**, 595–611 (1969)
49. Galligioni, F., Bernardi, R., Pellone, M., Iraci, G.: The veins of the posterior cranial fossa: an angiographic study under pathologic conditions. Amer. J. Roentgenol. **1**, 110, 39–49 (1970)
50. Galloway, J.R., Greitz, T.: The medial and lateral choroïd arteries. An anatomic and roentgenographic study. Acta radiol. (Stockh.) **53**, 353 (1960)
51. Gillot, Cl., Aaron, Cl., Chassaing, P., Delmas, A.: Les veines cérébrales et les sinus crâniens. Presse méd. **73**, 229–234 (1965)
52. Giudicelli, G., Salamon, G.: The veins of the thalamus. Neuroradiology **1**, 92–98 (1970)
53. Goetzen, B.: Use of latex of synthetic rubbers in anatomical investigations of vessels and ducts. [In Polish]. Pol. Tyg. lek. **12**, 8, 296 (1957)
54. Goetzen, B.: Vascularization of arteries of the interior capsule of the brain and the core of the telencephalon. [In Polish]. (Warszawa) **10**, 4, 411 (1959)
55. Goetzen, B.: Vascularisation artérielle de la capsule interne du cerveau. Arch. Anat. Path. **7**, 2, 155 (1959)
56. Goetzen, B.: Use of chlorinated polyrinylchloride in anatomical investigations of vessels and ducts. [In Polish]. (Warszawa) **14**, 1, 77 (1963)
57. Goetzen, B.: Use of the latex Geon 265 in anatomical investigations of vessels and ducts. [In Polish]. (Warszawa) **25**, 1, 147 (1966)
58. Gomez Oliveros, L.: Venas del cerebelo. Arch. esp. Morph. **30**, 251–282 (1950)
59. Gozet, G., Clarisse, J., Franck, J.P., Bonte, G., Delandsherr, J.M.: Anatomie radiologique de la veine uncinée. Sem. Hôp. Ann. Radiol. (Paris) **17**, 3, 253–257 (1974)
60. Guicheney, A.: Thrombophlébites cérébrales postpuerpérales. Contribution à leur étude. Efficacité et particularités d'action de l'héparine. Sem. Hôp. Paris, SP, 521–526 (1960)
61. Gund, A.: Über die Beeinflussung der Hirnvenen durch raumbeengende Prozesse im schnellen Serienangiogramm. Acta neurochir. (Wien) **15**, 202–211 (1966)
62. Habel, J.: Über die Erkennbarkeit von Seitenventrikeldachdepressionen durch parasagittale Raumforderungen im seitlichen Phlebogramm. Radiologe **11**, 1–4 (1971)
63. Hacker, H.: Abflußwege der Sylvischen Venengruppe. Radiologe **8**, 383–387 (1968a)
64. Hacker, H.: Venenabflüsse des Gehirns. Deutscher Röntgenkongreß 1967, Berichtsheft Teil A. Stuttgart: G. Thieme 1968b
65. Hassler, O.: Deep cerebral venous system in man. Neurology **16**, 505–511 (1966)
66. Hedon, C.: Étude anatomique sur la circulation veineuse de l'encéphale. Bordeaux, Thèse **79**, 1–96 (1888)
67. Huang, Y.P., Wolf, B.S.: Veins of the posterior fossa: superior or Galenic draining group. Amer. J. Roentgenol. **95**, 808 (1965)
68. Huang, Y.P., Wolf, B.S.: Precentral cerebellar vein in angiography. Acta radiol. [Diagn.] **5**, 250–262 (1966)
69. Huang, Y.P., Wolf, B.S.: The vein of the lateral recess of the fourth ventricle and its tributaries. Roentgen appearance and anatomic relationships. Amer. J. Roentgenol. **101**, 1–21 (1967)
70. Huang, Y.P., Wolf, B.S.: Angiographic features of fourth ventricle tumors with special reference to the posterior inferior cerebellar artery. Amer. J. Roentgenol. **107**, 543–564 (1969)
71. Huang, Y.P., Wolf, B.S.: Angiographic features of brain stem tumors and differential diagnosis from fourth ventricle tumors. Amer. J. Roentgenol. **110**, 1, 1–30 (1970)
72. Huang, Y.P., Wolf, B.S.: Differential diagnosis of fourth ventricle tumors from brain stem tumors in angiography. Neuroradiology **1**, 4–19 (1970)
73. Huang, Y.P., Wolf, B.S., Antin, S.P., Okudero, T.: The veins of the posterior fossa-anterior or petrosal draining group. Amer. J. Roentgenol. **104**, 36–56 (1968)
74. Huang, Y.P., Wolf, B.S., Antin, S.P., Okudero, T., Kim, I.H.: Angiographic features of aqueductal stenosis. Amer. J. Roentgenol. **104**, 90–108 (1968)
75. Huang, Y.P., Wolf, B.S., Okudero, T.: Angiographic anatomy of the inferior vermian vein of cerebellum. Acta radiol. [Diag.] **9**, 327–344 (1969)

76. Johanson, C.: The central veins and deep dural sinuses of the brain. Acta radiol. (Stockh.), Suppl. **107** (1954)
77. Kinal, M.E., Jackmann, J., Lechner, C.B., Nagle, D.B., Schlabach, D.: The radiographic value of dural sinus venography. Exhibit presented at the VII symposium neuroradiologicum, New York City, Sept. 1964, pp. 20–25
78. Koritke, J.G., Maillot, C., Tournade, A.: The superficial bulbar veins. Comparative anatomy. (Die oberflächlichen bulbären Venen. Vergleichende anatomische Untersuchungen). Neuroradiology **1**, 47–51 (1970)
79. Koritke, J.G., Tournade, A., Monnier, G., Maillot, C.: Les veins superficielles du bulbe: essai de systématisation. Bull. Ass. Anat. **149**, 791–801 (1970)
80. Krayenbuhl, H., Richter, H.R.: Die zerebrale Angiographie. Stuttgart: G. Thieme 1952
81. Krayenbuhl, H., Richter, H.R.: Die zerebrale Angiographie. Stuttgart: G. Thieme 1965
82. Krayenbuhl, H., Richter, H.R., Yasargil, M.G.: Die vaskulären Erkrankungen im Gebiet der Arteria vertebralis und Arteria basialis. Stuttgart: G. Thieme 1957
83. Laine, E., Delandsheer, J.M., Galibert, P., Delandsherr-Arnott, G.: Phlebography in tumors of the hemispheres and central grey matter. Acta radiol. (Stockh.) **46**, 203 (1953)
84. La Torre, E., Occhipinti, E., Pollicita, A.: Retrograde brachial angiography in the diagnosis of infratentorial tumor in infancy and childhood. Europ. Neurol. **4**, 77–99 (1970)
85. Lefebvre, J.C., Faure, C., Salamon, G.: Etude radiologique des gliomes infiltrants du tronc cérébral. Acta. radiol. (Stockh.) [Diag.] **1**, 343 (1963)
86. Lehmann, R.: Zur diagnostischen Relevanz des Vertebralisphlebogramms für Tumoren in der fossa posterior cranii. Berlin 1974
87. Lehmann, R., Szdzuy, D.: Das Vertebralisphlebogramm bei Akustikusneurinomen. Neurol. Neurochir. Pols. Symposium
88. Lehmann, R., Szudzuy, D., Porstmann, B.: Die Angiographie des kraniellen Venensystems. Radiol. diagn. allem. **14**, 4, 379–398 (1973)
89. Leman, P., Cohadon, F., Leifer, C.: Valeur de l'artériographie vertébrale dans les tumeurs de l'angle pontocérébelleux. Ann. Radiol. **10**, 791–802 (1967)
90. Lin, P.M., Mokrohisky, J.F., Stauffer, J.M., Scott, M.: The importance of the deep cerebral veins in deep cerebral angiography. J. Neurosurg. **12**, 256 (1955)
91. Lindgren, E.: Percutaneous angiography of the vertebral artery. Acta radiol. (Stockh.) **33**, 389 (1950)
92. Lindgren, E.: Central angiography des Herzens und der Gefäße. In: Roentgendiagnostik, p. 602. Berlin-Heidelberg-New York: Springer 1964
93. (a) Long, J.M., Kier, E.L., Schechter, M.M.: The radiology of epidermoid tumours of the cerebellopontine angle. Neuroradiology **6**, 188–192 (1973)
(b) Lopez, F.: Sistema venoso de la fossa posterior. Estudio radioanatomico. Clinica "Puerta de Hierro", Madrid (1971)
94. Megret, M.: A landmark for the choroidal arteries of the fourth ventricle – branches of the posterior inferior cerebellar artery. Neuroradiology **5**, 85–90 (1973)
95. Namin, P.: Angiographie vertébrale. Thèse de la Faculté de Médecine de Paris 1953, p. 479
96. Namin, P.: L'angiographie vertébrale. Paris: Doin & Cie 1955
97. Nolten, W.: Verbindungen zwischen intra- und extracraniellen Venen der Schädelbasis. Inaugural-Dissertation, München 1961
98. Padget, D.H.: Cranial venous system in man in reference to development of adult configuration and relation to arteries. Amer. J. Anat. **98**, 307–355 (1956)
99. Padget, D.H.: Development of cranial venous system in man from viewpoint of comparative anatomy. Contr. Embryol. Carneg. Instn. **611**, 36, 79–140 (1957)
100. Paillas, J.A., Bonnal, J., Payan, H., Pruvost, A.: Les thromboses puerpérales du sinus longitudinal supérieur. Presse méd. **1964**, 2265–2268
101. Papo, I., Salvolini, U.: Meningiomas of the free margin of the tentorium developing in the pineal region. Neuroradiology **7**, 4, 237–243 (1974)
102. Peeters, F.L.M.: The vertebral angiogram in patients with tumours in or near the midline. Neuroradiology **5**, 53–58 (1973)
103. Perese, D.M.: Superficial veins of brain from surgical point of view. J. Neurosurg. **17**, 402–412 (1960)
104. Pernkopf, E.: Topographische Anatomie, Vol. IV/1 and 2. München: Urban & Schwarzenberg 1957
105. Pernkopf, E.: Topographische Anatomie des Menschen. München-Berlin: Urban & Schwarzenberg 1960
106. Picard, L., Renard, M., Eloit, J., Brice, M., Roland, J.: Phlébogramme vertébro-basilaire. Etude de la radioanatomie normale du point copulaire: intérêt dans la localisation des processus expansifs de la fosse postérieure. J. Radiol. Électrol. **55**, 3, 211–216 (1974)
107. Poole, G., Kane, R.: Posterior fossa midline venography. Neuroradiology **3**, 53 (1971)
108. Potts, K.G., Taveras, J.M.: Differential diagnosis of space-occupying lesions in the region of the thalamus by cerebral angiography. Acta radiol. [Diagn.] **1**, 373 (1963)
109. Raimondi, A.J.: Pediatric Neuroradiology. Philadelphia: W.B. Saunders 1972
110. Rauber-Kopsch, F.: Lehrbuch und Atlas der Anatomie des Menschen, Vol. II. Leipzig: Thieme 1955
111. Richter, H.S.R.: Phlebogram in brainstem tumors. Acta radiol. (Stockh.) **40**, 182 (1953)
112. Rodriguez-Carbajal, J., Palacios, E.: Intraventricular meningiomas of the fourth ventricle. Amer. J. Roentgenol. **120**, 1, 27–31 (1974)
113. Rosa, M., Viale, G.L.: Studio angiografico del sistema venoso vertebro-basilare. II. La vena sopraculminare (vermiana superiore). Sist. nerv. **20**, 143–146 (1968)
114. Rosa, M., Viale, G.L.: Studio angiografico del sistema venoso vertebro-basilare. IV. La vena cerebellare precentrale. Sist. nerv. **20**, 286–294 (1968)
115. Rosa, M., Viale, G.L.: Studio angiografico del sistema venoso vertebro-basilare. VI. La vena cerebellare antero-superiore (grande vena cerebellare). Sist. nerv. **20**, 332–337 (1968)
116. Rosa, M., Viale, G.L.: Studio angiografico del sistema venoso vertebro-basilare. VIII. Le vene occipitali. Sist. nerv. **21**, 129–135 (1969)
117. Rosa, M., Viale, G.L.: Diagnostic value of vertebral phlebogram. Neuroradiology **1**, 147–150 (1970)

118. Rosa, M., Viale, G.L.: Subependymal veins in the angiographic diagnosis of brain tumours. Neuroradiology **2**, 90–93 (1971)
119. Rosa, M., Viale, G.L.: Vertebrobasilar venous system in the angiographic diagnosis of hydrocephalus. Acta radiol. [Diagn.] **13**, 2, 590–594 (1972)
120. Rosa, M., Viale, G.L.: Studio angiografico del sistema venose vertebro-basilare. X. Le confluenti nella vena cerebrale interne. Sist. nerv. (in press)
121. Ruggiero, R., Dettori, P.: La soustraction d'image en angiographie cérébrale bidirectionnelle simultanée. Neuchâtel: Delachaux & Niestlé 1970
122. Saldino, R.M., di Chiro, G.: Tentorial calcification. Radiology **111**, 1, 207–210 (1974)
123. Savoiardo, M., Lemay, M.: Diagnosis of posterior fossa lesion by angiography. Brit. J. Radiol. **43**, 291–302 (1970)
124. Savoiardo, M., Vaghi, M.A. (in Raimondi): Angiography in brain stem tumours. Neuroradiology **8**, 99–112 (1974)
125. Scotti, L.N., Goldman, R.L., Hardman, D.R., Heinz, E.R.: Venous thrombosis in infants and children. Radiology **112**, 2, 393–399 (1974)
126. (a) Seeger, J.F., Gabrielsen, T.O.: Angiography of eccentric brain stem tumors. Radiology **105**, 343–351 (1972)
(b) Signargout, Y., Simon, J., Duval, J.M.: La veine basilaire. Variations et semiologie. J. Radiol. Electrol. **53**, 317–328 (1972)
127. Sobotta, J., Becher, H.: Atlas der deskriptiven Anatomie des Menschen, Vol. III. München-Berlin: Urban & Schwarzenberg 1960
128. Spettowa, S., Chrzanowski, R.: Changes in the superficial venous system in cases of tumours of the brain. Acta med. pol. **4**, 4, 377–397 (1963)
129. Sutton, D.: Radiologic aspects of pontine gliomata. Acta radiol. (Stockh.) **40**, 234 (1953)
130. Takahashi, M., Okudera, T., Tomanaga, M., Kitamura, K.: Angiographic diagnosis of acoustic neurinomas: analysis of 30 lesions. Neuroradiology **2**, 191–200 (1971)
131. Takahashi, M., Odudera, T., Tanaka, M., Kitamura, K., Yonemasu, Y.: Angiographic diagnosis of cerebellar medulloblastomas. Evaluation of pre and post-operative vertebralangiographies. Amer. J. Roentgenol. **118**, 3, 622–632 (1973)
132. Takahashi, M., Wilson, G., Hanafee, W.: Significance of petrosal vein in diagnosis of cerebellopontine angle tumors. Radiology **89**, 834–840 (1967)
133. Takahashi, M., Wilson, G., Hanafee, W.: Catheter vertebral angiography: a review of 300 examinations. J. Neurosurg. **30**, 722–731 (1969)
134. Taveras, J.M., Wood, E.H.: Diagnostic neuroradiology. Baltimore: Williams & Wilkins Co. 1964
135. Tovi, D., Schisano, G., Liliequist, B.: Primary tumors of the region of the thalamus. J. Neurosurg. **18**, 730 (1961)
136. Turtas, S.: Pineal veins in vertebral angiography. Neuroradiology **10**, 107–110 (1975)
137. Viale, G.L., Rosa, M.: La vena sopra-culminare (vermiana superiore). Ibid. **20**, 143 (1968)
138. Viale, G.L., Rosa, M.: La vena vermiane inferiore. Ibid. **20**, 280 (1968)
139. Viale, G.L., Rosa, M.: La vena cerebellare precentrale. Ibid. **20**, 286 (1968)
140. Viale, G.L., Rosa, M.: La vena cerebellare marginale. Ibid. **20**, 328 (1968)
141. Viale, G.L., Rosa, M.: La vena cerebellare anteroposteriore (grande vena cerebellare). Ibid. **20**, 332 (1968)
142. Viale, G.L., Rosa, M.: Le vene cerebellari emisferiche. Ibid. **20**, 338 (1968)
143. Viale, G.L., Rosa, M.: Studio angiografico del sistema venoso vertebro-basilare. Le vene ponto-mesencefaliche. Sist. Nerv. **20**, 131–142 (1968)
144. Viale, G.L., Rosa, M.: Studio angiografico del sistema venoso vertebro-basilare. III. La vena vermiana inferiore. Sist. nerv. **20**, 280–285 (1968)
145. Viale, G.L., Rosa, M.: Studio angiografico del sistema venoso vertebro-basilare. V. La vena cerebellare marginale. Sist. Nerv. **20**, 328–331 (1968)
146. Viale, G.L., Rosa, M.: Studio angiografico del sistema venoso vertebro-basilare. VII. Le vene cerebellari emisferiche. Sist. Nerv. **20**, 338–343 (1968)
147. Viale, G.L., Rosa, M.: Studio angiografico del sistema venoso vertebro-basilare. IX. La vena del recesso laterale del VI Ventriculo. Sist. Nerv. **21**, 136–142 (1969)
148. Vuia, O., Alexianu, M.: Insuffisance circulatoire veineuse cérébrale. J. neurol. Sci. **7**, 495–511 (1968)
149. Wackenheim, A.: Some views regarding the diagnostic value of the veins of the posterior fossa. Neuroradiology **3**, 75–76 (1971)
150. Wackenheim, A.: Schéma des veines de la fosse postérieure visibles sur les clichés d'une angiographie vertébrale de routine. J. Radiol. Électrol. **4**, 694 (1973)
151. Wackenheim, A.: Visualization of the chorod plexuses of the lateral ventricles in the frontal projection in vertebral angiography. Neuroradiology **6**, 231–232 (1973)
152. Wackenheim, A.: The herniation of the superior vermis. Neuroradiology **7**, 221–227 (1974)
153. Wackenheim, A.: L'engagement du vermis superieur dans l'espace sous-dural sus-vermien ou engagement vermien supérieur. J. Radiol. Électrol. **55**, 41–43 (1974)
154. (a) Wackenheim, A., Braun, J.P.: Angiography of the Mesencephalon. Berlin-Heidelberg-New York: Springer 1970
(b) Wackenheim, A.: Roentgen Diagnosis of the Craniovertebral Region. Berlin-Heidelberg-New York: Springer 1974
155. Wackenheim, A., Ben Amor, M.: Phlebographic location of the inferior vermis. Neuroradiology **3**, 12–15 (1971)
156. Wackenheim, A., Ben Amor, M.: Arteriovenous separation of the prepontine vessels as a sign of intrapontine tumour. Neuroradiology **3**, 77–79 (1971)
157. Wackenheim, A., Braun, J.P., Babin, E., Ludwiczak, R., Dupuis, M., Capesius, P., Ammerich, H.: Analyse du phlébogramme cérébral de l'hydrocéphalie infantile. J. Méd. (Strasbourg) **7**, 14, 753–763 (1976)
158. Wackenheim, A., Braun, J.P., Bradac, G.B.: Angiographie der Tumoren des Mittelhirnes und seiner Nachbarschaft. Radiologe **8**, 354 (1968)
159. Wackenheim, A., Heldt, N., Ben Amor, M.: Variations in the drainage of the lateral mesencephalic vein. Neuroradiology **2**, 154–161 (1971)
160. Wackenheim, A., Nakayama, N., Wende, S.: Magnification of the veins in vertebral angiography. Neuroradiology **6**, 56–59 (1973)

161. Wackenheim, A., Stoffels, C.: La localisation antéro-postérieure du point de jonction de la veine du récessus latéral du IVème ventricule et de la veine bulbo-pontique latérale. Neuroradiology (1978) (in press)
162. Wilner, H.I., Navarro, E., Eisenbrey, A.B., Gracias, V.: The inferior vermian veins as a useful adjunct in the differentiation of brain stem tumors from midline cerebellar masses. Amer. J. Roentgenol. **118**, 3, 605–616 (1973)
163. Wolf, B.S., Huang, Y.P.: The subependymal veins of the lateral ventricles. Amer. J. Roentgenol. **91**, 406 (1964)
164. Wolf, B.S., Huang, Y.P., Newman, C.M.: The lateral anastomotic mesencephalic vein and other variations in drainage of the basal cerebral vein. Amer. J. Roentgenol. **89**, 411–422 (1963)
165. Wolpert, S.M.: Angiography in posterior fossa tumors of infancy and childhood. Amer. J. Roentgenol. **112**, 296–305 (1971)

Subject Index

Clinical Computer Tomography

Head and Trunk

Editors: A. Baert, L. Jeanmart, A. Wackenheim
1978. 414 figures, 2 tables. 261 pages
ISBN 3-540-08458-4

Contents: Introduction to the Technology of Computer Tomography. – Head: Seller Region: Normal and Pathologic Conditions. Empty Sella and Pituitary Gland. Midline Lesions. Ventriculocisternal Pathology in Children. Endocranial Calcification. Area of Maximal Density in Extracerebral Tumors. – Tumora Masses of the Posterior Fossa. Cerebellopontile Expansive Lesions. The Cervical Medullar Canal and Its Content. Cerebral Ischemia. Cerebral Metastases. Inflammatory Diseases of the Brain. Epidermoid Cyst. Meningiomas en Plaque. Neuro-ophthalmology. Arterial and Arteriovenous Malformations. – Trunk: Chest. Liver. Liver and Pancreas. Pancreatic Disease. Kidneys. Retroperitoneal Region. Pelvis. Abdominal Computer Tomography and Contrast Media.

A. Wackenheim, J. P. Braun

Angiography of the Mesencephalon

Normal and Pathological Findings

1970. 128 figures. XI, 154 pages
ISBN 3-540-05266-6

Contents: Introduction. – Review of Literature. – Normal Findings: Normal Radioanatomy of the Arteries. – Normal Radioanatomy of the Veins. – Pathological Findings: Tumours of the Pineal Region. Tumours of the Posterior Thalamus. Tumours of the Cerebral Peduncles.

A. Wackenheim

Roentgen Diagnosis of the Craniovertebral Region

1974. 500 figures. XXII, 601 pages
ISBN 3-540-06615-2
Distribution rights for Japan:
Igaku Shoin Ltd., Tokyo

Contents: Generalities: Historical Review. The Craniovertebral Region. Biotypology. Facial Biotypology as Seen by Famous Portrait Artists and Sculptors. Radioanatomical Evolution in Vertebrates. Clinical Examination. Normal Findings: Techniques of Radiological Examination of the Craniovertebral Angle. – Artefacts of Radiological Examination. Normal Radiographic Anatomy and Its Variations. Embryology. – Pathological Findings: Basic Radiological Abnormalities. Traumatic Lesions. Neoplasms. Abnormalities of the Soft Tissues of the Neck, in the Blood Vessels, and of the Bones in the Petrospheno-occipital Region. Bone Malformations in the Craniovertebral Region. Diseases of the Cervicooccipital Ligaments. Torticollis. Rheumatoid Arthritis and Allied Diseases. Cervicooccipital Involvement in Various Diseases. Chordopathy. Malformations of the Cerebellum and Spinal Cord.

Springer-Verlag
Berlin
Heidelberg
New York

GPSR Compliance

The European Union's (EU) General Product Safety Regulation (GPSR) is a set of rules that requires consumer products to be safe and our obligations to ensure this.

If you have any concerns about our products, you can contact us on ProductSafety@springernature.com

In case Publisher is established outside the EU, the EU authorized representative is:

Springer Nature Customer Service Center GmbH
Europaplatz 3
69115 Heidelberg, Germany

Batch number: 10615099

Printed by Printforce, the Netherlands